The Computational Infant

The Computational Infant
Looking For Developmental Cognitive Science

Julie C. Rutkowska

School of Cognitive and Computing Sciences
University of Sussex

HARVESTER WHEATSHEAF

New York London Toronto Sydney Tokyo Singapore

First published 1993 by
Harvester Wheatsheaf
Campus 400, Maylands Avenue
Hemel Hempstead
Hertfordshire, HP2 7EZ
A division of
Simon & Schuster International Group

Typeset in 10/12pt Times
by Hands Fotoset, Leicester

Printed and bound in Great Britain by
T.J. Press, Padstow, Cornwall

British Library Cataloguing in Publication Data

A catalogue record for this book is available from the British Library
ISBN 0-7450-1414-3 (hbk)
ISBN 0-7450-1415-1 (pbk)

1 2 3 4 5 97 96 95 94 93

Contents

Preface

This book is about four main things: infancy, action, computation and cognitive science. First and foremost, it addresses the problem of how we can best conceptualize the abilities of young human infants. My solution is a cognitivist one, but I do not view the infant primarily as a scientist with the wrong theory, a hypothesis testing model builder or a naive theoretician ready equipped with concepts about the world. Rather, the infant of this book is more appropriately viewed as a situated agent. S/he is a computational infant because I consider action to be the key to understanding infant ability, and computational concepts from cognitive science to be the best route to understanding action systems.

One of my central aims is to convey how a computational way of thinking about the infant can support a cognitive treatment of action, and a pragmatic treatment of cognition and of the knowledge of the world with which cognitivist positions generally associate it. An equally relevant aim is that of trying to make the exchange between developmental psychology and cognitive science a two-way process. My analysis is motivated by the assumption that the developmental psychology of infancy can offer unique insights for cognitive science and a way of evaluating its explanatory possibilities. Overall, this book is an exploration of what a computational developmental psychology or a developmental cognitive science might look like.

Given the potential scope of this endeavour, the material featured is necessarily selective. As far as computational work is concerned, I have concentrated on discussing foundational ideas from cognitive science, and on examples of programmed models whose underlying organizational principles have found a significant and enduring place in the development of the discipline. In the same vein, infancy material is featured that is representative of types of data and styles of interpretation that have proven significant and continue to be so. On neither of these fronts is there any overall consensus concerning the optimal concepts and style of explanation for handling psychological phenomena, which contributes greatly to the interest of seeing what can be done by trying to bring the approaches together. I have, therefore, chosen to concentrate on areas and issues that I believe should be of mutual concern to developmental psychologists of infancy and to cognitive scientists, and for which the overall aim of deepening understanding may have the most direct reciprocal benefits. The introductory chapter lays out the general structure of the landscape to which these selection criteria give rise, and to whose exploration the following chapters are dedicated.

Julie C. Rutkowska
School of Cognitive and Computing Sciences
University of Sussex

Acknowledgements

I am grateful to the late Max Clowes for inviting me to join the group at the University of Sussex that became the School of Cognitive and Computing Sciences, where (past and present) colleagues have provided the environment and the enthusiasm that make interdisciplinary ventures such as this book possible. For taking seriously my questions about artificial intelligence, I am indebted to Steve Draper, Geoff Hinton, Tom Khabaza, Aaron Sloman and David Young. I am also indebted to those people whose work, and/or feedback on my own work, has led me to change my mind in significant ways about issues that are featured in this book, in particular: Margaret Boden, Annette Karmiloff-Smith, Frank Keil and Pierre Mounoud. Thanks to Mike Scaife for encouragement to forefront my own ideas, and for helpful suggestions about ways of organizing material. And special thanks to Gerald Gazdar for contributing unlimited unconditional support throughout the duration of this book's preparation.

Early versions of some of the arguments that are developed here have appeared in the following journal articles or book chapters:

'Developmental psychology's contribution to cognitive science', in K. S. Gill (ed.) *Artificial Intelligence for Society*, Chichester: Wiley, 1986.

'Computational models and developmental psychology', in J. C. Rutkowska and C. Crook (eds) *Computers, Cognition and Development*, Chichester: Wiley, 1987.

'Action, connectionism and enaction: a developmental perspective', *AI & Society*, 4 (1990), pp. 96–114.

'Looking for "constraints" in infants' perceptual-cognitive development', *Mind and Language*, 6 (1991), pp. 215–38.

'A computational alternative to the Piagetian infant', *Infancia y Apprendizaje*, 57 (1992), pp. 99–120.

Chapter 1

Infancy, cognitivism and computation

Cognitive science has yet to take any significant account of the past 25 years of research into human infancy – ought it to do so? The psychology of infancy, for its part, has engaged only minimally with the computational concepts that provide a focus for the constituent disciplines of cognitive science – is this inhibiting its progress? In my view, 'yes' should be the answer to both of these questions. 'Should be', rather than simply 'is', because successful *rapprochement* between these two areas of inquiry is currently a hope rather than a certainty. We do not yet know what reciprocal, interdisciplinary possibilites exist for them, nor even if they can be brought together successfully as part of a genuine *developmental cognitive science*. Justifying the attempt to bring them together, and clarifying the kinds of issues and phenomena that such a science will be forced to face, is more straightforward; it is the task of this chapter, which provides a foundation for the exploratory chapters that follow.

Why a computational psychology of infancy?

The new wave of infancy research dates back to the 1960s, the period that saw the crystallization of a cognitive psychology of the adult. Against this background, theorizing about infancy predominantly adopted an orientation that shares some of the general features of the *cognitivist* style of explanation that has been considered characteristic of cognitive science (Haugeland, 1978). The mind became a legitimate focus of study – subjects not only behaved but understood their world, and were active in constructing this understanding by using their knowledge of the world to determine the intake and interpretation of information from the environment. The first major review of the new research phrased the key issue as: 'How does the world of the adult, physical and social, come to be represented in the mind of the child?' (Kessen, Haith and Salapatek, 1970, p. 287).

Since that time, an increasingly impressive array of empirical findings relating to the abilities of newborn and young infants has been assembled (for overviews, see Banks and Salapatek, 1983; Butterworth, 1981a; E. J. Gibson, 1987; E. J. Gibson and Spelke, 1983; Haith and Campos, 1977; Harris, 1983;

Osofsky, 1987; Slater and Bremner, 1989).[1] As Stone, Smith and Murphy's (1974) substantial compilation, *The Competent Infant*, illustrates, there is general agreement that infants are active – not simply reactive – and far more competent than was previously suspected. But there is no theoretical consensus concerning how this early competence is to be understood. Research reached what has been called a 'climax of fact finding' (E. J. Gibson, 1979, p. 148). A major departure has been waning enthusiasm for cognitive questions and explanations. In the 1970s, workers could be described as 'striving for prototheories which utilize internal constructs' (Haith and Campos, 1977, p. 252). Now, even the key issue of representation, noted above, has become controversial with the emergence of ecological psychology, which argues forcefully that notions of cognitive processes and representations are unnecessary to explain either infant abilities or their development (e.g. E. J. Gibson and Spelke, 1983; Wilcox and Katz, 1981).

In *The Computational Infant*, I propose a new form of cognitivist solution to the problem of how we can best conceptualize the nature and development of infant abilities, one which aims to overcome the deficiencies of precomputational cognitivist attempts to explain infancy, such as Piaget's (1953, 1955) seminal constructivist account. An improved explanation will need to do three things: deepen understanding of what infants are doing, and of how and why they do it; bring increased coherence to what seem to be disparate data and theoretical insights; and generate new questions to clarify what kinds of empirical and theoretical directions are worth pursuing.

The standard arguments for the benefits of a computational approach to cognitive psychology in general have been well rehearsed and need be recapped only briefly here. They characterize traditional psychology as adhering too single-mindedly to the correspondence theory of truth (Johnson-Laird, 1983), and overreliant on a *theory-demonstration* methodology (Miller, 1978), with its assumption that plausible theories of psychological mechanisms exist and that subjects' behaviour on appropriately constructed tasks can decide between them. In fact, the constructs of such theories often pose questions (Boden, 1981), or are simply redescriptions of essentially the same character as their accounts of what they purport to explain (Pylyshyn, 1984). Even when empirical observations appear to confirm a particular theory, the psychological processes that the theory proposes may not actually 'do' anything at all in the absence of a sympathetic audience adding many unspecified, missing assumptions (Sloman, 1978). Notably, Piaget and Vygotsky, both seminal developmental thinkers, provide two of Johnson-Laird's (1983) three examples of theories whose empirical 'predictions' follow only with the assistance of a good deal of the reader's intervening intuition.

In contrast, the computational perspective of cognitive science and artificial intelligence (AI) starts from the belief that it is very hard to construct psychological theories – or, indeed, that there are no plausible candidates yet. Its primary commitment is to the coherence theory of truth, and it endorses attempts to articulate theories in terms of working models and effective procedures (Johnson-Laird, 1983). Unlike virtually any precomputational theory, this *theory-development* methodology holds out the promise of a *sufficient explanation*: one which proposes a set of mechanisms that are physically possible; states the rules governing interactions between these mechanisms; and shows that these interactions will perform the function in question (Miller, 1978).

A good deal in this general line of argument can (and in the following sections of this chapter will) be mapped onto the prevailing correspondence-demonstration orientation of mainstream infancy research, whose often startling data are not matched by compelling theory. But there remains a considerable gap between the 'evangelical', PR side of computational thinking and its reality. As far as the psychology of infancy *in particular* is concerned, the explanatory possibilities of grounding theory in a computational approach are unclear. Not only is there very little by way of a computational psychology of infancy, there are not even ready-made answers about how to proceed.

For example, what has been called *genetic AI*, by analogy with Piaget's genetic epistemology, attempts to provide a detailed understanding of knowledge by specifying the changing computational mechanisms that are assumed to underlie it. This perspective has made contact with infancy through Drescher's (1986) attempt to translate Piaget into the high-level programming language LISP, outlining a possible 'schema mechanism' for some of Piaget's (1953, 1955) infancy data. Welcome as this attempt most certainly is, it seems remarkably *ad hoc* and unprincipled. Why the particular computational ideas invoked are employed is not justified: they do not seem to be constrained by prevailing computational thinking about how sensory and/or motor processes are likely to be organized; and it is not self-evident that translation into LISP is any more interesting than translation into Chinese or Italian. Also, the endeavour focusses solely on Piaget's observations, specifically dismissing the relevance of contemporary work and thus failing to engage with the numerous issues raised by demonstrations of early competence.

The traditional AI route to establishing a theory's sufficiency has been to model it in a working computer program, supposedly escaping the experimental psychologist's need for replication and statistical analysis (Newell and Simon, 1976). But, as far as infancy is concerned, it would be naive to equate a computational approach with writing and running programs (and there are now important arguments that the same conclusion should apply within AI itself (e.g. Marr, 1977)).

The traditional analogy between program–computer and mind–brain relations on which this alternative to behavioural techniques of theory validation is based seems straightforward enough. The computer's program, which determines the order and/or circumstances in which its operations are performed, can be thought of as a *virtual machine*: it 'models' or 'imitates' a set of relationships that may be quite unlike those expressed in the operation of the physical machine in which it is implemented, as evidenced by the possibility of running the same program on a range of computers with different physical structures. Cognitive processes can be conceptualized at the same computational level as the program. Therefore, the computations underlying cognition can be adequately modelled by a computer program, despite physical differences between the (brain) 'wetware' and (computer) 'hardware' that normally support them.

The scarce examples of developmentalists' adopting this approach to test theories of infant cognition have been interesting, instructive and disappointing (Rutkowska, 1987). They make the traditional view that a single successful run of a program can establish a theory's sufficiency look remarkably quaint.[2] For example, Luger, Bower and Wishart's (1983) programmed model of infant performance on a number of visual tracking tasks improves little on the original verbally stated theory (Bower, 1974, 1982). The theory's claim that behaviour is mediated by a series of conceptual rules appears superficial because the rules look like mere redescriptions of infants' general behaviour patterns, or what Pylyshyn (1984) calls 'generalizations stated over properties of the represented domain'. For instance, infants who look for objects in places or on paths of movement are attributed with a conceptual rule that defines objects in terms of places or paths of movement. Expressed in the logic programming language PROLOG, the contribution of these conceptual rules to generating 'behaviour' remains obscure. Theoretically important details of how the computational model works cannot be distinguished from the account of the program and the AI language in which it is implemented; active, processing aspects of the program's operation seem to rely on general PROLOG processes, rather than on processes that have been formulated explicitly in the theory and subsequently encoded in the program.

In view of the current status of computational psychology, it is likely that our understanding of most areas is, as yet, too incomplete for programming to offer great benefits. Certainly, there are plenty of programs in AI itself that appear to do, at some level, things that infants do. But grafting these together would yield only a Rube Goldberg/Heath Robinson creation. It would not give us a *principled*, internally coherent computational framework for infancy any more than theoretical advance could be achieved through the *ad hoc* eclecticism of merging or juxtaposing concepts from opposing psychological theories into a 'comprehensive' account of the infant.

At present, I believe, it is more important to relate developmental and computational approaches at a conceptual level. Emphasis should be on *foundational* ideas from cognitive science, and on AI systems that illustrate *robust* ideas from the discipline about ways in which computation might be organized, having proven their worth across time and over a range of systems, not on the details of many one-off programmed models that will be rapidly superseded. Notions about ways in which programs can be structured may prove more useful as a source of ideas to further understanding than as formalisms for actually writing programs.

This conceptual emphasis locates my starting point for a computational psychology of infancy at the weaker end of Newell's (1970) hierarchy of relationships that might hold between AI and cognitive psychology, where computational ideas are said to provide: a *metaphor* to aid *attention focussing*; a form of mental hygiene by forcing *operationality* of fuzzy or mentalistic theoretical constructs; and an *appropriate language* for discussing processes and behaviour. The use of selected AI systems as *base* (*ideal*) models against which to compare infant performance will illustrate a stronger relationship from Newell's hierarchy. For those infancy phenomena where there are no particularly relevant principled, robust models suitable for such a comparison, the next best approach will be to begin formulating outline process models in computational terms. And, although I am not advocating programming, bearing in mind the goal of *sufficiency analysis* helps to highlight vagueness and keep in mind the importance of trying to express ideas in terms of precisely defined structures and processes.

'Appropriate language' proves particularly important because many unresolved theoretical conflicts in the infancy literature stem from the kind of everyday language in which they are posed: do infants see objects as we do, or in a qualitatively different way? To what extent does infants' knowledge of the world depend upon action rather than upon perceptual, conceptual or representational abilities? When young infants exhibit what looks like appropriate behaviour, do they really understand what they are doing? At what point in development are infants' activities controlled by intentions and goals? Do infants have expectations about how the world works, and are they capable of anticipating events that occur within it? Is there a radical reconstruction of psychological processes during infancy, with the development of concepts and representations of the world? Does representation develop during the first or second year of life, or do even newborn infants have the capacity to represent objects and events?

Developmental psychologists generally accept that such questions can be asked and that there are answers to them, but I am going to dispute this because I believe that such questions are unlikely to be answered. It is not just that they are hard to answer, or that it is difficult to discover the answers – the vocabulary used to express them is simply not finely tuned enough to support a scientific analysis of infancy. Not only is it too vague to map unambiguously onto empirical findings, it is also potentially misleading in establishing what may be inappropriate dichotomies or choices, and it is likely to obscure important properties of the phenomena we wish to explain.

Trying to apply this kind of computational approach to infancy proves equally important for computational cognitivism. Critiques of its success in explaining the abilities of naturally intelligent systems continue to accumulate, some of them explicitly incorporating the author's intuitions about human infants – though notably lacking relevant psychological data or insights. For example, Dreyfus's (e.g. 1981) attack on AI systems that model performance in microworlds argues that they are unable to approach the flexibility even of a 6-month-old child. These computational systems, he claims, 'may in no way resemble general-purpose human and animal intelligence' (1981, p. 173).

A similar perspective emerges from Winograd and Flores's (1986) desire to break away from AI's rationalistic model of human ability, where everything needed for task solution is attributed to the subject in the form of an exhaustive representation of the world. They note that a typical AI explanation of a neonate's ability to get food would probably include: a set of goals such as 'drink milk'; a set of operators such as 'cry' and 'suck'; a model of the world setting these in correspondence; and possibly even a model of the caregiver's goals and plans, with the infant's cry conceptualized as an attempt to activate an appropriate plan on their part. Yet infants, they assert, do not really know anything – their abilities do not involve representations, planning or analysis.

Putting this state of affairs together with the fact that cognitive science itself now features internecine conflicts concerning the nature of computational explanation gets us into a particularly interesting situation. We can exploit this opportunity for some back-to-basics theorizing in both infancy and cognitive science. The aim becomes to assess whether or not a computational psychology is possible; and, if so, what form it should take. This sets up the possibility of validating a computational approach by seeing which of the available concepts prove useful in an explanation of infancy.

Three overarching issues chart the common territory that this approach establishes for infancy and cognitive science, and they form an approximate hierarchy, albeit with considerable overlap between them. Most fundamental is the need to clarify the *mechanisms* that could underlie the infant's abilities. Decisions about mechanisms have implications for the issue of *intentionality*, of how we should view the psychological status of the infant. And decisions about intentionality impose restrictions on the forms that *development* might take.

Mechanisms

The mechanism question essentially asks *how* infant abilities are possible, but 'what' we think our subjects are doing is intricately bound up with 'how' we believe they are doing it. To draw on a familiar example from AI, we would be unlikely to describe the performance of a computer as reasoning, even if it succeeded on a behavioural Turing-test, if we discovered that its program was no more than a gigantic look-up table listing independent responses to each and every question that it might be asked.

Because the mechanisms proposed in mainstream infancy theorizing are often underspecified, its 'what–how' mapping is generally poor. The same theoretical terminology is used with sometimes glaringly different meanings, so there is no real agreement, beyond a superficial level, on 'what' infants are doing in the first place. Numerous debates feature the key notions of 'perception', 'action' and 'representation', without there being any real consensus on what these terms mean.

Within developmental psychology, Piaget's (1953, 1955) work has been taken as the prototype for cognitivist approaches to the relevant issues. His questions and concepts still provide the background against which numerous researchers interpret the meaning of their work, even when they reject his view of the infant: devoid of psychological organization; relying on action to bring order and meaning to chaotic sensory experience; and taking the first 18 months or so of life to differentiate the self and action from the world through the construction of a notion of object permanence.

Since Fantz's (1958, 1965, 1966) seminal studies of infants' 'visual preferences' demonstrated what he originally believed was an innate sensitivity to pattern, few have concurred with Piaget's assumption that infant perception consists of literal, static, unrelated images of sensory inputs which 'only constitute spots which appear, move, and disappear without solidity or volume' (Piaget, 1953, pp. 64–5).[3] However, consensus *that* early organization exists does not preclude the continuing coexistence of distinct approaches to what the nature and role of that organization is.

Each development in neuroanatomically or neurophysiologically grounded theories of perception has tended to produce a new explanation of the infant's attention and discrimination. These range from invocation of the notion of dual visual systems (e.g. Bronson, 1974, 1982), together with the well-known findings of cortical cells that are thought to act as feature analyzers (Hubel and Wiesel, 1959), to the infant visual system's capacity for performing spatial frequency analysis on the image, with its associated linear systems theory (Banks and Ginsburg, 1985; Banks and Salapatek, 1981; Gayl, Roberts and Werner, 1983).[4]

It would be reassuring to think of such neurally informed explanations as compatible with claims about the psychological organization of infant perception, but there are significant cases where they continue to be presented as alternatives. The long-running debate over face perception provides a good example of how big the gap between them can be.

Exploiting data from infants' visual following, rather than visual preference for stationary paired stimuli, Johnson (1988, 1990) and Morton and Johnson (1991) conclude that human infants possess two modular mechanisms dedicated to perception of the human face. CONSPEC is thought to be innate and subcortically mediated; to predispose infants to attend to roughly facelike stimuli (for example three high-contrast blobs on an oval ground); and to drop out after it has established the appropriate learning space for CONLERN. CONLERN is thought to be cortically mediated and devoted to learning the details of specific faces. This distinction is used to account for why faces have not always been preferred in earlier studies, and why stimuli need to be increasingly facelike to attract attention as the infant matures.

Banks and Ginsburg (1985), on the other hand, argue that a 'sensory' interpretation of the neonatal data in terms of their linear systems model is perfectly sufficient. In support, Kleiner and Banks (1987) demonstrate that neonates 'prefer' stimuli that share the amplitude spectrum of a schematic human face to stimuli that share its phase spectrum and look far more facelike to adult observers; the latter stimuli are 'preferred' by 2-month-old infants. Thus, they reject the core Johnson–Morton claim that the data require a 'social' interpretation in terms of innate face knowledge.

If such conflicts are to be resolved, some way of relating neurally and psychologically grounded accounts of mechanisms is going to be needed. It seems unlikely that any 'sensory' interpretation could account for the full range of phenomena supporting claims that the neonate's vision provides – *at some level* – an appreciation of many of the more general properties of objects. For example, some of the most striking studies suggest that by about 3 months of age or even younger infants already segment scenes composed of stationary and/or moving entities in many ways as adults do (reviews of their own and their colleagues' work are provided by Bower (1974, 1979, 1982) and Spelke (1982, 1985, 1988, 1990)). In particular, the three-dimensional properties of a scene have a significant impact on how it is visually segmented. Though there are certainly important differences between adult and infant performance, as we shall see when perception is discussed further in the following chapters, the findings appear to support Spelke's (e.g. 1985, 1990) conclusion that infants start out by perceiving scenes in terms of *unitary*, *bounded*, *persisting objects*.

The main contender for an overall theory of infant perception has been ecological psychology, with its foundations in Gibson's (1950, 1968, 1979) realist theory of *direct perception*. It challenges not only Piaget's uncompromisingly deconstructed view of the infant but also all cognitivist theories that assume that the input does not provide a sufficient basis for an organism's knowing its environment; and the organism provides the necessary 'embellishment' (Michaels and Carello, 1981).[5]

The foundational claim of ecological psychology hopes to redefine the task of the infancy theorist. It assumes that a sufficient basis for perception is provided by unambiguous *information* that is 'available', in the environment, in the form of *invariant* relations and transformations in the structure of the ambient optic array surrounding the subject's position. We should view infants as 'preattuned' to 'pick-up' information that uniquely specifies the objects, events and surface layouts that structure light so as to 'broadcast' that information. And 'information pick-up' should be regarded as a primitive notion that is not amenable to psychologically meaningful decomposition in terms of the kind of interacting 'information processing' components that cognitivists have proposed. So, the psychologist's task should be the discovery of what things infants perceive and of what information specifies those things (E. J. Gibson and Spelke, 1983). Elsewhere, I have evaluated this position in detail (Rutkowksa, 1988), but here I shall only note briefly that its rejection of the mechanism issue is far from compelling.

A notion like 'tuning', as Scriven (1977) points out, cannot provide either testable content or useful insights unless it is considerably clarified. Too often, however, perception is said to be unproblematic for human infants and adults because the 'biological design' and 'optimal tuning' of the brain enable information to 'resonate' with 'highly evolved neural structures' in 'an elegant and simple way' (e.g. Olson, Yonas and Cooper, 1980; Shaw and Bransford, 1977). But this simply confounds levels of explanation and has no genuinely explanatory status. At best it is a plea to the inadequate criterion of *material instantiation* (Fodor, 1968), similar to a programmer who can get a computer to perform some function but who can explain how it does so only by pointing to the existence of the working program: the machine exists, and it works, but how?

In general, the direct perception approach has found far greater credence in theoretical accounts of infancy than it has in cognitive science, probably because its nativist stance gives the illusion of fitting the infancy data so well, whereas Piagetian theory most certainly does not. On the one hand, I agree with the cognitive science assessment by seeing this enthusiasm as largely misplaced. Disproving Piaget cannot, by default, establish the validity of the Gibsonian alternative – evidence of preadapted ability greatly underdetermines what mechanisms underlie it. On the other hand, in ecological psychology's favour is its concern with the nature, role and validity of *action* in subjects' understanding of the world. Infancy research highlights the fact that perception has evolved to exploit the world for action. So, from the infancy perspective, one is inevitably going to pay greater heed than most of cognitive science at least to certain metatheoretical aspects of this framework.

The infancy literature now contains increasing evidence for the *functional coordination* of perceptual and behavioural components directed at objects, people and events (cf. Bruner and Koslowski, 1972; Butterworth, 1981a, 1989; Jones, Spelke and Alley, 1985; Mounoud and Vinter, 1981; von Hofsten, 1980). The specificity of perceptual and behavioural components is seen in both neonatal activities and those that emerge predictably during the course of infancy, and their selective coordination appears to be established prior to the kind of feedback from practical experience that Piaget's traditional theory deemed necessary.

A key example is provided by prehension and manipulation of objects. Neonates show a form of visually elicited reaching which is sometimes accompanied by grasping (Alt, summarized in Trevarthen, 1984; Bower, 1972; Bower, Broughton and Moore, 1970b; von Hofsten, 1982, 1984). This 'neonatal reaching' or 'prereaching' (Trevarthen, 1974; von Hofsten, 1982, 1984) generally declines, to be succeeded at around 5 months by a new emergence of prehension, in which a wider range of manipulatory behaviours occur. In both forms, several specific properties of 'real' objects have been claimed to influence the behaviours observed, such as distance, solidity, size and movement (Bower, 1972; Bruner, 1968; Bruner and Koslowski, 1972; von Hofsten, 1977, 1980; von Hofsten and Spelke, 1985; Yonas, Granrud and Pettersen, 1985).

This functional preadaptation extends to the social domain. When faced with people, very young infants have been shown to engage in what have been described as attempts at communication: repeated patterns of facial and body gestures, and distinctive mouth and tongue movements (Brazelton, Koslowski and Main, 1974; Legerstee, 1992; Stern, 1985; Trevarthen, 1974, 1975, 1980, 1984; Trevarthen, Hubley and Sheeran, 1975). Trevarthen terms these 'prespeech' and contrasts them with the 'prereaching' behaviours characteristically shown to inanimate physical objects.

Further studies reveal functional responses to events involving the infant. For example, some studies have found what could be seen as defensive behaviour towards an approaching object on a collision course in young infants. Backward head rotation occurs, which may be accompanied by the hands being raised in front of the face (Ball and Tronick, 1971; Ball and Vurpillot, 1976; Bower, Broughton and Moore, 1970a). The developmental pattern of performance on the well-known visual cliff proves to be complex, as we shall later see, but there certainly are conditions that support replication of the original finding that crawling infants will cross to their caregivers only across the 'shallow' side (E. J. Gibson and Walk, 1960). And, at the onset of sitting and standing, infants lose their stability if the walls and ceiling of the room surrounding them are moved. The specificity of the direction of fall (backwards if the room appears to move towards the infant; forwards if it appears to move away), suggests that the infants generally exploit vision to monitor their position and are attempting to compensate for a perceived – albeit unreal – loss of balance, not suffering from shock evoked by the room suddenly moving (Butterworth and Cicchetti, 1978; Butterworth and Hicks, 1977; Lee and Aronson, 1974).

To the extent that certain behaviours are linked from the outset with visually provided information, the (Piagetian) problem of how vision and behaviour get linked up in the first place would seem to be eliminated. 'Reciprocal assimilation' of perceptual and behavioural schemes may not be the foundation for the earliest organization, which is no bad thing since it is difficult to make the notion carry the theoretical burden that Piaget places upon it (Rutkowska, 1988, 1992a). It seems that prior motor activity is inessential for initial perceptual organization (e.g. Butterworth, 1981a; E. J. Gibson and Spelke, 1983). But this conclusion leaves open the equally significant question of how such selective coordination might work – a question that is vital for phrasing developmental issues.

Piaget's view that behaviour reveals control by underlying sensorimotor *schemes* is notorious for its insufficiency, and has contributed immensely to a jaundiced view of the explanatory power of 'action'. The scheme notion is vague; is handled inconsistently by Piaget; and says remarkably little about the sensory-perceptual contribution to sensory-motor knowing (Rutkowska, 1988, 1992a). This has led many to reject the notion as circular and mentalistic (e.g. Wilcox and Katz, 1981), or to equate Piaget's view of action simplistically with overt behaviour or motor response (e.g. Brainerd, 1978; Kagan, 1972). While this is probably an unfair reading of what Piaget 'meant', it is a reasonable interpretation of what he 'said', since the nature of schemes is never spelt out beyond description of the behaviours they are assumed to control, such as looking or grasping.

Ecological psychology highlights perception's use in action, but developmentalists subscribing to direct perception theory tend to be satisfied with the conclusion that infants perceive objects accurately in terms of the adaptive activities they afford, not in terms of meaningless physical properties, and 'coordinate their actions to realize those affordances' (Jones, Spelke and Alley, 1985, p. 262).

The notion of direct pick-up of 'affordances' serves only to confirm doubts about the explanatory power of direct perception theory. Whether or not activity plays a role in the initial organization of perception, how far can prehension be explained by saying that infants 'just see' that an object is solid or graspable? Failure to cross the visual cliff has been attributed to detection of an invariant specifying a 'falling-off place' (Gibson, 1979) and to absence of an invariant specifying a supporting surface (Rader, Bausano and Richards, 1980) – distinct yet behaviourally indistinguishable proposals. The claim that direct perception of the appropriate meshing between the nature, direction, timing and intensity of one person's actions and those of another underlies immediate awareness of other persons, the interpersonal self and the relationship between them (Neisser, 1988) appears only to explain away the early communication findings and the fraught notion of 'intersubjectivity'. And the transformational invariant of 'looming' (symmetrical expansion of a closed optical contour centered in the midpoint of the visual field) may accompany an object's approach (Schiff, Caviness and Gibson, 1962), but how adequately can it explain infants' avoidance behaviour?

Ecological psychologists go further by suggesting that the appropriate level of analysis for a psychological account of perceiving and knowing the world should be in terms of naturalistic or causal *ecological laws* relating environmental 'affordances' to the subject's 'effectivities' (Turvey, Shaw, Reed and Mace, 1981). This leaves the explanation of selective perceptual-behavioural coordination with notions like 'attention' (Gibson, 1979), or a relation of 'mutual constraint' or 'dual complementation' between affordances and effectivities (Turvey, Shaw, Reed and Mace, 1981). While capturing the intimate match between behaviour and environment, the extent to which such notions are explanatory remains debatable. The notion of 'action', receiving no independent definition, becomes identified with the subject's effectivities, and comes close to looking like Piaget's seeming restriction of action to the behavioural-motor side of activity.

Despite their differences, the positions considered so far concur in separating the infant's perceptual-behavioural achievements from the arena of cognitive-representational activity. More intriguing, to my mind, is the direction taken by (scarcer) hybrid positions which do not partition ability in this way (such as those viewing infants' developing ability in terms of modes of action (Trevarthen, 1974, 1980); conceptual rules for controlling behaviour (Bower, 1974, 1979, 1982; Wishart and Bower, 1985); or distinctive coding capacities for representing experience (Mounoud, 1981, 1982, 1986)). For example, Bruner's (1968, 1969, 1970, 1973) account of *skilled action* anticipates, in precomputational form, two general themes that are sympathetic to my version of a computational perspective.

One is the essential compatibility of action-based and cognitive views of the infant. Bruner attributes a highly cognitive status to the sensory-motor skills that he considers the foundation of infant competence. The organization of activities such as object manipulation, he suggests, is best seen as a kind of problem solving: preadapted constituent behaviours are selected, ordered and executed to attain predetermined objectives. Equally important is Bruner's emphasis on the line of continuity between infant and adult activity, with the skilled action of the infant seen as based on internal 'competence' that is the same as or *homologous* to that underlying adult representational abilities such as language, thought and problem solving. Though they serve different functions, important parallels are to be found between the ordering principles, strategies or rules for combining constituents that he believes underlie their flexibility. He appositely notes that usefully precise ways of specifying deep parallels are not going to be easy to come by.

These proposals contrast with the mainstream infancy treatment of representation, where Piaget's philosophically motivated assumptions dominate. These consider the ultimate goal of infant development to be liberation from the limitations imposed by perception and action, through a radical reconstruction of action to an 'objective', conceptual-representational level of functioning. 'Representation' is restricted to *substitution*, and is grounded in a circular operational definition: the existence of re-presentation is demonstrated by phenomena that are deemed to require re-presentation for their explanation; that is, by behaviour that suggests some appreciation of an 'out of sight' object/event, suggesting that it is not 'out of mind' for the infant.

Examples include the dramatic findings of infants as young as 0.7 hours of age imitating gestures such as mouth opening and tongue protrusion of an adult model (Maratos, 1982; Meltzoff, 1981; Meltzoff and Moore, 1977, 1983, 1985), with special attention being drawn to the temporal delay that exists between presentation of the model and the infant's post-test response to it (Meltzoff, e.g. 1981). Evidence that young infants appear to perceive an object as continuing behind another which partially occludes it is also pertinent (Kellman and Spelke, 1983), as is the finding that by 5 to 6 months of age infants who have habituated to a screen which rotates back and forth will dishabituate when it rotates back after an object has moved behind it without emerging (Baillargeon, 1986; Baillargeon, Spelke and Wasserman, 1985). And, although infants in at least the first half-year of life generally appear untroubled by changes in the features of an object that moves out of view then reappears (Bower, Broughton and Moore, 1971; Gratch, 1982; Muller and Aslin, 1978), by a few weeks of age they will sometimes anticipate a moving object's reappearance by turning to look to the opposite side of a screen behind which it moves from sight (e.g. Bower, Broughton and Moore, 1971; Gardner, 1971).

The most important aspect of such findings is generally taken to be their apparent demonstration of representational ability *earlier* than Piaget's theory allows. Typical conclusions are that the capacity to represent absent objects and events is the very starting point for infant development, not its end-point (Meltzoff, 1981); by 6 months of age infants believe that unseen objects continue to exist (Baillargeon, 1986; Bower, 1974, 1982); even the youngest infants have an 'object concept' (Spelke, 1985, 1990); and by 9 to 10 months of age infants develop cognitive 'hypotheses' and can 'think' (Kagan, 1972). While the timing issue is important for assessment of the Piagetian framework, from the mechanism perspective it remains crucial to ask: what is 'representation'? How does it work? How is it related to the kinds of phenomena generally associated with 'perception' and 'action'? And what are the developmental implications of answers to these questions?.

My route into these issues will forefront three dimensions. First, the perspective that I adopt is unashamedly *functionalist* in claiming and emphasizing that the mind is for acting in the world.[6] Next, it aims at a reinstatement of the explanatory power of action, considering infant abilities and their development to be best understood as having their basis in *action systems*. Finally, it argues that a *computational* account of action systems can supersede other treatments of action, rendering obsolete many disputes about the import of action and supporting a unified perspective on abilities in the first year of life.

Just as developmental psychology lacks consensus over the explanation of infant ability, cognitive science currently features a number of debates about what precisely ought to be meant by a 'computational' explanation. My focus will be on exploring how far the possibilities offered by what have recently come to be called 'classical' computational concepts of the *physical symbol system hypothesis* (PSSH) can take the analysis of infancy. I do not intend to exclude other approaches, such as connectionism, from my definition of computation, but, for infancy purposes, I do believe that they prove to be further away from offering a new paradigm for computational psychology. From a psychological perspective, PSSH remains attractive for a number of reasons which are also relevant to considering its (frequently controversial) fit to more recent approaches.

The major force of PSSH is that humans (and computers) are members of a class of physical systems that are capable of having and manipulating symbols that can correspond to aspects of their internal and external environments (Newell, 1980; Newell and Simon, 1976). Such a system's 'intelligent' functions are based on computation, which is generally defined as rule-governed symbol manipulation (e.g. Sloman, 1982). PSSH thus purports to offer an account of the processes that underlie our subjects' understanding of their world, and a valid alternative to the inappropriate quest for nomological explanations in the form of lawful, causal relations that predominates in the physical sciences. The PSSH style of explanation can be described as *systematic* (Haugeland, 1978). It is a type of morphological explanation in which the phenomena to be explained result from properties of the system's functional components, their structure and, most distinctively, the *organized, cooperative interaction* between interdependent components which could not achieve their functional role independently of each other; in fact, one of these components' main properties *is* their ability to interact in this way. In principle, PSSH holds out the possibility of a fine-grained analysis of ability, enabling us to get behind everyday yet mysterious phenomena like 'perceiving an object' or 'having a concept' by clarifying the kinds of symbol structure construction and manipulation that underlie them.

Newell (1980, p. 156) characterizes the general relation between the symbols of a computational system and the entities to which they refer in terms of *designation*, which he considers the most fundamental PSSH concept: 'An entity X designates an entity Y relative to a process P, if, when P takes X as input, its behaviour depends on Y.' The relation between computational symbol structures and processes clearly has non-trivial implications for how this notion works in the PSSH context, and for what it can do, and these will be taken up in chapter 4's discussion of information. In the current context, it is important only to focus on how this basic idea opens up a less restrictive perspective on representation.

As far as PSSH is concerned, representation is generally thought of as another way of looking at symbol structures that designate. Here the definitions of 'representation' and 'symbol' are internal to PSSH and distinct from their meanings in most infancy-related theorizing. In Piagetian theory, for example, the equation of representation with substitution is based on a definition of a symbol as a material event (a word, image or imitative gesture) which is differentiated from, and takes the place of, some other event that is known but not present. In contrast with this, computational symbols are more appropriately seen as the building blocks for many kinds of intelligent system. Symbols can be input to many kinds of process, and the implications of a particular symbol structure cannot be understood without knowing with what process(es) it is associated – that is, understanding its role in the *system* of which it is a component. In the general case, their representational role comes from mediating the subject's *selective correspondence* with the world, and a wide range of such correspondences is possible. This subsumes the kinds of correspondences established by a perceiving-behaving system or by a system whose performance is conceptual-representational in Piaget's 'strict' sense. This notion of representation also ties in with the subject's 'knowledge' of the world, since the selective correspondences that it mediates permit the subject to interact with or 'know' certain aspects of the world and not others; and this has the potential to support a notion of knowledge as an active process of computation, in place of the reified, molar knowledge structures so often associated with cognitivist explanations.[7]

The most widely felt disputes concerning the status and relative utility of PSSH feature the *parallel distributed processing* (PDP) dominated ideas of the new *connectionism*. Much of the force of connectionist challenges to PSSH derives from the idea that the collections of massively interconnected simple units that make up a PDP system function at a *subsymbolic* level and cannot be conceptualized appropriately in terms of symbol manipulation (Smolensky, 1987, 1988). In place of PSSH's concern with programmed rules and with symbols, PDP explanations concentrate on networks of units and their behaviour: the activation values of units; connection strengths and activation passing; and the sequence of patterns of activation into which the network as a whole settles (e.g. Rumelhart and McClelland, 1986; Smolensky, 1987, 1988). Rule-like behaviour is said to emerge from specific patterns of activity recurring within a network.

Connectionist networks have successfully implemented a number of functions that look relevant to understanding infancy, with their outstanding success being the performance exhibited by trained but unprogrammed self-organizing networks in pattern learning and classification.[8] In Rumelhart and McClelland's (1986) view, connectionism is uniquely well suited to account for how infants could have come to have innate knowledge, though agnostic with regard to innate/acquired controversies. A system's memory can be thought of in terms of functionally persistent trained networks, so any behaviours that prove useful in the course of human evolution can simply be built into the pattern of prewired interconnections between the infant's input units and their connections to the output units that control effector activity. Rumelhart and McClelland consider it unlikcly that rules of the kind proposed by cognitivist accounts of linguistic competence could 'get into the head' of the newborn, and 'implausible that the newborn possesses elaborate symbol systems and the systems for interpreting them required to put these explicit, inaccessible rules to use in guiding behaviour' (1986, p. 142).

As Fodor and Pylyshyn (1988) point out, making a priori assumptions about what newborns do and do not seem likely to possess is an ineffectual way of doing cognitive psychology. The subsymbolic focus of PDP may make it look more congenial than PSSH as far as explanations of early infancy are concerned, but the difficulty of seeing how PSSH could help to clarify the content and functional organization of infant knowledge probably owes more to the way the notions of 'symbol' and 'symbol manipulation' have been (ab)used by AI than to the potential of the concepts themselves (Rutkowska, 1986, 1990).

The connectionist emphasis on explanations that are fine-grained, forefront processes and get behind the molar phenomena of cognition is certainly welcome, but we have to note that these properties equally describe what PSSH promised for computational psychology. In practice, the outcome of a good deal of mainstream AI work that aimed at immediate implementation was very different, and its limitations for psychology are evident without connectionist prompting. 'Symbols' became far removed from the fine-grained building blocks that one was led to expect. Often they simply mirrored the very terms of common-sense 'folk psychology' that they were supposed to help unravel. A typical example comes from applications of PSSH to expert systems, where Partridge (1987, p. 107) notes that the symbol structures employed 'tend to be objects such as propositions – e.g. JOHN THINKS THAT HE IS SICK'. Unlike subsymbolic connectionist units, these kinds of PSSH symbols imply 'something that we can put a conceptually meaningful label on, such as "apple", "dog", etc.'. However, not all AI has followed this route, a significant exception being computational work on vision that will be featured in chapter 2, and there is no necessity for the conceptual use of PSSH notions in relation to infancy to do so.

In conciliatory mode, some theorists consider PSSH notions of symbol structures and processes to offer a *macro-theory* (Rumelhart and McClelland, 1986) or rough approximation (Smolensky, 1987, 1988) to the performance of a PDP system. This is broadly compatible with seeing symbolic functioning as an emergent property of connectionist networks, as some PSSH theorists consider the phenomena described by the everyday language of folk psychology to emerge from symbol structure manipulation. Interestingly, the kinds of process-oriented (procedural) classical concepts that prove useful in analyzing infant action, 'pattern matching' for example, turn out to be ones that are increasingly being called upon in connectionist implementations of ability.

At the other extreme is the view that neither everyday terminology nor classical computational notions can provide a valid or useful explanation of cognition. For example, Churchland (e.g. 1986a, 1986b, 1989b) opts for a radical (eliminativist) *neurocomputational* approach whose explanations are to be couched in terms of connectionist notions that he attempts to map directly onto the neuroanatomical mechanisms underlying behaviour. Many of his examples feature sensory-motor abilities that are also prominent in the infant's repertoire, providing a ready-made opportunity for some pertinent mappings between infancy data and different styles of computational analysis. Significantly, such abilities are increasingly becoming the focus of computational explorations of intelligence, and provide a basis for a number of approaches that reject the necessity of *any* notion of symbol manipulation, emergent or otherwise. For example, from the perspective of recent work in 'mobotics' (i.e. with mobile robots), Brooks (1991b) argues that much AI, preoccupied as it is with supposed components of the 'central' system (e.g. planning, knowledge representation, or learning), suffers from self-deception about its 'successes'. These rely on factoring out all aspects of perceptual and motor skills, which are the hard problems solved by intelligent systems, and solutions to them will seriously constrain solutions to the remaining components of natural intelligence.

In general, dichotomies like 'PSSH versus PDP' are likely to turn out to be simplistic and misleading. For example, it may prove to be the case that the kinds of massively parallel machines stressed by the PDP style of computation will be necessary for effective implementation of some aspects of psychological theories. But, as Minsky and Papert (1988) suggest, a single computational system may turn out to combine properties from 'serial versus parallel', 'hierarchical versus heterarchical' and 'localized versus distributed' dichotomies, and it would be misguided to think of them as supporting a principled distinction between PSSH and PDP.

The primary task for *any* computational analysis of infant action is going to be to determine the functional decomposition of the system's components, their properties, and how they are organized and work together. Provided the pitfalls of past AI treatment of 'symbols' and 'representation' are avoided, 'classical' computational notions should prove best suited to cast light on the *complete* range of phenomena and questions with which the analysis will have to be concerned.

Intentionality

During no other period of development is there a more intimate link between psychological processes and what some might dismiss as 'merely' functional behaviour or motor skill. This gives infancy a special status in the attempt to clarify what it takes – and what it means – to perceive, act in and understand the world. And, under various terminological guises, it raises the (frequently unacknowledged) problem of *intentionality*. Philosophers consider intentionality to be the property that is uniquely characteristic of the mind of a psychological subject and of actions that are guided by the mind. Intentional phenomena are said to be 'representational' or 'meaningful', to have 'semantic content' in the sense that they are 'directed at' or 'about' objects or states of affairs (Boden, 1972, 1981 ch. 2; Dennett, 1978a, 1983; Haugeland, 1978).

In philosophy, the notion of intentionality has often been limited to the language-using subject and to the special logical properties of linguistic accounts of intentional phenomena. From this starting point, intentionality becomes synonymous with the so-called propositional attitudes ('believing that *x*', 'intending to *x*', etc.), with mental states whose representation is characterized in sentential-propositional terms. From my perspective on infancy, however, I shall be thinking about 'intentionality' in a sense that is less restricted but, I believe, equally valid: to distinguish between non-intentional explanations, which might be couched in the causal language of neurophysiology or reflexes, and intentional explanations, in which it is necessary to invoke mental constructs – not necessarily the linguistically/syntactically represented mental states that typify many philosophical accounts – to make sense of subjects' abilities. This is consistent with the idea that we require a theory that can handle the apparently cognitive activities of infra-verbal infants (Churchland, 1980); and that '*where there is language, there must be meaning*' does not imply that 'where there is meaning, there must be language' (Barwise and Perry, 1984, p. 32).

'Intentionality' casts the problem of interpreting infant performance into especially sharp relief. Typical infancy issues introduced so far in this chapter acknowledge a number of traditional intentional phenomena such as 'expecting', 'intending', 'knowing', 'seeing or recognizing as', 'understanding', 'conceiving of as', 'believing' and 'thinking'. Even among cognitive approaches to infancy, there is (implicit or explicit) disagreement as to whether or not such psychological constructs are appropriate or necessary in explanations of young infants' abilities. The infant of this book may be accused of being a monadic creature (there are limits to the number of degrees of freedom that can be handled), but it is interesting to note that the same issues arise in relation to the infant's appreciation of both physical and social worlds, and they may prove susceptible to a similar treatment. This is despite these two forms of understanding traditionally having been considered to involve quite different orders of intentionality: most of the infant's activities would be seen as candidates for 'zero-order' versus 'first-order' intentional analysis (for example, reflex grasping movements versus 'infant *intends to* grasp X'), while genuinely communicative intentions are assumed to involve at least 'third-order' intentionality (for example, 'infant *believes that* adult *knows that* infant *intends to* grasp X').

Piaget's (1955) commitment to traditional philosophical distinctions leads him unequivocally to deny the existence of intentionality until the infant meets his criteria for conceptual-representational status. Some agree there is no reason for explanations of the infant's earliest competences to invoke more than the organization of the nervous system (Mounoud, 1982) or 'scattered reflexes, or evolutionary vestiges of pre-constructivist intelligence' (Drescher, 1986, p. 377).

Opposed to this is the claim that the attribution of intentionality to even the youngest infants is evidently justified. Bower (1977), for example, asserts that infants are not just 'little machines' that exhibit appropriate behaviour; they 'know', 'believe' and 'think' in a way related to what adults do when they do such things.[9] Functional perceptual-behavioural coordinations, such as prehension and communication (protoconversation), are said to exhibit intention on the part of the infant *ab initio* (Bower, Broughton and Moore, 1970b; Trevarthen,1974), with Trevarthen emphasizing his point with the (redundant) description '*psychological* action'. And psychological states such as desires and intentions are deemed by ecological psychologists to be essential and primitive constructs, their inclusion in scientific accounts of perception and action being 'no less scientific than talking about liver cells' because they are a consequence of evolution (Michaels and Carello, 1981, p. 75).

Such conflicts are not simply due to the changing data-base from Piaget's seminal studies to the present, for the performance of infants in contemporary studies is just as likely to be interpreted as lacking intentionality. A particularly important approach emphasizes the adult's structuring role in the progressive organization of the infant's apparently appropriate behaviours. On this view, young infants are biological rather than psychological beings. Their behaviour is not driven by mental states (for example, crying may be caused by physiological changes but is not initially motivated by the infant 'knowing' that s/he is hungry or 'desiring' food, much less 'intending' to 'communicate' this to an adult in order to summon their help) but it becomes so through adults treating them *as if* it is (e.g. Kaye, 1982; Richards, 1974; Valsiner, 1987). In a related though more complex position, Bruner (1973) talks of 'skilled action' to highlight that 'intention' is vital for even initial sensory-motor coordinations, however philosophically debatable, while simultaneously maintaining the necessity for a developmental path between 'reflexly' or 'instinctually' determined patterned behaviour and intentional action.

For the developmental psychologist of infancy even to begin touching on issues that fall within the scope of intentionality, two things must be made explicit from the outset: is this kind of terminology going to be used in explanations, and, if so, when? And onto what mechanisms, if any, are such terms to be mapped?

The most straightforward way of trying to handle the phenomena would be to adopt the *intentional stance* (Dennett, 1971, 1978a, 1983, 1988). This provisionally decouples the two issues by holding that the pertinent terminology is appropriately used for explanations whenever it has predictive power, no matter what the 'innards' of the system under consideration. The attempt to predict subjects' behaviour should start from an assumption of their 'rationality' and tendency to behave in accordance with the beliefs, desires and intentions that we would expect them to exhibit under their particular circumstances.

Isolated perceptual-behavioural coordinations might be attributed to 'first-order intention' (e.g. Butterworth and Hopkins, 1988). If, however, prediction and the rationality assumption are brought into play, over a range of infancy data, just how much leverage might be gained on the explanation of developmental phenomena appears questionable. For example, it was noted earlier that very young infants exhibit visual search for moving objects that disappear from view behind a screen. Adopting the intentional stance, this could be explained by their 'belief' that the object continues to exist and move while out of sight, together with their 'desire' to re-establish contact with it. (Cf. Bower's (1974, 1982) original view of what the task was testing.) With these beliefs and desires, equal success would be predicted for infants on the ubiquitous manual search tasks, but there appears to be a non-trivial lag between these achievements, even after the requisite motor capacity is in place.

Perhaps infants' beliefs and so forth are being characterized far too inaccurately to achieve predictive success across the situations. The search for greater specificity is unlikely to succeed with a terminology that may simply be too *molar* to offer useful explanatory primitives for the kinds of phenomena with which we are concerned. Developmental psychology characteristically handles behaviours that change over weeks, months or even years. The subtle and complex path involved in the acquisition of search for hidden objects, for example, could only crudely be captured by considering the infant in zero- versus first-order intentionality terms, as 'having' or 'not having' an 'object concept', 'belief' in the object's continued existence and/or an 'intention' to find it.

A more differentiated language for discussing intentional phenomena is needed. Without this, Valsiner's (1987) belief that considering the development of purposiveness and intentionality may get us out of the impasse engendered by either positing or denying their existence in the infant seems overoptimistic. The most plausible route is to locate the discussion of infant intentionality, from the outset, in the context of possible mechanisms underlying changing performance. Dennett (1983) has characterized the intentional stance as offering an 'intermediate descriptive method' (neither behaviourist nor neurophysiological), without entailing particular cognitive mechanisms. He hopes that it can offer a way-station to the further goal of explicating cognitive processes, and that describing behavioural and cognitive competence in this kind of (everyday) language will constrain strongly what underlying mechanisms might be involved. However, it seems to me unlikely that it is going to constrain them adequately or in the right way for infancy purposes.

Turning to mechanisms may also help to tease out some of the theoretical underpinnings of intentional phenomena in general. For instance, Dennett (1978a) himself notes the imprecision of the core notion of 'semantic information', which is the object of many discussions of intentional phenomena and on which the intentional stance is predicated. While ecological psychology addresses this with the controversial notion of 'affordances', cognitively oriented infancy theorists have yet to clarify alternative solutions to the pertinent questions of: *what* information a signal carries, which requires a qualitative answer, in terms of content and structure (Dretske 1981, 1983); *where* it is – to what extent mind-dependent or mind-independent; and *how* some states or events can carry information about others for the infant? More 'sensory', neurophysiologically motivated interpretations of infant performance, in terms such as contour density or spatial frequency analysis, have actually considered their insensitivity to the meaning or significance of stimuli as a theoretical advantage (e.g. Banks and Salapatek, 1981), which leaves open the question of how such inputs ever could attain semantic status for the subject.

Classical computational treatments of the subjective and purposive aspects of ability have relied on the mind–program analogy and on the in-principle separation of virtual and physical machine levels of explanation. The functioning of the program level is susceptible to a teleological or purposive analysis, but the program is implemented in the computer, which is a causal machine. This identifies the attempt to understand intentionality with the exploration of computational mechanisms, and a combination of two of the explanatory stances that it supports offers a useful route into the analysis of intentionality in infancy.

First is the *design stance* (see, for example, Dennett, 1978a). This does not ask whether we should or should not attribute intentionality to infants. Instead it aims to explore what type of mechanism could exhibit the functions with whose presence intentionality has generally been associated. The implications of using the mind–program analogy to analyze intentionality phenomena in the specific case of the infant will depend on what kinds of program structure prove necessary to explain their performance.

The other has been called the *revisionist stance* (Sloman, 1987a), and holds that the knowledge, expectations, intentions and so forth of any intelligent system cannot be treated as primitive constructs. Minsky (1980) refers to such notions as 'prescientific idea germs'. They are, he argues, too coarse to support powerful theories: 'they are not (like milk and sugar) objective things our theories must accept and explain; they are only first steps toward better concepts' (Minsky, 1980, p. 439). It is better to think of them as intuitive, global labels that we apply to the properties of computational systems and their operation.

Allied to this is the *space of possible minds* proposal, which combines behaviourist concern with classifying the types of things that different kinds of behaving systems do with exploration of potential kinds of virtual machine structure(s) (Sloman, 1984). Instead of drawing a line between things with/without minds, thus fostering inevitable boundary disputes due to different decision criteria, this accepts that there are likely to be numerous divisions in the space of possible behaving systems, rather than a continuum or dichotomy. Some of the divisions and distinctions that are so prominent in the macro-language of intentionality disputes may map onto virtual machine distinctions; many may not; and significant novel distinctions may be uncovered.

The key issue raised by this approach in relation to infancy involves the implications for our understanding of intentionality of viewing infant ability in terms of action and of computation. In this, *embodiment* proves to be a key theme.

For the purposes of cognitive modelling by computer, the in-principle separation of program and machine may be essential. For psychological theory, however, the relationship between them is equally vital, if not more so. The notion of an action system's *instantiation* is important for our understanding of the nature and development of abilities and of what underlies them. Unlike the computer used as a modelling tool, a real-world action system is embodied in a specific way, and understanding its actual and potential functioning requires us to take seriously reciprocal constraints that are likely to hold between cognitive and physical-motor mechanisms. A key concern of the analysis of infant abilities must be the extent to which computational cognitivism can take on board and help to develop such ideas. Certainly, the analysis will neither get very far nor go in the right direction if it holds to the traditional cognitivist view that functionalist attempts to understand cognitive processes need not go beyond the program level (e.g. Johnson-Laird, 1983).

An equally important concern becomes definition of the boundaries of computation and cognition, especially with regard to the relation between subject and environment. The embodiment of a naturally intelligent, real-world computational system underwrites its embedding in the environment, which is essential to our understanding of action and of its relation to the mind, biological preadaptations and behaviour. This suggests an important direction for trying to extend the notion of a systematic cognitivist explanation beyond its usual strictly intra-subject cognitive/program focus. To put it crudely: you have to get out of the head to understand the mind.

This emphasis challenges the kind of computational approach that has often contributed to the view that cognitivism offers an implausible framework for psychological and developmental issues (e.g. Shaw and Turvey, 1980; Sinha, 1985). For example, Fodor's influential version of the 'computational theory of mind' (see his 1980) does for 'computation' much what Piaget's 'schemes' have done for 'action'. Seeing no possibility of specifying lawful relations between an organism's mental states and objects in its environment, Fodor insists that a computational system's connections to the world cannot provide a basis for a science of cognition. But this 'methodological solipsism' is contentious, and, as we shall see, not only from the standpoint of infant action.

Emphasizing notions such as action and embodiment might seem to take the exploration of infancy far from this section's primary concern with the mind and its mechanisms, but in actuality it does not. What it does is open up a different approach to understanding what the mind is.

Development

Here and elsewhere, 'development/al' has often been used in a fairly cavalier way to denote phenomena and issues that are of concern to developmental psychologists, even when purely *synchronic* processes are at stake. Serious proposals addressing *diachronic* processes of psychological change are more elusive, although they are self-evidently crucial to a developmental psychology of infancy, and equally so for any attempt to arrive at a non-trivial conceptualization of human ability. Contemporary work in cognitive science acknowledges that a psychologically valid model of mind must be malleable and accommodate the fact that learning and development across time are fundamental human characteristics; it needs to be capable of acquiring its capabilities through development, learning from the environment, coping with new environmental demands and learning by doing (Newell, 1980; Norman, 1980; Pylyshyn, 1978, 1979, 1984). On the face of it, human infants appear to spend a good deal of time acquiring abilities that look as if they could simply have been 'wired in'. If we could understand better the nature of the acquisition process, this should tell us a good deal about what development is and why it is so essential.

Developmental psychologists, however, do not agree about even the broadest characteristics of the 'development' that they are aiming to explain. Even whether development will turn out to differ in any principled way from learning is unclear (Boden, 1988; Newell, 1980). Some kind of *qualitative discontinuity* has often been taken to be the hallmark of development as opposed to learning. Most prominently, this has been taken to imply across-the-board changes in mental processing, such as those thought to result from the global, repeated *reconstructions* of the subject's knowledge to distinct 'higher' levels of functioning that characterize traditional stage theories (e.g. Piaget, 1970, 1978a, 1978b; Piaget and Inhelder, 1969); or from a more *additive* acquisition of novel coding mechanisms due to maturation (e.g. Mounoud, 1981, 1982, 1986) or of 'higher' mental processes through social interchange (e.g. Vygotsky, 1978, 1981). Yet the very feasibility of radical changes or discontinuities in natural conceptual development is increasingly being challenged by arguments for essential *continuity* in significant properties of initial knowledge structures and their mature counterparts (e.g. Keil, 1981a, 1981b, 1985, 1990).

Clearly, many questions about the development of the infant's understanding of the world remain to be asked and answered. Some appropriate directions are provided by issues for infancy that parallel concepts and debates that are informed by the cognitive science literature.

At the most general level of discussion, the question 'What are the *constraints* on development?' currently aims to supersede obsolete 'innate versus acquired' disputes. It maintains the issue of predetermination through its concern with the determinants of regular developmental outcomes in the face of environmental variation, but additionally it supports some promising novel issues. Keil (1990) offers a heuristic classification of constraints in terms of a 2 × 2 matrix featuring the categories 'innate versus acquired' and 'domain-specific versus domain-general' (acknowledging that the scope of what counts as a knowledge domain varies widely from one theorist to another). On this analysis, the extreme empiricism that relies on domain-general acquired constraints finds little place among infancy theorists. The most nativist perspectives are reflected in proposals that forefront domain-specific innate constraints, a position close to Spelke's (1982, 1985, 1988, 1990) views on the principles underlying infants' understanding of objects.

Taking up some of the issues dealing with constraints should help to provide the clarification that the notion of 'constraints' itself urgently needs and merits. The most evident thing about it at present is that constraints are thought necessary, and that putative explanations are proposing many kinds at many levels (Gelman, 1990; Keil, 1990). If more or less any component of a model of development comes to be viewed as a constraint, the notion will inevitably lose its explanatory power.

One issue that must be of concern is the extent to which *modularity* characterizes the infant mind (cf. Mehler, 1985). Fodor's (1983) 'modularity of mind' thesis suggests a possible functional architecture for the initial infant system by proposing a principled distinction in mental processes between fixed input systems and 'higher' central ones. The former are to be viewed as autonomous, hard-wired, inflexible, domain-specific processing modules, the latter as domain-neutral and far less restricted in what they can do (indeed, this assumed flexibility leads Fodor to conclude that their precise nature is liable to remain opaque). This proposal is angled primarily at synchronic analysis, but, to the extent that it proves applicable to the infant, it contributes one way of trying to clarify the nature of domain-specific innate constraints on functioning and their possible role in any subsequent development.

As we shall see in chapter 2, work on perception provides a good route into exploring the potential for infancy of the modularity notion. Whatever the conclusions, they are independent of whether or not it is appropriate to reinstate the dichotomy between perceptual and cognitive processes, which cognitive science has predominantly abandoned, and/or to conceptualize it in the way that Fodor does. And we shall need also to consider to what extent any apparently prewired components really are fixed. A suggested developmental advantage of connectionist models is that built-in interconnections that prove relevant to the subject's activities could be maintained, while those that prove irrelevant might be (re)used for new functions (Rumelhart and McClelland, 1986). Locating such flexibility, if it indeed exists, is clearly important for a full understanding of factors that control the developmental path.

A second major issue more directly highlights diachronic concerns: the extent to which constraints, whether thought of as facilitators or as restrictors of development, are to be found primarily in *structure* or in *process*. Keil (1981a, 1981b, 1985) has marshalled evidence to support the view that Chomsky's (1965, 1980) approach to language development can effectively be extended to cognitive development in general. For 'natural' conceptual domains, the *formal constraints* proposal holds that initial and outcome knowledge structures can be shown to share abstract, general properties, and it is these properties that ensure the convergent and predictable developmental outcomes that developmentalists repeatedly seem to encounter. What we are witnessing is said to be *developmental differentiation* rather than any *qualitative restructuring* of knowledge, an emphasis in keeping with Chomsky's (1980) assumptions of innate endowment and the predetermined form of the language system's growth along an intrinsically determined course. Object knowledge has been considered as a candidate for this approach, with the implication that unique, *domain-specific innate* structures should play the significant role in the development of the infant's understanding of the physical world and the impact of any *general-purpose* (domain-general) developmental process(es) should be negligible.

Low enthusiasm for general-purpose developmental processes is evidently justified if we consider examples like Piaget's (1953, 1955, 1978a) notion of 'equilibration', which is notorious for failing to move beyond redescription to a sufficient explanation (Rutkowska, 1985). At best it can be seen as marking an important question that remains obscure (Boden, 1982), at worst as 'inchoate' (Brown, 1979) or as 'surplus baggage' (Bruner, 1959). Like his 'schemes', Piaget's 'equilibration' is an easy target, but it seems to me unlikely that we can abandon the notion of a significant role for general-purpose developmental processes in our explanations so easily. Structuralist arguments for continuity and differentiation of structure are, by definition, based on a highly abstract level of description of ability, and how this relates to process models of the mechanisms that subjects may be using is far from clear-cut. By way of contrast with this, approaches that share a functionalist concern with the details of changing behaviour have tended to steer closer to mechanism proposals and to forefront the role of the (potentially domain-general) processes involved.

In this vein, Bruner's (e.g. 1969, 1970, 1975) assumption that different functional systems share processes of operation and change leads him to query domain-specific and modular views of development. Such commonalities suggest to him that acquired skill in one system may actually be the key constraint on acquisition of another, as when the organization of cooperative action with caregivers is proposed as the infant's route into acquiring both pragmatic and syntactic aspects of language (Bruner, 1975). For Mounoud (e.g. 1981, 1986), development involves the construction of representations in a series of distinct codes ('perceptivomotor' during infancy, then 'conceptuomotor' and 'semioticomotor'). Representations are thought of as 'internal organizations of contents', comparable to a notion of domain-specific knowledge mechanisms. The construction of 'content(s)' in these internal codes is attributed to exchanges with the environment mediated by a process that is common across contents in whatever codes they are represented. And Karmiloff-Smith (e.g. 1990, 1991) provides behavioural data from a wide range of tasks to support the claim that an endogenous process of representational redescription, operating across the many forms of representation that she believes subjects use (from kinaesthetic to linguistic), is a key constraint on development.

These approaches begin to show that *how* domain knowledge is represented is as significant as *what* domain knowledge is represented. *Systems* (*codes* or *modes*) *of representation* and their properties offer a significant source of potential insights into the nature of developmental constraints.

From this perspective, my approach to infant development is based on the role of perceptual-behavioural action in representing and knowing the world. In the main, as noted earlier, the infancy literature does not now consider this to be much of a role at all, with action and representation being viewed as independent kinds of ability. For example, the '-motor' suffix sometimes ascribed to Mounoud's codes for representing domain knowledge constructed on the basis of experience is a distinctly unusual amalgamation, forefronting the physical basis of subject's exchanges with the environment, that has received scant attention. So, Trevarthen (1974, 1980) may talk of early ability as revealing two *modes of action*: 'doing' and 'communicating', directed at objects and persons respectively. But Carey (1985) interprets similar data as evidence for two *innate theories*: a naive physics, based on the concept 'object'; and a naive psychology, based on the concept 'person' (cf. Spelke, 1988). This illustrates a significant contemporary trend, even though the attribution of theories to the infant is contentious (Hobson, 1991; Keil, 1990). In my view, action and theory approaches are not simply offering us ways of talking about different (for example, 'overt behavioural' and 'internal representational') aspects of infant ability. They have distinct implications for where and at what level we should seek development, and for the range of solutions that becomes possible.

One way of tackling infant development has been to attempt something like a developmental transition between such approaches. For example, a *rapprochement* between Gibsonian and Piagetian theories has been proposed that conceptualizes development as a transition from 'precursors of knowledge in perception and action' to objective understanding through representational mechanisms (Butterworth, 1981a, p. ix).

This type of proposal faces a number of difficulties, not least that the demarcation between phenomena amenable to interpretation in terms of perception and action as opposed to representation is nowhere as clear-cut as it needs to be. As we have seen, young infants' precocious successes on a range of tasks that involve time delays (e.g. imitation) or out-of-sight objects (e.g. visual search) are widely interpreted as evidence for ready functioning representation, concepts and even theories rather than for perceptual or action-based mechanisms. Either there is no clear discontinuity between the types of task performance suited to the two kinds of explanation, or it is unclear where/how to establish it.

Even if we agree that the direct perception approach is best seen as an explanation of everyday behaviours such as locomotion and grasping (Turvey, Shaw, Reed and Mace, 1981), it proves difficult to draw a clear demarcation between these and supposedly more cognitive abilities. Such 'basic' abilities undergo complex changes in the direction of greater anticipation and changing use of information. Such developments may be better conceptualized as changes in control and organization rather than as an addition of, or switch to, re-presentational processes (cf. Bruner, 1968, 1970; Mounoud and Hauert, 1982a, 1982b). Neither perspective fares well at this task: the effectivity style concepts underlying ecological psychology's account of action prove too molar and inflexible to handle the phenomena; and the control of posture, locomotion and prehension is not the kind of issue to which the Piagetian treatment of representation can be applied readily, if at all.

The key locus for my exploration of developmental change will be with this type of *restructuring* of action. (Cf. Mounoud's (1986) proposals for appropriate performance phenomena on which to base an account of natural representational development.) Viewing early perceptual-behavioural coordinations in this light should prove equally relevant to understanding what is generally considered as infant cognitive development, since these coordinations fit closely Keil's (1981b) four criteria for strongly constrained knowledge structures: the apparently predictable development of many perceptual-behavioural coordinations parallels the requirement for a wide range of experience to be mapped into a narrow range of (mental) structures; these acquisitions seem to be effortless, rapid and universal, without evidence of formal tutoring; the psychological mechanisms that underlie them are relatively closed to introspection; and violations result in a strong sense of anomaly (*eventually* – a qualification which will contribute towards understanding the changes that are taking place).

This brings us back to the original concern with mechanisms, since the restructuring issue obviously cannot be addressed without a detailed account of the infant's initial state. In fact, computational concepts probably can contribute more directly to a synchronic analysis appropriate to initial mechanisms than they can to a transformational account. PSSH may stress the significance of process models, but opting to explore this kind of approach does not automatically buy you into any form of developmental model. There is a genuine issue as to whether or not this kind of computational reductionism can provide mechanisms to account for development without sleight-of-hand.

The best starting point is with the attempt to characterize the infant's initial state, allied with the general question: 'What restrictions does a computational conceptualization of the infant's initial mechanisms impose on the possible form and range of learning and development?' Of particular interest is the way a computational focus on (changing) action mechanisms will affect the way we carve up developmental issues.

Regarding the way mechanisms along the developmental path are viewed, for example, a major concern lies in exploring the import for the action–representation relationship of extending the notion of 'representation' to its 'selective correspondence' reading. Rather than assuming the distinctiveness of action and representation and attempting to define criteria that distinguish them, the aim is to develop a way of discussing action that supports a more open comparison of possible mechanisms underlying the functioning of infant action and of what are generally thought of as *naturalistic representational systems*. Similarly, the requirement for an extended systematic explanation of the mind, which arises from the centrality of an action system's embedding in its environment, gives the restructuring of action a new scale at which to seek out development.

Two AI-related computational areas offer leads for the exploration of change, albeit scarce direct explicit links to developmentalists' work. Most obvious is research dealing with machine learning, and the potential mutual benefits of comparing its underlying assumptions with those made by a developmental psychological analysis (Draper, 1987). Less obvious but equally important is Marr's (1977, 1980, 1982) work on vision and the explanation of information processing systems in general. Handling vision primarily at the synchronic level, his work offers a direct route into clarifying the infant's initial state. But it also generates links with developmental issues through its methodological links with Chomsky's 'programme for cognitive science' and Fodor's speculations on the 'modularity of mind'.

Notes

1. To avoid confusion, J. J. Gibson is referred to throughout as Gibson; references to other Gibsons include their initials.
2. Cf. Ritchie and Hannah's (1984) important critique of Lenat's (1979) machine learning program, the 'Automatic Mathematician'.
3. Two empirical methods underlie many contemporary infancy findings. *Visual preference* uses differential fixation time as an index of discrimination between paired stimuli. *Habituation* or *familiarization* involves repeated presentation of a stimulus, until a decrease is observed in whatever behaviour is being used as an index of attention; for example, fixation time. Recovery of the behaviour ('dishabituation') when a new stimulus is introduced, indicates its discrimination.
4. The dual visual systems hypothesis (e.g. Humphrey, 1972, 1974; cf. Trevarthen, 1968) proposes that subcortical regions mediate perception that there is something there and *where*; but the visual cortex is required for pattern recognition and perception of *what* the thing is. The linear systems approach suggests infant pattern preferences can be explained in terms of the 'contrast sensitivity function' (an index of the spatial frequencies to which the visual system is responsive and the amount of contrast that it requires to detect them) by assuming that, after stimuli have been filtered, infants preferentially attend to the one with maximum contrast. (See Aslin, 1987 or Banks and Salapatek, 1983 for overviews.)
5. Its direct (unmediated) version of 'realism' assumes there is an organized world of objects and events independent of our mental activity; that this is the world we perceive and know; and that we need not be sceptical of the validity of this knowledge.
6. In psychology, functionalism is generally characterized by emphasis on action, adaptation and the practical consequences of mind in the world (e.g. Schultz, 1981). In philosophy, functionalism's focus has primarily been on the definition of mental states, the adaptive, practical dimension receiving less attention. It holds that mental states cannot be identified with physical states; they are functional states, to be defined in terms of the network of relations they hold to one another and the causal role they perform in mediating between the system's inputs and outputs (e.g. Block, 1980).
7. This sense of 'knowledge of the world' is increasingly used by cognitive psychologists, and it is adopted in Newell's (1980) statement of PSSH. However, some philosophers would distinguish between belief and knowledge, arguing that knowledge must involve the ability to hold and to justify beliefs (cf. Pylyshyn, 1984).
8. 'Unprogrammed' means that a network is not equipped with rules determining what it should look for in the input domain or how it should respond to it. Generally, the network starts out without predetermined connection strengths between its input, output and intervening – 'hidden' – units; and changes in its connection strengths (and 'behavioural' output) are achieved by 'training' on a range of input examples according to a learning rule. For example, the *Hebbian learning rule* changes connection strengths on the basis of inputs to the system, without benefit of feedback, in what is sometimes identified as unsupervised learning. In *backward propagation*, the network is trained through feedback about the discrepancy between its actual and 'desired' output. (Rumelhart, McClelland *et al.* (1986, vol. 1) provide useful technical discussion.)
9. Even the innocent expression '*appropriate* behaviour' connotes a degree of subjective selectivity and purpose incompatible with viewing infant performance in terms of causally determined reflex movements. As Dretske (1981) emphasizes, it is intentionality which determines the appropriateness or otherwise of behaviour.

Chapter 2

From low-level computer vision to early infant vision

In this chapter, I explore two reciprocal questions that bear on the mapping between data and theory in infancy and Marr's (1976, 1977, 1980, 1982) AI-based computational account of vision: how can Marr's framework for low-level vision open up our thinking about early infant vision? Does research on early infant vision support the psychological validity of Marr's computational model of low-level vision?

In part, the grounds for beginning this book's infancy–computation mapping with the domain of vision are purely pragmatic. It is not my intention to single out perception as the primary contributor to infant abilities in general or to infant action in particular, however hard it is to get away from considering it as the 'front-end' of the system. Indeed, I shall argue later that a similar analysis is needed, but currently much further off, with regard to the behavioural component of the infant system.

In the case of vision, researchers in infancy and in AI have already come to address closely related issues about how it is possible for us to see the world. Infancy work wants and needs to understand the perceptual abilities of newborn and young infants, especially the nature of underlying preadaptations and their implications for development. AI seeks principles for the design of computer vision systems that can function effectively in our particular world, and has increasingly come to acknowledge the importance of potential insights from the evolutionary design of their biological predecessors. As a consequence of this, a great deal of material on infant perceptual organization has accumulated, mostly on vision; and vision proves to be the domain where the most precise, immediately relevant AI work is to be found, with the level of detail offered by vision models providing a rare exception to the disappointingly molar models of much mainstream AI.

Over and above this, however, there are principled reasons, in the sense that I outlined in the introductory chapter, for starting out with vision. Marr's vision scheme proves to be an excellent candidate for a base (ideal) model with which to compare infant performance. The general framework in which it is embedded consolidates and extends a range of significant ideas from cognitive science about the nature of computational explanation; and the model itself features robust ideas from AI about specific kinds of computation that might be employed by the visual system.

Starting off by considering Marr's framework and infant vision introduces conceptual tools of analysis that will continue to be useful when we move from the confines of early/low-level vision to locate perception in the context of action, and from the synchronic focus of Marr's model to broader diachronic concerns. This certainly does not mean that all of Marr's ideas will be wholeheartedly endorsed throughout. It does mean that they can play a valuable *heuristic* role in charting the territory. The first task is to ground this framework in the kind of infancy phenomena that have preoccupied developmental psychologists, by sketching the background to Marr's model and then comparing it in some detail with a representative sample of specific work from infancy research.

Vision as computation: Marr's perspective

Many introductions to Marr's framework are now available.[1] Here, I shall recap briefly only the main ideas that underwrite his own model of low-level human vision: his view of vision as computation; assumptions about its underlying design principles; and the tripartite explanatory scheme that he proposes for information-processing systems in general.

In line with PSSH, Marr conceptualizes vision as a process involving the construction and manipulation of symbol structures that designate, ('model' or 'refer to') some particular aspect(s) of the subject's environment. Vision is a problem in information processing because it involves mapping between representations, each of which can be thought of as offering a distinctive 'vocabulary' within which things may be described. Since any representation is defined in terms of its primitive elements, their properties and the types of relations that can exist between them, it will be capable of conveying some information explicitly (whatever is available without recourse to further processing), some implicitly (whatever could be made explicit provided the right additional processes are available to operate on it) and some not at all.

In the case of vision, the starting point is an 'image', which can be thought of as a type of representation whose capacity for describing the world is limited to making explicit information about light-intensity values at each point on its surface. The task of vision is to get from this to an efficient description in a representation that makes explicit information better suited to the subject's uses and purposes, such as information about physical properties of the world.[2]

Marr (e.g. 1976, 1982) articulates a number of principles concerning the organization of complex symbolic processes such as recognition, which suggest how a visual system – biological or otherwise – needs to be designed to fulfil its role(s) effectively. Three of these are of particular relevance to infancy concerns. The most fundamental principle – *explicit naming* – falls out of the view of vision described so far, and captures the computational significance of being able to describe things in an appropriate representation. Once a collection of data has been given a 'name', it can be manipulated as an entity in its own right, have properties assigned to it, and be referred to by other, external structures and processes (cf. Newell's (1980) notion of 'designation'). Thus, certain processes may be readily applicable to a description in one representation but applicable only with difficulty, or not at all, to another which contains the same information in an implicit form.

The centrality of explicit, appropriate descriptions sets the stage for the principle of *modular design*. An efficient information processing system should split up large computations into relatively independent parts, so that a change at one location does not entail changes at numerous others. This makes it much easier for either a human designer or biological evolution, as appropriate, to 'debug' or to improve the system. As we shall see, Marr's vision model follows this principle by dividing low-level vision into a number of distinct stages, each involving a range of modular processes aimed at producing a different kind of description. This provides us with the basis for a much more fine-grained comparison of computer, adult and infant vision than is generally possible.

The system should also follow the principle of *least commitment*: attempting to render explicit all of the useful information in the image without prejudice to higher levels of structure, so that things done early on do not later need to be undone. This leads to the proposal that low-level vision must be *data-driven*, its processing proceeding in a bottom-up direction, uninfluenced by downward-flowing information which might be equated with 'hypotheses' or 'expectations' concerning, for example, an object giving rise to the image.

As far as biological visual systems are concerned, such basic processes are taken to be an outcome of evolution. Marr therefore purports to provide a principled analysis of what information the visual system can make available without any individual experience of particular objects. And he offers us a model, based on these principles, of computational structures and processes that could achieve this in the human case. Effectively, we are equipped with *explicit predictions* of what we should expect the organization of the young infant's visual system to look like. Additionally, Marr offers a third conceptual tool for our analysis of infant vision: three levels of explanation at which any information processing system can be explained and understood.

The 'top-level' explanation or *computational theory* should provide a universal, precise specification of the task, and of its principles of solution as it confronts any system, animal or machine (Marr, 1977, 1982). *What* does the system have to compute, and *why* is that the appropriate thing for it to be doing in terms of *constraints* in the domain concerned? At the most general level, we can think of constraints simply as relations or patterns that are established by the interaction of entities in a domain. For vision, they result from the cohesion and continuity of physical matter in the world. Uncovering them requires a formal (preferably, to Marr's mind, mathematical) account of the physical principles that determine how 'things' in the world interact with one another. Clarifying physical constraints on the organization of the world, and on the image formation process, can prove that unequivocal interpretation of the image is possible, and show why a particular computation is the sensible one for the vision system to be attempting.

Marr (1977) emphasizes the comparison between his computational theory and Chomsky's (e.g. 1965) abstract 'competence' level of analysis for linguistics. Other notable precursors are Clowes's (1971) and Huffman's (1971) formal analysis of the properties of trihedral objects, which were the focus of many early AI scene-analysis programs, and Newell and Simon's (1972) notion of 'task invariants', properties of a task environment that determine behaviour yet are independent of the properties of the problem solver.

This abstract proof that a computation is, in principle, possible is seen by Marr as the most vital part of any explanation, and as the most neglected by psychology, neurophysiology and the process models of traditional AI. Understandably, he and Chomsky (1980) mutually endorse one another's methodology for doing science. Certainly, much experimental infancy research shares the difficulty of lacking a deep appreciation of the structure of the task(s) with which the infant is being faced. Often, for example, there is only a vague or intuitive appreciation of dimensions of similarity and difference in the experimental stimuli used, casting doubt on the clarity of the questions being put to infants and on the validity of proposed answers.

Marr's own vision model, based on his design principles and explanatory prescriptions, suggests some new ways of looking at the kinds of questions that have often figured in theoretical discussions of infant vision. Its overall task is to arrive at a description that can support the efficient recognition of three-dimensional objects. However, Marr argues that low-level vision forfeits its generality if it attempts to see 'objects' as immediately as possible (not least because of the variability of what can count as an object – the features of a person's face, the person or a group of which s/he is a member each has the potential to be viewed as an 'object'). It is doubtful that any of the modular computations involved correspond to an obvious partitioning such as 'image to object'; even (apparently fundamental) figure–ground separation, Marr (1977) suggests, is unlikely to have a single underlying computational theory. Instead, he proposes the construction of descriptions in three distinct representational formats, each with its own associated computational theory: the *primal sketch* and *2½-D sketch* support viewer-centered descriptions of image discontinuities and surfaces respectively; and the *3-D model representation* supports object-centered description of an object's overall shape.

This bears on the key problem of infant *object understanding*. The central question has often been posed as some version of: 'Does the infant see *it* as an object or not?' Investigation is then directed towards demonstrating the critical dimension(s) of 'objectness' for the infant. We should, however, be cautious about adopting premature or ill-defined assumptions about what it is to know something 'as an object'. A more differentiated starting point is provided by Marr's model, which offers the intriguing possibility of replacing a quest for *the* object with the notion that infant (and adult) vision probably provides *multiple descriptions* of objects and events. This idea supports new questions:

what representational formats are available to infants? What descriptions can infants construct within those formats? And what functions can those descriptions serve? Especially persistent in the infancy literature is the notion that shape or form is the vital dimension of 'real' object understanding. However, we shall see that conflicts regarding the existence and/or significance of shape perception turn out to be more apparent than real once they are contextualized within the computational framework.

Marr's second, *algorithmic* level of explanation highlights an area in which I have already suggested that developmentalists' explanations are frequently weak: the specification of sufficient information processing mechanisms. It concerns *how* a system could perform its task by means of processes that make use of the relevant domain constraints. The construction of a description in an effective, useful representation has to be explained in terms of the input and output representations employed and the algorithm(s) that map between them.

It is at this level that Marr locates the role of the empirical psychologist, in uncovering which of the range of possible representations and algorithms are actually used by a particular subject. He maintains, characteristically, that their process models are unlikely to offer 'true understanding' unless they are informed by a prior top-level theory. If the computational theory is wrong, or lacking, speculations about processing mechanisms will most likely be in error too. Because the algorithmic proposals of each stage of Marr's vision model *are* backed by computational theory, exploring correspondences between them and the process models of infancy research provides a useful route to establishing whether the developmentalists' models really are sufficient for the information processing task they claim or hope to perform. Marr's specific suggestions are certainly not the only or ultimate possibilities; they will undoubtedly be superseded, but they retain heuristic value for the comparison process.[3] Discovering correspondences can be mutually beneficial, since positive parallels help to establish the psychological validity of the representational formats and algorithms that comprise Marr's computational model.

Marr's definition of an algorithmic explanation for vision also contributes directly to one of the broad theoretical concerns that figured in the introductory chapter by suggesting a way of beginning to link the somewhat vague notion of 'constraints on knowledge' with specific processing mechanisms. In line with earlier AI vision work and the outcome of his own top-level computational theory, Marr argues that the input alone is inadequate to explain how the data-driven visual system can get to an output with more of the explicit information that it needs. Input–output mapping becomes possible if additional 'assumptions' about the domain concerned can be brought to bear on restricting the interpretation process, and the processes of the visual system are equated with this *innate, implicit knowledge* of constraints in the task domain (cf. Chomsky's (e.g. 1965) argument that a strong language-specific biological basis is essential for the acquisition of syntax).

Saying that these processes 'embody' or 'make use of' domain constraints in no way implies that the system explicitly represents these assumptions; they are not, for instance, stored in data structures to be called up or consulted by visual processes. In vision, the processes themselves provide an implicit (procedural or process-based) representation of 'facts' about images that derive, through evolution, from the organization of the physical world. The processes do not 'look like' the assumptions they model, their operation simply satisfies the constraints imposed by the world.

The third and final level of explanation, *hardware implementation*, must seek the physical realization of computations in neurophysiology or hardware. Here, we can expect to find the greatest variation because we know that a single algorithm can be executed by quite different physical machines, and this shows why it is so difficult to get directly from neuroanatomy or neurophysiology to an effective explanation of psychological abilities such as perception. Support from higher levels of explanation is needed. Without a top-level computational theory, we may fail to understand *what* the neurophysiology is doing. And many findings (e.g. from single-cell recording) suggest *that* the brain has solved some problem in information processing, but they do not tell us *how*.

These ideas bear directly on influential suggestions for exploiting *neuro-anatomy* as a framework for explaining infant perceptual abilities and their development. In particular, a dichotomy between cortical and sub cortical processes has long proved attractive. For example, a general model proposed by K. R. Gibson (1981) argues that ontogeny recapitulates phylogeny in the order of maturation of neural processing mechanisms. Many early infant behaviours (e.g. auditory-visual coordination, avoidance of looming objects, cross-modal matching, and neonatal imitation and reaching) are attributed to mechanisms located in the midbrain colliculus and its connections with the brain stem. Development of these abilities is then attributed to an addition of cortical processes, and an attempt is made to establish correlations between the maturation of specific cortical regions and the emergence of abilities that are significant in Piagetian theory. Abilities whose physical and psychological foundations are obscure, such as neonatal imitation, are allocated to either subcortical or cortical regions, as though this were explanatory.

We cannot expect simplistic ideas of localization of function to inform us about the mechanisms of early abilities. Such ideas are, at best, contentious (Luria, 1973). A subcortical versus cortical distinction cannot handle either the psychological or the neurophysiological complexity involved, no matter how tempting it may be to view direct perception phenomena as 'lower', hard-wired subcortical processes, and mediated perception and object conception as 'higher' cortical ones. The pertinent question is not '*Where* are the relevant brain regions?' but '*What* are they doing?'

An important implication of Marr's tripartite explanatory scheme is that the significance of some of the phenomena in which we are interested may well be restricted to one particular level of explanation. Ideas about the physical substrates of behaviour may be of relevance only at the implementation level, and not be indicators of the top-level task the system is performing, or of the algorithms it is executing. Behavioural/psychophysical data may be due to specifics of neurophysiology, not evidence of processes usually involved in executing vision algorithms (e.g. afterimages); or illustrate the different algorithms necessitated by different types of representation, not any difference in what is being computed and why (for example, calculating the larger of two numbers expressed in prisoner's tally or roman numerals). Mapping between infancy phenomena and Marr's three representational formats in the light of his explanatory levels may help to tease out such effects in infant vision.

The primal sketch – infant attention to 'high information areas'

What type of information does the image contain, and how can it all be made explicit? Marr's main contribution to our understanding of the nature of this problem, and of how it might be solved, comes from a mathematical perspective (Marr, 1982; Marr and Hildreth, 1980). He uses this to support the argument that physical phenomena (shadows, object boundaries, textures and so forth) give rise to discontinuities in image intensities. So, unlike absolute intensity values, these discontinuities can be informative about the 'things' of the physical world. The first task of vision is to get from the image's analog (continuous) representation of intensities to a primitive but rich digital (discrete) symbolic description that makes explicit the location, direction, magnitude and spatial extent of important *types of intensity change* in the image.

We have little detailed knowledge of the neurophysiology underlying the visual information processing covered by Marr's model, but most of the neurophysiology–computation links that there are concern this initial representational format – the *primal sketch*. This provides a valuable area for interchange with the extensive literature on infant attention and discriminative 'preferences', especially with a variety of neurophysiologically motivated attempts to explain infant performance. Typical approaches use notions of *feature detection* (e.g. Bronson, 1974, 1982; Kagan, 1972; Karmel and Maisel, 1975); of the relative maturity of *X- and Y-channels* projecting to the cortex from retinal ganglion cells that are thought to analyze 'pattern' and 'movement' respectively (Maurer and Lewis, 1979); and *spatial frequency analysis* (Banks and Ginsburg, 1985; Banks and Salapatek, 1981; Gayl, Roberts and Werner, 1983). Much of this work contends that the neonate's perception is qualitatively different from that of the older infant or adult, based on response to 'salient' stimulus characteristics but not necessarily to objects (e.g. Bronson, 1974, 1982; Fantz, 1966; Kagan, 1972; Salapatek, 1975). This then raises the question of just how the infant's apparent fascination with contour-like stimuli could operate as a biological prerequisite for perceiving, recognizing and classifying objects. Convincing answers are not forthcoming because, in the main, such theories have drawn on research that provides good examples of trying to describe what the brain is doing without explaining why.

Most studies use two-dimensional, geometric stimuli, and correspondingly characterize determinants of the infant's preferences in terms of quantitative variables: amount of contour, area, degree of complexity (e.g. number of squares on a checkerboard), brightness and so forth. Where semantic considerations are introduced, apparently straightforward disagreement results.

For example, in view of young infants' frequent failure to discriminate between different configurations of the same contours, Bronson (1974) considers their responses to involve only low-level mechanisms for enhancing spatial and temporal discontinuities in illumination. The so-called cortical 'feature detectors' that he assumes necessary for pattern analysis, and thus for object recognition, are deemed inoperative. Related models propose that contour processing in the subcortical secondary visual system predicts infants' visual preferences for particular contour densities in the early weeks of life (Haith, 1980; Karmel and Maisel, 1975). And habituation studies have been taken to suggest that the neonate can discriminate the orientation of line segments but not integrate them into angles or forms (Cohen, 1988). The X-Y-channel hypothesis also suggests immaturity of the supposedly pattern-relevant X-channel, in the face of infants' interest in moving stimuli.

Yet some researchers believe habituation techniques have demonstrated that even neonates exhibit *orientational selectivity*, suggestive of cortical function, and can make discriminations on the basis of ('simple') patterns, such as crosses versus circles or acute versus obtuse angles (Slater, Mattock and Brown, 1991; Slater, Morison and Rose, 1983; Slater, Morison and Somers, 1988; Slater, Morison, Town and Rose, 1985). Slater *et al.* (1991) are faced with alternatives to their claim that infants perceive the form of two lines joined at an angle: given neonates' poor acuity, perhaps their discrimination is simply based on what looks to them to be one line (an obtuse angle) rather than two (an acute angle); or on 'blobs' of different sizes, since the apex of the acute angle will give rise to lower spatial frequency information than the obtuse.

Such theoretical choices cannot be made independently of a precise model of the goals of visual processing; without this it is not clear even to what extent they are alternatives. Even a 'blob'-level description may contribute towards the infant's psychological understanding, once we grasp its role in the visual (and action) system. A striking property of this work is convergence between the language of the explanations and the vocabulary of Marr's independently motivated primal sketch. The primitives of this representation include edge segments, bars, blobs and terminations (or discontinuities), together with their orientation, length, width, contrast and position; the full primal sketch also aims to combine these and to use virtual lines to describe the geometry of their more global (two-dimensional) spatial structure. (Thus, descriptions based on a close-up of a mackerel-tabby cat's fur might successively yield the density and orientation of individual hairs, then stripes and finally parallel markings.)

Looking at Marr's computational task analysis for the primal sketch, and at the sketch's role in subsequent processing, may begin to offer the possibility of *rapprochement* between infancy approaches, instead of fruitless opposition. Indeed, Marr's model would predict that we ought to find evidence of discrimination along all of the lines suggested, and that each is relevant to seeing physical objects (albeit not necessarily in the way their proponents believe). It also offers a viable alternative to the otherwise appealing direct-perception claim that 'feature analyzers' and so forth are simply artifacts due to misguided assumptions and empirical methodologies, revealing nothing but incidental properties of the physical machine (e.g. Michaels and Carello, 1981). The explanation of primal sketch construction begins to bring such phenomena within the scope of a valid psychological explanation (which is not to say that there is not still a long way to go, as the following chapters will show).

Viewed mathematically, the discovery of intensity changes requires a *differential spatial operator*. (Essentially, an operator is a local calculation applied at each point of an array, such as an image, and making use of the intensity values there and in the immediate vicinity.) Marr uses a priori mathematical arguments to establish that a second-order differential operator is best suited to the task. (This analyzes the rate of change in the derivative of the first-order operator, which delivers the rate of change in intensity across the array.) There is a role here for the retinal *X-cells*, whose sustained response to stationary areas of contrast has linked them with infant 'pattern' and object perception. Their receptive fields prove to have mathematical properties very close to those of the particular operator that Marr proposes.[4] This is why two kinds of X-cells are found: on-center cells detect positive values of the second derivative; off-center cells detect negative values; and pooled outputs from adjacent on- and off-center cells detect 'zero-crossings', which implement the digitalization of intensity information by indicating where the value of the continuous intensity change function changes from positive to negative.

Marr is also able to offer an interpretation of the psychological function of *Y-cells* by establishing that the time derivative of his differential spatial operator can provide a single source of information about movement: its *direction* (Marr, 1982; Marr and Ullman, 1981). He substantiates the psychological validity of this analysis through (adult) psychophysical experiments and neurophysiological studies. For example, the predicted behaviour of his time derivative for edges and bars matches closely that recorded in neurophysiological studies of the Y-cells. This gets beyond the vague notion that Y-cells must be 'movement detectors' because they show a transient response except with moving stimuli. Instead, it provides a theory of how what they are doing is implicated in making explicit movement information, and in the subsequent processing that will be making use of this information.

Marr's analysis establishes *spatial frequency analysis*, which the infant visual system undoubtedly executes in some form, as insurance against the loss of potentially important information during image processing. Changes or gradients of intensity are sometimes sharp (for example, at the boundary of a well-illuminated dark object), and sometimes gradual (for example, shading on a surface inclined towards the source of illumination). So, intensity changes actually need to be measured at different *scales* of analysis. This is achieved by first passing the 'raw' image through a number of filters, each averaging over different numbers of receptors to provide a description of image intensities at a particular scale of analysis. Marr views neurophysiological findings of single cells that respond selectively to stimulation at different spatial frequencies as support for the view that the brain is, in fact, making explicit this information.

Although many find uncontroversial the notion that so-called feature analyzers perform some function relevant to perception, precisely what that function might be is more opaque. To achieve the primal sketch's higher-level description of image properties, the individual intensity changes that are found must be combined to make explicit the length and orientation of *continuous intensity changes*. Evidence that the brain performs this function is provided, Marr argues, by the characteristics of Hubel and Wiesel's (e.g. 1959) well-known *simple feature analyzer cells*, whose optimal stimuli include bars and edges, and their (more or less) specific orientation and location in the retina.

Developmental theorizing has often failed to acknowledge the implausibility of a direct route from features to patterns (considered as unique configurations of features) to object perception. No simple one–one correspondence can be assumed between what goes on in the retina and the shape, orientation and movement of objects. A contour in the visual field could correspond to numerous features of an object, or indeed to none at all – it could be evidence for a discontinuity in illumination or for a texture boundary. And precomputational 'schema' theories, which propose matching of input to models of previously encountered feature configurations (e.g. Jeffrey, 1968, 1976; Kagan, 1972), walk straight into the infinite regress of the homunculus problem: if the unlearned, intrinsic stimulation characteristics to which infants are attracted do not allow them to perceive things, then a copy of those same characteristics cannot help, unless some additional interpretive process is invoked.

Marr's task analysis for this stage of perceptual organization helps to clarify the complexity of what might be meant by detecting features in the image. It proposes that combining continuous intensity changes reveals the system exploiting a constraint on the way the physical world works: most intensity changes in an image can be shown to be produced by physical phenomena that are *spatially localized*. If continuous intensity changes with the same orientation and position are found at different scales of analysis, they are then combined by a grouping process which is viewed as the system's (implicit) assumption that they indicate the presence of an intensity change associated with a single physical phenomenon – a change in reflectance, illumination, depth or surface orientation.

This account replaces ill-defined features or patterns of dubious semantic status with a sequence of (symbolic) descriptions constructed in a precisely defined representation. The hierarchical grouping processes of primal sketch construction begin to illustrate how features (or 'forms' in the sense of geometric feature configurations) are to be found at many levels and scales of processing. Analysis of what the system is doing and why, at the level of the computational theory, establishes that these descriptions really can make explicit information about the physical world and what that information is. The analysis of how changes in physical phenomena are associated with discontinuities in image intensities explains *why* infants preference for contour-like stimuli can, in fact, be viewed as adaptive attention to 'high information areas' (Mendelson and Haith, 1976).

The kinds of two-dimensional stimuli that are often used to study infant vision might usefully be considered in the same way as line-drawings in adult perception. These may be easy for humans to interpret because they correspond to a form of 'intermediate representation' that is normally constructed during visual analysis (Draper, 1980). In terms of Marr's model, they may contain information similar to that usually made explicit in the primal sketch. On this view, line drawings do not provide the same stimulus as what they depict, but they do contain (at least some of) the same information since they can be described within the same representation. Along the same lines, the contour displays that attract infant attention may do so because they contain some of the information that is provided by real objects during visual processing.

Primal sketch information is certainly a long way from the (quick) object-specifying information that infancy theorists often assume or hope for from image processing. It is viewer-centered (retinocentric), tied to its input modality, and there is no attempt at this level to make explicit objects, or even all aspects of figure–ground separation. The term 'edge', for example, is used to describe a type of intensity change due to *some* physical phenomenon – it could be a texture boundary or a shadow.[5] Nevertheless, it does make explicit some of the most *general* properties that are shared by *all* physical objects. And it is central to making explicit other types of information – in particular, information about the existence and structure of surfaces.

The 2½-D sketch – parsing scenes and seeing surfaces

Marr's formal analysis of how physical phenomena give rise to the image establishes that image intensities implicitly embody information about local *surface dispositions* (e.g. orientation, depth) and *material properties* (e.g. texture, colour). This information cannot be obtained directly from the image, but it can be made explicit by processes operating on the arrangements and motions of primal sketch elements. For example, *stereopsis* can now come into play because objective mapping of corresponding positions in the two images, which allows calculation of the disparity values needed to assess distance and orientation, becomes possible with the primal sketch (Marr and Poggio, 1976). (At the image level, only intensity values would be available as a basis for deciding which points to compare, and many of those points might have identical intensity values.)

These processes output a description in a representation which Marr calls the *2½-D sketch* because it makes explicit information about the depth and orientation of *visible surfaces*, and about *discontinuities* in these properties, within a viewer-centered framework (Marr, 1980, 1982; Marr and Poggio, 1976; Marr and Ullman, 1981). Since it does not attempt to segment the input into regions that are meaningful as physical objects, the 2½-D sketch can be produced without a priori knowledge about the shapes of viewed objects.

As far as Marr was concerned, by far the hardest part of this stage of his investigation, and the biggest breakthrough, was identification of the constraints involved, hence the precise nature of the task attempted by this level of visual processing. He noted, however, that there was little (adult) psychophysical work on the 2½-D sketch; it was not even known whether it existed in the form he described. This opens up an important role for infancy data in the psychological validation of this level of Marr's model, while simultaneously assessing the extent to which Marr's ideas can sharpen up explanations of the infancy data.

A particularly pertinent area for computational–psychological comparison is the substantial body of work due to Spelke and her colleagues (e.g. Kellman and Spelke, 1983; Kellman, Gleitman and Spelke, 1987; Kellman, Spelke and Short, 1986; Kestenbaum, Termine and Spelke, 1987; Spelke, 1982, 1985, 1988, 1990; Spelke and Born, 1982). Their studies are usually presented as exploring the conditions under which young infants (up to about 12 to 16 weeks of age) perceive *unitary objects*. But, to avoid adopting premature assumptions about what it is for an infant (or adult) to know something 'as an object', the problem with which their tasks face the infant can provisionally be expressed more neutrally as: how do infants parse scenes composed of stationary and/or moving entities in various spatial arrangements?

Infant 'successes' and 'errors' in these studies offer support for the view that a key goal of low-level vision is to describe things in terms of a representational format like the 2½-D sketch. The data reveal that infant performance is not influenced by sources of information that are *not* employed in 2½-D sketch construction, and suggest a straightforward hypothesis about what infants *are* doing: initial parsing is determined by the 2½-D sketch's description of surfaces. Following the initial familiarization period, a new scene will be discriminated to the extent that it generates a novel description in 2½-D sketch terms.

Spelke *et al.*'s discussions of the information that infants might exploit for their tasks have drawn on Gestalt principles of perceptual organization. Two typical kinds of study can illustrate their general conclusion that infants do not appear to parse scenes on the basis of: *similarity* of colour or texture; *simplicity* of shape or form; or *good continuation*.

One series of experiments assumed that infants would be 'surprised' if the boundary of what they perceived as a unitary object was violated. When a stationary cylinder was viewed suspended before a flat background, infants did not appear 'surprised' if the whole object subsequently began to move back and forth. Nor were they 'surprised' when they viewed an area similar in (projective) size and shape to the original cylinder, surrounded (in the same plane) by an area of similar (projective) size and shape to the cylinder's original background, if half of that display moved back and forth, even if the two areas differed in both colour and texture. However, 'surprise' was exhibited by infants viewing the stationary cylinder suspended before the flat surface if half of the cylinder and its background moved back and forth together, even when the cylinder and its background were the same colour and texture.

In a second kind of study, dishabituation measures suggested that infants parsed two adjacent, touching objects as a single entity, even when each had a simple, continuous boundary and their edges were unaligned (for example, adjacent horizontal and vertical rectangular blocks that are described as two objects by adults). They would parse them as two entities if one was in front of the other and partly occluded it. Studies with suspended objects confirmed the importance of adjacency and contact.

Discontinuities in depth are at the heart of such infant discriminations, and in several ways the performance pattern converges with ideas about the 2½-D sketch. According to Marr's computational theory of the 2½-D sketch, the constraints that permit its construction, and the algorithms that make use of them, cannot be based on the information that infants also appear unable to use or fail to use. *Colour* cannot be used to deliver information about discontinuities in relative depth and orientation, and Marr's model proposes that it is relatively peripheral to object recognition in general. *Texture* discontinuities are made explicit in the earlier primal sketch, which can describe only two-dimensional image properties, not three-dimensional properties of surfaces. (It seems feasible that texture *per se* may prove more relevant to perception of surface properties such as hardness, smoothness and so forth.[6]) And Marr argues that low-level vision – to ensure its generality and efficiency – should not exploit a priori information about the possible *shapes* of objects. The type of representation involved offers a suitable format for any of the many physical objects that an infant may encounter.

This means that 'object' figure–ground discrimination may be incomplete at low levels of visual processing. Marr's example of an image of two superimposed leaves shows how similarities in intensity can lead to the construction of a primal sketch that fails to separate objects, despite each of them having a 'simple', continuous form (Marr, 1976, p. 504; 1982, p. 273). Separating entities that are not distinguished in the primal sketch can be done by using *top-down knowledge* (in this case that the two leaf segments do not match); and it may be that the Gestalt law of good continuation implicates processes that rely on individual experience of forms. Alternatively, and more appealingly to Marr's mind, information from *motion* or *stereopsis* can be used.

These processes are central to Marr's theory of how the 2½-D sketch can be constructed from the primal sketch without benefit of top-down information, and either process will result in adjacent regions that are identically oriented in relation to the viewer being grouped together at this level of description. If infants' perception of object arrangements in these studies is mediated by 2½-D sketch construction, we should expect them to fail to separate two objects under adjacency conditions – as they do.

Under certain conditions, infants' ability to parse spatial arrangements extends to partly occluded objects. For example, infants who were habituated to movements of a rod whose center was occluded by a stationary block subsequently showed recovery of looking time to a split rod, but not to a complete rod exhibiting the same motion. This implies that the two visible rod segments in the initial display were perceived as a single entity; the broken rod stimulated dishabituation because it appeared novel in terms of the infant's initial description. In fact, infants grouped together physically unconnected areas differing in colour, texture and shape, and whose contours and major axes were not aligned, provided those areas moved together (for example, a rod and an irregular polygon on opposite sides of the block).

This kind of ability ('perceiving the unity of a partly occluded object' as Kellman and Spelke (1983) termed it) may also be mediated by 2½-D sketch construction. Marr (1982) argues that *filling in* and *interpolation* of missing contours and surfaces occurs first in the 2½-D sketch, being essential to the internal consistency of this format's representation of depth and orientation. Smooth continuation of discontinuities, possibly including continuation behind occluding planes, may be an outcome of stereopsis and the *continuity* constraint that it satisfies. Stereopsis embodies the assumption that disparity varies smoothly almost everywhere in an image, resulting in only a relatively small part of an image being composed of discontinuities in depth.

Different versions of Spelke's theory of the infancy data share the same general form; all focus on information about surfaces, depth and motion (Spelke, 1982, 1985, 1988, 1990). Initially (e.g. Spelke, 1982), infant performance was interpreted in terms of two principles dealing with *connected surfaces* and *common movement*, which were assumed to stem from an innate 'notion' or 'unlearned conception' that the physical world is composed of things that are coherent or cohesive.[7] More recently, it proposes that perception directly delivers a representation of the arrangement and motion of visible surfaces in terms of *surface points*; then, a central cognitive system employs a range of *conceptual principles* (the initial 'object concept') to 'carve up' this continuous array into 'unitary, bounded, persisting objects'. Successive versions of the theory are open to similar analysis and evaluation from the perspective of the 2½-D sketch hypothesis.

Consider first the *cohesion* and *boundedness* principles (Spelke, 1990). These are related to the earlier connected surface principle, and like it they confound 'what' subjects may be doing with 'how'. They are closer to a redescription of the fundamental *task* of making explicit surface information that infant vision must tackle in these experimental situations than to a processing solution to how unitary objects are perceived. They assume that points lie on the same object if a path of surface points connects them; and lie on different objects if no such path connects them. The cohesion principle will not apply if there is spatial separation or relative motion of surfaces; and the boundedness principle will not apply if there is no spatial separation or relative motion of surfaces.

This part of the theory is predicated on the assumption that perceiving surfaces or surface points and their distance, orientation and motion in three-dimensional space is unproblematic and directly 'given' for the infant, unlike object perception. But the explanation of infant vision cannot take this for granted. Viewed computationally, applying principles to surface points presupposes prior solution of the task of computing which primal sketch elements are evidence for a surface, rather than for some other physical phenomenon. And the processes that solve that task (which is the construction of the 2½-D sketch) will deliver connections and discontinuities between surfaces – like the cohesion and boundedness principles, these two properties are interdependent, and both emerge from processes for making explicit surfaces and their spatial arrangement.

Two further principles highlight *rigidity* and *no action at a distance* (Spelke, 1990). They assume that infants are predisposed to interpret object motion as rigid; and separated objects as moving independently of one another. Surfaces that undergo different rigid motions will be treated as distinct bodies; and surfaces that undergo the same rigid motion will be treated as connected. Like their predecessor the common movement principle, but unlike 'cohesion' and 'boundedness', these principles can appropriately be viewed at the process-based *algorithmic* level. However, like 'cohesion' and 'boundedness', the task they are tackling may be misphrased.

Computationally speaking, infants in these tasks do not perceive surfaces then group them into 'objects' on the basis of assumptions about their movement; rather, they exploit motion to recover three-dimensional surfaces, and the segmentation this achieves offers a sufficient account of their discriminative performance. Processes exploiting motion are significant for 2½-D sketch construction and rely on an assumption of rigidity, a constraint which is embodied in Ullman's (1979) procedure for getting three-dimensional *structure from motion*. This exploits the full primal sketch, and involves matching successive views. The force of the rigidity assumption comes from the underlying (mathematical) computational theory, which establishes that this method will succeed in finding an interpretation only if the motion is due to a rigid body. Given the way the physical world works, the possibility of arriving at an interpretation in the absence of a rigid body is negligible.[8]

Spelke *et al.* have viewed their organizational principles as 'constraints' on object behaviour in space which infants can use to 'find objects'. But although notions such as 'coherence' (Kellman and Spelke, 1983) or 'cohesion' and 'boundedness' (Spelke, 1988) feel intuitively plausible, they are not constraints on infant knowledge in the sense exemplified by 'continuity' or 'rigidity' within the computational framework. They *may* prove to be constraints, but currently fail to address two central questions involving the conversion of principles into mechanisms.

First, establishing that 'cohesion' and so forth are constraints requires a formal account of how they work, and, central to psychological concerns, clarification of what parsing process(es) operating on the input of surface points (or whatever) could satisfy them; otherwise, such principles need to be read *metaphorically*, not as statements about mechanisms. As Marr (1982) notes, Gestalt psychologists formulated rules that described what perception did, but they failed to embody them in grouping *processes*. Gestalt laws, and by analogy Spelke *et al.*'s principles, 'work' when they embody useful assumptions about the world, but it takes a computational analysis to clarify what those assumptions are and why they are valid.

Second, in what kind of representational format is the output of this 'carving up' process to be described? The notion of 'a general representation of object unity and boundedness' (Spelke, 1990) does not specify the necessary primitives, properties or relations. Neither is it precise enough to generate performance predictions different from those of the 2½-D sketch hypothesis, nor would it seem effective for its purported purpose since often it clearly fails to discriminate between cases where there is or is not a single object at a level of processing suitable for three-dimensional object recognition.

These objections notwithstanding, the infancy data raise some useful new questions for Marr's model. There is, for example, a need to clarify which motion-based algorithms infants are using, and their relationship with other processes, especially stereopsis. Spelke *et al.*'s studies highlight the role of motion of object surfaces in relation to each other and to the infant. This might implicate either low-level *directional selectivity* information, obtained during primal sketch construction, which embodies an assumption of 'continuity of direction of flow' and can be used to separate the boundary of a moving area roughly from its background (cf. the role of Y-cells at the primal sketch level); or Ullman's (1979) higher-level *structure from motion*, which is favoured by Marr due to its mathematical 'cleanliness', although he acknowledges a lack of psychophysical evidence for its use.

Neither of these could account for how stationary scenes are parsed as they are. It would be straightforward enough to check the contribution of stereopsis, if any, by comparing binocular with monocular viewing conditions. A further possibility is that these infancy findings may contribute towards questioning Marr's (1982) conclusion that *optical flow* is probably insignificant in the recovery of surface structure.

An interesting question is whether the properties of each of these three types of motion computation can be specified in a way that supports unique behavioural predictions for infants. For directional selectivity, it proves possible to exploit Braddick's (1973, 1974) experimental designs, which used random dot/line displays to specify the parameters that yield segregation and apparent motion of part(s) of a display. For instance, if central dots in a display move in the same direction, adults perceive them as a group (for example, a square), segmented from surrounding dots that are moving in random directions, even if the central dots are moving at different speeds from one another. Using this method, Wattam-Bell (1992) has found no evidence for infant sensitivity to this source of motion information prior to 1 month of age.

The sufficient information for structure from motion and for motion parallax, a form of optical flow due to the subject's motion, has also been isolated in computer-based demonstrations. In both cases, adult viewers see these as random dot arrays, until motion is added. In the former case, the dots' movement results in perception of two coaxial, counter-rotating cylinders (Ullman, 1979); in the latter case, displacement of the dots with the viewer's lateral movements simulates the patterns of relative motion that would be produced by a real three-dimensional corrugated surface at the distance of the screen (Rogers and Graham, 1979). 5-month-old infants (somewhat older than those in the studies under discussion here) seem able to exploit the Rogers and Graham demonstration to specify the location of a convex surface, as indicated by their reaching behaviour (Yonas, Smith and Granrud, 1983, cited in Yonas and Granrud, 1985, p. 120). It should be possible to adapt either demonstration for younger infants by using habituation and/or preference techniques.

The interpolation of missing boundaries or surfaces highlights a different type of ability, and the picture here gets more confusing. Marr's consideration of psychophysical data from adults leads him to stress the role of stereopsis and the continuity constraint in filling in surfaces and smooth continuation of discontinuities, possibly including continuation behind occluding planes (Marr, 1982, p. 287). Some support for the stereopsis possibility might come from the fact that the 3- to 4-month-old infants of these studies are moving into the age range when stereopsis comes into play; evidence considered by Atkinson and Braddick (1989) suggests 12 to 13 weeks of age as the lower boundary. Neonates, who certainly lack stereopsis, have recently been shown to exhibit no sign of the filling-in characteristic of their older counterparts on the partial occlusion task (Slater, Morison, Somers, Mattock, Brown and Taylor, 1990).

The infancy data are particularly interesting since they appear to challenge Marr's account by suggesting a central role for motion in filling in with partly occluded displays. The experimenters argue that '*real object motion*' is the sufficient stimulus, and that infants are sensitive to relations in optical flow patterns. Motion of the partly occluded object laterally, vertically or in depth seems effective, but not subject motion around a stationary display (Kellman, Gleitman and Spelke, 1987). Of course, the 'real motion' here is simply one defined by elements of the rod and block moving relative to one another in the visual field. It need not imply that the infant's perception is based on the subsequent motion of previously perceived surfaces; it could be based on computations at the primal and 2½-D sketch levels.

Although Marr does not consider the possibilities of motion processes for arriving at continuation, perhaps this could (and should) be accommodated in his model? Structure from motion, which is a form of apparent motion, obviously involves a type of filling in of surfaces. And, on the face of it, computational theories of optical flow are similar to structure from motion. They involve a type of successive matching (probably operating on the full primal sketch), and rely on an assumption of rigid surfaces to guarantee a unique interpretation of a scene (Longuett-Higgins and Prazdny, 1980; Prazdny, 1981). Alternatively, perhaps the performance pattern across ages and conditions reveals an interaction between processes based on motion and stereopsis; for instance, motion might be required to get surfaces in the first place, then stereopsis to achieve filling in. Additional studies to specify type(s) of motion used and the availability/use of stereopsis should clarify this issue.

Even if the 2½-D sketch can account sufficiently for the infant discriminations discussed above, Spelke is certainly right that some kind of further segmentation would be required to provide a suitable input to the object-recognition process. Spelke's notion of 'surface points' is equivalent to a pointillist depth map, whereas the 2½-D sketch also yields boundary information in the form of contours that make explicit discontinuities in orientation and depth. Even so, the contour primitives of a 2½-D sketch type of representational format will fail to individuate 'objects' in certain kinds of arrangement.

Marr's solution is sketched in the next section, but the need for an intermediate level between the 2½-D sketch and his object-based 3-D model description has been suggested. Fisher (1986) proposes an intervening representation – *surface clusters* – which supports an identity-independent description of 3-D bodies in static scenes. Looking for surface connections will not solve the problem; indeed, they often *are* the problem, as when surfaces of different objects are touching one another (cf. limitations of the 'cohesion' and 'boundedness' principles discussed above). Fisher's solution exploits insights of early AI scene-analysis work to determine which boundaries between surfaces are segmenting boundaries. For example, a concave boundary is provisional evidence for segmentation because, when one object is in front of another, the boundary that intervenes between them will always be either concave or occluding. The developmental theory raises the possibility that different kinds of motion information might be usable for arriving at the 2½-D sketch and then for further segmentation. While this should not be discounted, what algorithm(s) might be involved are, as yet, unclear.

The 3-D model description – shape for object recognition

Since the 2½-D sketch is viewer-centered, it is unsuitable for recognizing any particular three-dimensional object as 'the same as one seen earlier'. Marr considers a more efficient representation for this purpose to be one which makes explicit readily identifiable geometric features of an object's *overall* shape and is *object-centered*; that is, it can specify those features and their dispositions relative to the shape itself. This would make it possible to construct a *canonical* description that is unaffected by whether one saw the object from the front, side, or whatever. The final type of representation that is proposed within Marr's vision scheme – the *3-D model representation* – is intended to satisfy these requirements (Marr and Nishihara, 1978), and provides the first description in which objects are individuated.

Marr argues for the utility of a representation whose primitives are based on *volumetric* as opposed to surface properties. The geometric properties of the volume occupied by a shape can be used to construct a description of the *relative* spatial arrangement of the components of the object involved. A suitable primitive is Binford's (cited by Marr and Nishihara, 1978, p. 277) notion of a *generalized cone* (a surface that is created when a cross-section is moved along a given straight axis, the cross-section varying smoothly in size but not shape). Having an axis as an integral part of their structure, generalized cones can be used to define a local coordinate system.

Complex shapes can be described in terms of *3-D models*. For example, the complete 3-D model description for the human body consists of a hierarchy of 3-D models, each comprising a model axis for the shape involved together with its component axes. This permits the body to be described at differing levels of detail. The most general 3-D model in the hierarchy is a model axis corresponding to the gross size and orientation of the body, together with six component axes (for the torso, head, arms and legs). A more detailed description is achieved by developing 3-D models for each of these components, and so on.

The 3-D model description may seem much farther removed from the image than either the primal or 2½-D sketches. However, Marr was able to establish that if a physical surface is a generalized cone there are specific relationships between the points on its surface that give rise to occluding or bounding contours in the image and inflections (convexities and concavities) of that contour. So, there is a natural link between generalized cones and the imaging process. This means that data-driven processes making use of these relationships offer one method of recovering the natural axes of a shape from its image to provide the basis for a 3-D model description. No a priori knowledge is required, other than the (implicit) assumption that the three-dimensional shape involved is composed of generalized cones.

Nevertheless, as far as early infant performance is concerned, this type of representational format proves far more elusive than the primal or 2½-D sketches. Relevant tasks are those testing the infant's ability to maintain or generalize an object description in the face of variation in viewer-centered properties like orientation, such as traditional explorations of *shape constancy*. However, no consensus has emerged from studies along these lines with newborn and young infants (up to about 4 months of age); conclusions about the infant's capacity for *orientation-independent description* vary according to the experimental methods and stimulus materials used.

Different theorists have explained discrepancies in a similar way. Cohen (1979) proposes a distinction between 'simple' patterns or forms, which even the youngest infants treat as compounds and as equivalent across orientational change, and 'complex' ones, such as faces and coloured forms, which they do not. Along the same lines, E. J. Gibson and Spelke (1983) conclude that young infants have the ability to discriminate and recognize 'simple' shapes but not 'complex, embedded' ones. But such *post hoc* dichotomies seem merely to redescribe (the experimenter's intuitions about) stimulus materials associated with infants' 'successes' and 'failures'. It is important to look more closely at why it may be difficult to demonstrate the unequivocal presence of this kind of description.

To begin with, we need to consider whether infants really are capable of describing shape in a viewpoint-independent way in typical tasks, let alone in Marr's particular 3-D model description terms. Even the type of data that supposedly reveals precocious shape constancy with simple cases is, in fact, more problematic than is generally thought.

For example, very young infants who habituated to one shape, a square, appeared to 'classify' it as the same when it was rotated through 45 degrees (Schwartz, cited by Day and McKenzie, 1977, p. 310). But rather than infants displaying a higher, more abstract level of description with this simple stimulus, the opposite may be the case: for adults the square and the diamond are *not* equivalent. This may be linked to adults' representational formats including 'significant directions' (e.g. Hinton, 1979), a computational notion that is compatible with the axes that are central to Marr's 3-D model scheme.

It may be that infants are unable to generate such a description with the kinds of two-dimensional black and white stimuli that were used in Schwartz's and similar studies, and which seem ideally suited to description in the primal sketch terms discussed earlier. In Marr's scheme, silhouettes can be recognized in terms of the relationship between generalized cones, contours and image formation. It may not be possible to recover the volumetric information that could form the basis for an object-centered representation from the type of stimuli that Schwartz used, due to the nature of their bounding contours and the consequent absence of the necessary correspondence between contour inflections and potential points on the surface of a three-dimensional object. It would be instructive to explore this issue empirically using two-dimensional contour figures similar to Picasso's *Rites of Spring*, which Marr has used to illustrate properties of his 3-D model representation and whose formal properties are better understood.

A well-known shape constancy phenomenon involving apparently simple three-dimensional objects also proves problematic in an interesting and different way within the context of Marr's model. Very young infants have been found to generalize an operant head-turning response from a solid rectangle in the fronto-parallel plane to the same rectangle oriented at an angle of 45 degrees to their line of sight (Bower, 1966). Similarly, they have shown identical patterns of habituation with either repeated presentations of a single cube or a series of presentations of the cube in different orientations (McKenzie and Day, 1973). In both sets of studies, photographic stimuli failed to produce the generalization pattern characteristic of solid objects. Reviewing these and other shape constancy studies, Day and McKenzie (1977) concluded that such early capacities for establishing equivalence have their basis in *slant* information derived from *motion parallax* (cf. Bower, 1966).

Marr's tripartite explanatory system proves highly relevant here. It suggests that these conclusions concerning either *what* infants are achieving or *how* they are achieving it – or maybe both – may be in error. Motion parallax due to subjects' lateral head movement might be useful for specifying the slant of stationary surfaces in a 2½-D sketch-like description, but different views are not going to be equivalent under that kind of description. It is not clear how infants are relating different views of the objects involved across trials, nor how they might produce a single, unifying shape description or sequence of equivalent descriptions. The information stressed by the developmentalists is simply inadequate for object shape constancy under the conditions in which they have tested it and in the sense in which they generally define it.[9]

Although the kind of motion highlighted in the structure from motion process also results in a viewpoint-specific description (or descriptions), it involves a form of apparent motion and so can specify (albeit within narrow spatio-temporal limits) that a *single* object is involved. Ullman (1979) himself suggests that autonomous *correspondence through motion* processes may be of significance for infant object understanding. For different views of an object to be linked through this process, however, it is essential that they are related by *smooth motion*. There appears to be no way in which such processes could establish equivalence across *discrete presentations* of a static object in different orientations, unless the change between trials is rapid enough to give rise to apparent motion – an unlikely possibility.

This analysis concurs with Kellman's conclusion, from studies that varied both orientation and motion, that spatio-temporal continuity is central to whatever kind of constancy infants are exhibiting here (Kellman, 1984; Kellman and Short, 1987). When this was eliminated by exposing infants to multiple, static views of an object undergoing rotation, with intertrial intervals that were brief but not so brief as to support apparent motion effects, generalization of habituation to that object on a new axis of rotation was also eliminated. (This is quite unlike the adult pattern, where a single stationary view is often enough for this kind of recognition.) Kellman believes this shows *continuous rotation in depth*, whether obtained by rotating the object or by moving the subject around the object, to be the sufficient stimulus for *three-dimensional form perception* by (4-month-old) infants.

This illustrates a key difficulty for the attempt to map between Marr's ideas on the 3-D model description and infancy research. Developmentalists' vocabulary differs most markedly from that of the computational account with regard to 'shape' (in contrast with the close parallels at the level of primal sketch considerations). Shape description based on volumetric primitives is a long way from the notion of equating object form with (two-dimensional) patterns or feature configurations which characterizes much of the infancy work that was highlighted in discussing the primal sketch. Spelke's research leads her to conclude that shape or form is not even an 'essential' property of objects, although the studies concerned attempted to explore only the role of 'simplicity' or 'goodness' of bounding contours in parsing scenes.

As far as Kellman's studies are concerned, he argues that *perspective transformations* of an object's bounding contours *directly* specify its three-dimensional form for infants. However, although he is undoubtedly correct that infants are not responding in terms of proximal stimuli, no precise definition of the 'form' alternative is given. So, the task facing the infant, and the implications of these studies for Marr's model, are unclear. Informally, adults would most likely describe the two solid objects used as a wedge and an L-shaped block, but we must ask what kind of description is motion in depth allowing infants to construct in these studies? The objects have identical maximal extents but will differ in, for example, surface area. Therefore, although their main axes are probably different (their principal/component shape axes should not be confounded with the axes about which they rotated in the experiments), it remains possible that infants' recognition is based on surface information rather than on a canonical description of overall structure.

From a computational viewpoint, describing the shape even of something seemingly as simple as a cube is a far less banal task than it might intuitively seem. A structural description of its three-dimensional shape requires three orthogonal *axes* or *significant directions*. Adults, including vision researchers, find an imagery task that involves rotating a cube into a position where its new vertical has components in all three of these axes very hard indeed (Hinton, 1979). This type of shape analysis might profitably be turned to the task of exploring infant perception. For example, adults find that an imagery task that involves rotating a cube so that its new vertical has components in only two of its orthogonal axes is easier than the 'very hard' task where it has components in all three. If infants are exploiting a similar kind of description, this order of difficulty may correspond to their degree of generalization across different types of object-orientation change.

Clearly, the particular type of object used in the infancy studies is highly significant. This might be due not only to the appropriateness or otherwise of such objects' properties to description in 3-D model terms under the study conditions, but also to infant vision being specialized to describe some kinds of things more directly than others. Phenomena not usually associated with shape constancy may offer the most suggestive evidence of young infants employing a representational format along the lines of the 3-D model description. Paradoxically, as far as the infancy literature is concerned, the kinds of objects involved – hands and faces – would undoubtedly count as 'complex'.

Meltzoff (1981) argues that a mechanism involving very abstract representation must be implicated in the neonate's ability to imitate facial gestures such as tongue protrusion, and hand movements such as opening and closing. An adult hand, say, must somehow be 'matched' with the infant's, despite differences in size, texture and so forth. The volumetric primitives of the 3-D model representation, or some equivalent format, could fulfil this function, whereas neither the primal sketch nor the 2½-D sketch could do so. At the least differentiated level, a 3-D model description for a hand would consider it as a single volume, making explicit its main axis. At the next level, this would serve as the coordinate system for the fingers, each of which can be considered as an independent volume with its own axis. The axes of each 3-D model in the description of the human body provide a local coordinate system, which preserves modularity within the representation, enabling the shape of parts such as a hand to be described irrespective of their spatial relation to the body. This description would be abstract in the sense of being independent of the hand's location, orientation and size, and of the subject's viewpoint; and texture and colour would be peripheral components of the final object description.

By the same token, the volumetric 3-D model scheme might contribute towards an explanation of the infant's imitation 'errors' (for example, some, though not all, studies find that not only adult tongue protrusion but also other approach movements, such as those of a pen, will selectively elicit tongue protrusion in young infants; contrast Jacobson, 1979 and Legerstee, 1992); and the generalization pattern of infants' imitative 'near misses' which Meltzoff (1981) characterizes as some form of 'intelligent confusion'. It certainly offers a straightforward account of why it is so easy for adults to recognize massively 'simplified' objects, such as stick-figure animals, which share none of the surface structure of their 'real' counterparts: component sticks may correspond to the axes of generalized cones that approximate the shapes of component parts of the 'real' object. Whether infants would find this as easy is an open question that would be worthwhile addressing to them, again with stimuli whose properties are (at least better) understood.

These ideas may also have something to contribute towards 'social' interpretations of the literature on very young infants' reactions to facelike stimuli (Morton and Johnson, 1991). Marr (1982) suggests that volumetric primitives making explicit a main model axis and component axes could be used to describe certain kinds of two-dimensional configurations, including drawings of faces. Provided the configuration has a natural axis of symmetry, a canonical axis can be recovered, and component axes can be recovered for surface primitives such as eyes and mouth. This could certainly handle the 'oval ground plus three blobs' face-approximation preferred by the youngest infants.

Admittedly, this mapping is suggestive rather than compelling. However, approaching infant face understanding from this direction has the advantage of beginning to integrate it with the otherwise (methodologically and theoretically) independent imitation data. Its plausibility is increased somewhat if we consider Johnson's (e.g. 1988) attempt to define what constitutes 'henfulness' for a newly hatched chick. For example, what mechanisms underlie chicks' preference for a rotating stuffed hen over a rotating box of the same size? Absence of preference for a (somewhat grotesque) stimulus object composed of the jumbled limbs of a stuffed fowl suggests that surface properties such as rich contours or feather pattern are not implicated.

The key appears to lie in 'some group or groups of features, in their correct position relative to each other' (Johnson, 1988, p. 61); in particular, something about the configuration of head and neck. Whatever it is about this configuration that engenders preference, it is shared by a stuffed polecat, a natural predator to which a strong preference is also exhibited. Object-centered representation involving axis-based coordinates might explain this otherwise paradoxical finding in terms of the equivalence of shape descriptions for the head–body and head–neck regions of polecat and fowl respectively.

As Bruce (1988) points out, Marr's volumetric primitives describe *all* human faces as similar in terms of their overall three-dimensional structure. They are, therefore, an unlikely basis for the task with which Marr himself is concerned (recognition of particular faces), but may be suitable for the kind of 'universal face schema' required by the infancy data. Relevant here are Marr and Nishihara's (1978) proposals for interaction between 3-D model construction and a memory-based 'catalog' of *prototypes*, when the 3-D model description is used for *explicit* recognition.[10] The most important 'indexes' into this collection allow recognition to proceed from the general to the specific, and the two most general would match the infancy data: first, the most undifferentiated model, the main model axis, corresponding to gross size and orientation; next, the number of component axes and their distribution along the principle axis; only then, particular angles and lengths; and so on.

These links suggest a profitable direction that research might take, rather than firm conclusions.[11] For example, a key challenge to the Morton–Johnson model questions its assumption of the time-course of cortical learning because infants of no more than a couple of days of age appear able to discriminate their mother's face from that of a stranger on the basis of visual cues alone (Bushnell, Sai and Mullin, 1989). Viewed computationally, however, it does not seem useful to place localization of function claims at center-stage in this debate, unless they are accompanied by a corresponding clarification of mechanisms through clear dissociation of function. Bushnell *et al.* have made some progress on this difficult task, but the *visual* basis of the infant's recognition ability remains unclear. Further studies informed by computational ideas on face shape perception may further understanding of 'what' processing is going on rather than 'where'.

What happens next?

The preceding discussion illustrates how concepts and questions central to Marr's computational framework can suggest ways of integrating a wide range of otherwise disparate phenomena of infant vision, without resorting to *ad hoc* eclecticism. Correspondences between Marr's model of human vision and the infancy data begin to move us well beyond the notion that there is preadaptation on the infant's part, towards a deeper understanding of the nature of that preadaptation. Beyond this, new questions are raised for both sides of the comparison.

From the perspective of the computational theory underlying each of the levels of processing discussed above, it becomes clear just how complex are the tasks facing infant vision – often different from, and more numerous than, those envisaged by developmental theories. Considering the algorithms and representational formats that might be exploited by infant vision reveals support for the psychological validity of Marr's independently motivated model, although it also raises questions about some of his personal algorithm-representation preferences. This capacity to generate novel, precise questions of mutual interest can be viewed as a point in favour of the computational framework.

Where the computation–infancy mapping works at the algorithmic level, it not only supports the notion of domain-specific innate constraints on the infant's object knowledge, but also clarifies how these constraints are embodied in psychological mechanisms. The way that infant knowledge is grounded in mechanisms prevents us from falling into the trap of moving from the data to either over- or underattribution of competence to the infant. The multiple, process-embedded constraints that constitute the infant's 'visual knowledge' embody assumptions about some of the most general properties of the physical world. Such processes fit Fodor's (1983) idea that perceptual processes are candidates for his notion of modularity, in so far as they are *informationally encapsulated*. That is, unlike reflexes, they involve computation (inferences from data), but they are special-purpose devices, data-driven and operating only on a prespecified, restricted range of data (cf. Fodor and Pylyshyn's (1981) autonomous computational 'reflex').

Comparison at this level also demonstrates an important general sense in which a computational model has the potential to provide a powerful explanation. In the cognitive science literature, Brown and van Lehn (1980) suggest that linguistics' notions of grammatical competence and 'starred sentences' (deemed ungrammatical by native speakers of the language) can be extended to cognitive modelling. Because all sorts of process suggestions will be capable of generating a given outcome, a psychologically valid peformance model of a domain of ability should do more than generate all of the successful human performances in that domain. Additionally, it should generate all of the *bugs* (or errors) that humans actually produce and none of the 'star bugs'; that is, bugs that are not produced by human subjects, although they could, in principle, exist in the domain concerned. Marr's model goes some way to offering this kind of explanation for a range of infant visual performance, suggesting mechanisms that might generate 'appropriately' errorful or paradoxical behaviour.

It may be possible to go even further in this direction. Infants' interpolation of partly occluded surfaces, which was discussed in relation to the construction and properties of the 2½-D sketch, raises a non-trivial issue for Marr's computational model: the *maintenance* question. A genuinely functional view of the 2½-D sketch requires an explicit account of how the internal consistency of the surface descriptions that it generates are maintained over time. As we have seen, it is possible to begin discussing spatial interpolation in terms of the integration and internal consistency of the 2½-D sketch's representation of surfaces and depth. But what of such representations' *temporal* characteristics? To what extent might analogous mechanisms for integration/consistency over time yield temporal filling-in or *persistence* of surfaces? In Marr's model, 'memory' features explicitly only at the level of the 3-D model description. But the infancy tasks that have been shown to correspond closely to surface-based descriptions involve a short-term memory component (albeit of relatively brief duration, it is lengthy in comparison with the speed of visual processing).

Computational–developmental collaboration in this direction could well contribute to an explanation of some important findings with infants rather older than those considered in this chapter (Baillargeon, 1986; Baillargeon, Spelke and Wasserman, 1985). Following habituation to a screen rotating back and forth, they exhibit dishabituation if the screen rotates backwards after an object has moved behind it without emerging. Following habituation to an object that moves behind a screen to emerge at the opposite side, they look longer at an 'impossible' event, where the object continues to follow the same path although a box has been placed on top of the track behind the screen, than at a 'possible' one, where the box has been placed behind the track.

Baillargeon's (1986) conclusions are couched in the language of 'object permanence': by 6 months of age, infants believe that occluded objects continue to exist in space and time, and can represent their existence, location and trajectory. She acknowledges that 'belief in object permanence' is probably no isolated conceptual attainment; so, it is appropriate to ask to what extent the infant's early 'causal reasoning ability', as expressed in (dis)habituation behaviour, could be grounded in properties of the scene descriptions constructed by low-level vision. As with Spelke *et al.*'s studies, we are dealing here with infants' appreciation of surface arrangements in depth. An explicit characterization of the maintenance characteristics of the 2½-D sketch's description of surfaces – put another way, processes for establishing data-base consistency in this format – might offer a process-based explanation of infants' discrimination between 'impossible' events, where two objects appear to be in the same place at the same time, and 'possible' ones, where they do not.

As far as physical implementation is concerned, we might best consider Marr's work as highlighting the kind of complexity that will be needed by a successful model. Without assuming that his treatment of neurophysiology is complete or absolutely right, which would be premature, its importance as a serious attempt to integrate psychological and neurophysiological insights can be acknowledged. One of its implications, to which I have been sympathetic, is that the 'localization of function' methodology is unlikely to figure highly in increasing our understanding of the 'what', 'why' or 'how' of psychological processes. 'Dissociation of function', on the other hand, may offer clues for task analysis (for example, functional dissociation in brain-damaged adults informs Marr and Nishihara's (1978) ideas about the nature of object shape recognition).

In this vein, it may be mutually beneficial for the retina-to-cortex dominated computational model(s) of vision to consider ways of taking on board the (as yet far from clear) significance of many of the early infancy phenomena (from visual preferences to neonatal imitation) being associated with subcortical/midbrain areas. As yet, neither the functions of cortical and subcortical areas nor the interactions between them are well understood. Many options exist. For example, subcortical areas are often associated with mediating what have been thought of as 'reflex' as opposed to 'voluntary/guided' behaviours. One possibility, therefore, is that the two regions are not characterized so much by qualitatively different *perception* as by their processing (sometimes similar) information for different purposes and/or as a component of different forms of *action coordination*.

This points to an integration issue that is significantly broader than the maintenance question posed above: what happens to the descriptions constructed by low-level vision? How do they mesh with the infant system as a whole and with the knowledge that it supports? Spelke's (1990) reintroduction of a perception–conception distinction is compatible with Fodor's (1983) proposal for a distinction between input systems and central systems of belief fixation. But though vision's characterization as an input system seems apposite in some respects, I am sceptical of the value of accepting mysterious 'higher', central processes as the broader context for understanding it.

At best, taking that route would seem to lead to a problematic juxtaposition between, on the one hand, the detailed account of (some) computational mechanisms that was offered for vision and, on the other hand, an account of central processes in the very kind of molar folk psychology terminology that the computational concepts aim to replace. Re-evaluating the notion that the outputs of visual processing are used for the construction of explicit central beliefs about objects, and beginning to phrase an alternative, involves moving to a new scale of analysis. This must focus on expanding the context within which the mechanisms underlying infant ability are located.

Notes

1. Marr's (1977, 1980, 1982) own overviews provide a comprehensive introduction. Boden (1988) and Rutkowska (1988) link Marr's work and the concepts and methods of preceding AI work on scene-analysis. Links with neurophysiology are lucidly covered by Bruce and Green (1985) and Frisby (1979). And issues concerning explanatory levels for information-processing systems are introduced by Ullman (1980).

2. Marr's notion of 'a description within a representation' carries much the same import as the 'structural descriptions' of earlier AI vision work – both imply types of symbol structures that can designate certain kinds of entities in a certain kind of world.
3. Many extensions and challenges to Marr's framework are already under way, although they do not yet have clear implications for how we might exploit links between computational explanations and infancy. For example: different mathematical bases have been suggested for a formal computational theory (e.g. Pentland, 1985); additional levels for explanations of information processing systems (Peacocke, 1986; Rumelhart and McClelland, 1986); further stages and types of representation (Fisher, 1986; Sloman, 1987b); and different processes for recovering information with regard to exploiting motion in the image (Buxton and Buxton, 1983).
4. The response of X-cells to differences in intensity between their center and its surround has been characterized mathematically as taking the difference of two Gaussian distributions (Enroth-Cugell and Robson, 1966). Marr and Hildreth (1980) establish that this 'DOG' function is equivalent to their second-order differential operator.
5. This is important to the generality of Marr's model since it allows objects to be sought at different scales of analysis.
6. Spelke *et al.*'s claims do not cover *texture gradients*. In any case, Marr (1982) argues that getting from such gradients to the distance and orientation of surfaces is harder than Gibson (1950) originally believed.
7. These are discussed in Rutkowska (1988). Cf. also Bower's (1974, 1982) proposals for expressing the infant's object knowledge as a series of conceptual rules, with the infant's earliest rule featuring boundaries, volume, location and movement.
8. Which may make this constraint implausible. Many everyday motions, such as facial expressions, are non-rigid (Sloman, 1980). On the other hand, a biomechanical motion such as walking involves local rigidity, which infant vision seems able to use within limits set by the orientation of the (moving light-point) displays employed (Bertenthal, Proffitt and Kramer, 1987).
9. Some straightforward computational method for getting from the motion parallax parameters provided by an object at an angle to the line of sight to a description of its shape in the fronto-parallel plane may exist for objects such as rectangles with fairly simple surfaces (D. Young, personal communication), and offer a principled basis for a simple-complex distinction. In terms of the known mathematics, however, it is not like the relationship between motion and the 2½-D sketch, where the structure of the sketch is delivered as a 'result' by the process.
10. This 'catalog' returns us to the familiar territory of experience-based 'schemata'. How it could develop and operate computationally requires a great deal of additional clarification and will not be discussed further here.
11. See Humphreys (1987) for an introduction to some interesting developments in this direction with adult subjects; and Bruce (1988) ch. 6 for a discussion of face perception in this light.

Chapter 3

Getting perception into action

A predominantly centralized view of mind has led the cognitivist approaches most widely found in adult and developmental cognitive psychology, AI and cognitive science in general to share the tendency of viewing mind and brain as coextensive, and cognition as something that is localized strictly 'between the ears'. For example, figure 3.1 shows Pylyshyn's (1981) scheme for partitioning the world into sources of behavioural complexity. It does not aim to depict a mechanism for intelligent abilities, and its author insists that it is 'simple minded'. Nevertheless, it inadvertently succeeds in illustrating two features that are characteristic of most – though not all – computational models.

First, subject (device or organism) and environment are schematized as mutually exclusive sources of complexity. Indeed, Palmer's (1978) metatheory of cognitive representation is predicated on an assumption of two related but functionally separate worlds: a represented world and a representing world which 'stands for' it. Second, in keeping with the establishment of this internal–external dichotomy, behaviour rarely appears at all. Pylyshyn is not unique in taking 'behaviour' to mean something like 'what the subject can get done', directing attention towards task structure and away from the overt body movements that we can observe in the environment.

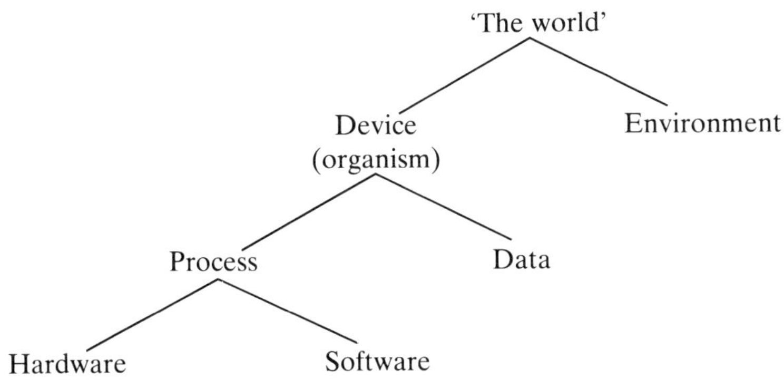

Figure 3.1 Sources of behavioural complexity. (After Pylyshyn 1981).

In the main, the developmental cognitive psychology of infancy has conformed to this vantage-point. A range of post-Piagetian theories start from the view that habituation or familiarization in the face of repeated exposure to stimuli, cues or patterns leads the infant to construct a memory-like representation of the input. Known variously as a 'model', 'expectancy' or 'schema', this is supposed to mediate future interactions between infant and environment, and to determine the likelihood of new information being incorporated into a modified internal model (e.g. Charlesworth, 1968, 1969; Jeffrey, 1968, 1976; Kagan, 1972, 1979; McVicker Hunt, 1961, 1965). A significant influence on these precomputational cognitivist theories has been the 'old connectionism' of Hebb (1946, 1949), with its notions of 'cell assemblies' and the motivating power of 'discrepancy' from 'habitual' perceptions (Rutkowska, 1990). In keeping with Hebb's seminal opposition to behaviourism, such theorists argue that the infant's 'memory' for static visual displays must go against Piaget's 'motor schematic' theory (Haith and Campos, 1977). This leads them into a blind alley as far as explaining the abilities of infants is concerned. Overcompensating for Piaget's apparent motor (over)emphasis, their centralist preoccupation ends up missing essential features of how those abilities work. Their infant turns out to be a peculiarly disembodied, disembedded and decontextualized creature.

In my view, the human infant is a uniquely suitable candidate for the turtle's role in Dennett's (1978b) 'Why not the whole iguana?' proposal. This asks how cognitive scientists could achieve theories of *whole* intelligent systems in place of models of partitioned-off subsystems such as perception, problem solving or reasoning. Putting all of these together will not result in a unified theory; how a subject's knowledge is brought to bear on a range of tasks needs to be understood. A model of a whole human being would be too vast to handle. The problem could, however, be simplified and scaled down by trying to model a less sophisticated whole creature such as a turtle, whose 'knowledge' poses ample issues relevant to how organization and control are directed at survival in real time in a real world. The drawbacks are that you would have to learn all about turtles; your findings would be unlikely to impress others who were not similarly preoccupied; and you might get bogged down in the eccentricities of turtle cognition, when what you are after are principles applicable to cognitive organization in humans.

To overcome this, Dennett suggests making up a whole cognitive creature – for example, a Martian three-wheeled iguana – and an environment for it to cope with. Given, however, that we lack a deep understanding of the principles governing not only whole cognitive creatures but also environments, the human infant seems a better candidate for the role of restricted whole system. We know a fair amount about infants, however hard our data are to interpret; and, whatever their eccentricities, infants predictably become adult humans. The nature and organization of their abilities can offer insights into one particularly relevant way in which a naturally intelligent system is designed to interact with and develop in the human physical and social environment.

Understanding the infant in this way involves defining an expanded scale of analysis on which to seek the mechanisms of infant ability, one in which behaviour can be grounded appropriately as an active interface between subject structures and processes and the ('external') structures and processes that comprise the environment. This means taking the perceiving *and* behaving subject *in* the environment as the unit of analysis, and renewing attention to the physical-motor mechanisms that make this embedding possible.

These considerations, and the range of infant perceptual-behavioural coordinations noted in chapter 1, suggest that the 'object' for explanation in an account of infant abilities is an *action system*. The central aim of this chapter is to get some leverage on what this notion means.

To most developmental psychologists, any move to 'get the subject into the environment' is virtually synonymous with endorsing ecological psychology, but this equation is by no means necessary. Functionalist considerations converge on a decentralized view of ability but cover a wide range of otherwise incompatible theoretical directions, ranging from ecological psychology's 'methodological realism' (Shaw and Turvey, 1980) to Varela's (1988) orthogonal commitment to the world's being 'brought forth' rather than pre-given. In and of itself, therefore, advocating a more mutual view of subject and environment does not have straightforward implications for how we should view the infant. Implications follow once specific mechanisms are proposed, and here ecological psychology and (any) computational approach inevitably diverge.

While computational and ecological approaches agree that different levels of explanation of a system have to be distinguished, they dispute the constructs that should be employed at the psychological level, and its relationship with the biological level. These differences make it clear that the two perspectives offer fundamentally different styles of explanation and are not compatible in the way some theorists believe (cf. Reed, 1986). Ecological psychology proposes that constructs concerning any organism and its environment should be phrased at three alternative 'grains of analysis', which cannot be causally related: psychology (for example, affordances and effectivities); biology, including the brain; and physics (Michaels and Carello, 1981). For the computational approach adopted here, psychology–biology interactions are at the heart of the virtual–physical machine relation. This supports a distinctive and powerful perspective on *functional perception*, the nature of *behaviour* and the *subject–environment relationship* in the organization of ability.

Taking behaviour seriously

Prior to mechanism considerations, it is necessary to frame the context in which the question 'What happens after low-level vision?' can usefully be asked. Central to my viewpoint is the assumption that we cannot understand naturally intelligent systems like the infant by decomposing their abilities into input systems (modular or not) and central systems. Output systems must also be included; in the human case this means behavioural processes which span subject and environment. Deciding to take behavioural processes seriously moves the attempt to understand perceiving beyond the confines of the descriptions constructed by vision to consider their *selective use* in *functional activity*. Evidence for modularity in low-level visual processing does not mean that perceiving as we generally think of it can be fully understood as an autonomous, data-driven, modular process; vision's coordination with behaviour must also be taken on board.

The type of complexity involved is illustrated by infant performance on the visual cliff. Recordings of infant heart rate reveal that they discriminate between the two sides of the cliff long before they can crawl. Their response to the deep side changes over the first year of life, and it is somewhat counterintuitive. In the youngest infants, heart rate on the deep side shows the pattern of change characteristic of interest or heightened attention; they actually cry less than they do on the shallow side. Only older infants show the pattern that is generally associated with fear (Campos, Hiatt, Ramsay, Henderson and Svejda, 1978; Campos, Langer and Krowitz, 1970).

The responses of younger and older infants suggest there is some sense in which they are 'interpreting' the same visually provided information about the cliff in different ways, and that difference lies in how the information is or is not used. Initially, infants can discriminate the available information for depth without 'recognizing' its conventional meaning or significance for adaptive activity; such recognition is associated with the use of that information in crawling (cf. E. J. Gibson and Rader, 1979). Information that has been made explicit by low-level vision may still be implicit or unavailable to the infant system at the level of action.

The distinction between information's availability to a subject – as indexed by, say, discrimination tasks – and its selective use in purposive activity has relevance beyond the functional coordinations of perception and behaviour found in young infants. Cognitive and cross-cultural studies of adults also point to the dynamic, selective aspect of perception and to its intricate connection with the goal-oriented aspect of activity. Furthermore, they illustrate relations between this and supposedly 'higher' cognitive processes such as evocation memory and classification.

Adult introspection may often suggest that knowledge of an object consists of constructing as complete and detailed as possible a percept or image of it (an exhaustive representation) but it is misleading. For example, most people 'know' what a penny is, but Nickerson and Adams (1979) found adults to be remarkably poor at drawing one unaided, drawing it from a list of features, choosing its features from a list of possibilities, indicating errors in a drawing, or even selecting the correct drawing from a set. The visual details of the object appear to be accessible through either recognition or evocation only to the extent that they are *useful* in everyday life.[1]

Deregowski's (1978) review of cross-cultural studies concludes that familiarization due to frequent encounters with phenomena is independent of their discriminability and recognition. Subjects are often unable to sort or classify objects on the basis of apparently obvious stimulus characteristics of commonly used, frequently encountered objects. Colour, for instance, may be unusable by people from a culture where a focal activity is flower arranging based on shape.

Similar studies feature in van Brakel's (e.g. 1991) arguments against the notion that a fixed number of domains of salient categories, such as those proposed by Rosch's 'prototype theory', provide the basic elements for all human cognition.[2] Whether biological taxa, colours or emotions are concerned, cultural '*"use" considerations*' regularly overrule supposedly universal perceptual gestalts. For example, the Cheyenne religionists' life-form *zeevseossomotomevo* has been translated as 'birds' and seen as convenient evidence for a conceptual domain shared with English speakers. But concentrating on the numerous bird names that do occur overlooks genera such as *vovetas*, which comprises not only vultures and nighthawks but also two kinds of insect and a meteorological event (tornado).

It is not the case that there is *no* perceptual basis for such behaviour; what *vovetas*' component 'species' share is a whirling movement pattern, a funnel-shaped configuration which is a significant Cheyenne religious symbol. But there seem to be virtually endless ways of seeing even directly obvious perceptual similarities, and they are use related. From this perspective, an apparently universal world view like that exhibited in the widespread existence of words for 'tree' may owe more to the universal utility of burning wood than to the form of a prototypical tree. And the same principle is likely to apply to apparently universal patterns of object understanding that are revealed in infant activities.

Functional perception

In keeping with the centralized view of computation and mind, most detailed AI analyses of vision have restricted themselves to considering it as a process in its own right. For example, Ullman's (1980) key computational critique of *The Ecological Approach to Visual Perception* (Gibson, 1979) does not acknowledge, let alone tackle, either Gibson's central proposal for the action-related nature of perception and knowledge or his attempt to handle it by introducing the 'affordance' concept. In principle, however, the issue of functional perception is in no way incompatible with cognitivist concerns. Even Ames, talking of the correspondence between perception and the objective world, stressed that: 'Whether or not correspondence occurs seems of no importance in and of itself. Correspondence matters only in relation to purposes' (1968, p. 107).

Few explicit insights about the kind of computational organization that might be involved are offered by Marr's framework. It characterizes an efficient representation as one that is useful for the subject's purposes, but generally treats the construction of descriptions within representations as a purpose in its own right. However, a way into considering the architecture of vision in relation to the generation of appropriate behaviour is provided by Marr's (1982) explicit equation of his model's levels of processing with *preattentive* vision. This is compatible with viewing low-level vision as providing automatically various types of information which may or may not be addressed by extra-perceptual processes. The latter would be implicated in the *control* of vision; that is, *how* the (potentially useful) information it provides is sought and used rather than *what* information vision can deliver.

Marr (1977) appositely noted the significance of what he termed 'the reference window problem': how might the description of an object provide the potential for its use in quite different contexts? How, for example, can a newspaper be perceived as suitable for reading, fire-lighting or squashing irritating insects against a brittle surface?

Neuropsychological studies of brain-damaged patients support the view that there are different and dissociable types of 'use knowledge'. For example, left parietal-lobe damage is associated with loss of the ability to provide or recognize the name, use, purpose and properties of an object, even though its shape can be correctly perceived (Warrington and Taylor, 1973). This suggests the existence of a representation which includes an explicit description of the activities an object might be useful for, and which is additional to visual description of the object: a hierarchical description, headed by the object and possibly tree-like, in which a range of properties are brought together and represented as applicable/relevant to the object concerned (i.e. 'X is 1, 2, 3, . . . and can be used for A, B, C, . . . ', etc.). This might be called an *integrated object description* (cf. the idea of describing knowledge in terms of a network of concepts and/or principles, which will be looked at in chapters 4 and 7).

Evidence for a different type of representation comes from subjects who have lost the ability to name an object, or to answer questions about its functions, but are quite able to 'recognize' it in the sense of engaging in appropriate activities with it (e.g. Humphreys and Riddoch's subject, J.B., cited by Caplan, 1987, pp. 186–7). An integrated object description is not essential here (independent representations of the form 'A can be done with/to X', 'B can be done with/to X', 'C can be done with/to X', etc., would do).

How might these types of use knowledge be analyzed computationally? Current ideas suggest two main possibilities for mapping these forms of 'recognition' onto the type of vision model that was outlined in chapter 2.

Computational approaches that are allied with traditional, schema-based constructivism propose a series of hierarchically structured descriptions or representations that become increasingly relevant to the control of action. Oatley (1978), for example, argues that the perceptual process will need to pass through a number of domains of description: image points, lines, surfaces, three-dimensional structures (e.g. cubes), three-dimensional objects with uses, and so on. Low levels of representation will not describe input in terms of the right entities, attributes and relations to be relevant to human behaviour.

To find something to sit on, for example, you would require a description in the three-dimensional object domain that specifies entities such as 'chair', attributes such as 'wooden' or 'comfortable' and relations such as 'provides support'. This could be centrally represented in, for example, a 'frame' of essential/universal and optional properties of chairs (cf. prototypes). The role of perception in sitting might be viewed as the activation of processes that attempt to match environmental objects with this abstract structural description of 'chairness'. To make explicit their relevance for human purposes, approaching objects would need to be described at an even higher level, in the four-dimensional domain of moving objects. An entity in this domain might be, say, a 'cricket ball', embodying relations such as 'dangerous' and 'approaching', and possessing attributes such as 'to be avoided'. The type of information provided by lower levels might contribute to the construction of these high-level descriptions, but it would not itself be considered relevant to generating or controlling behaviour.

It seems unlikely that such high-level integrated object descriptions are required for early infant activities, or indeed for all adult ones. An alternative approach seems more plausible. This involves rejecting the view that early descriptions are nothing but intermediate representations, progressively approximating to a description of the 'real object' towards which action must be directed. The type of object recognition that demands what I called an integrated object description is only one activity relying on perceptually provided information. Instead, *all* levels of description may provide potentially useful information for different purposes. In keeping with this, Marr (1982) suggested that each species will be capable of producing descriptions suited to its particular purposes; some descriptions may be dedicated to specialist purposes, while others are implicated in more general abilities.

Sloman (1983b, 1987b, 1989) begins to develop a proposal compatible with this line of argument. He stresses that an efficient vision system will need to construct many kinds of description, and that a range of processes may be able to access 'intermediate' descriptions. For instance, in cases of the continuous control of activity, such as keeping a line of motion central relative to the edges of a pathway, descriptions of relationships in the two-dimensional image structure might be most useful. It might be harder and slower to use a full description of the three-dimensional layout.

Simply thinking up descriptions that could support each individual function that comes to mind raises the danger of an unprincipled proliferation of unrelated, *ad hoc* proposals. I shall, therefore, restrict myself to considering some evidence for the functional possibilities of descriptions within the representational formats detailed in chapter 2.

Does it make sense to talk of descriptions in terms like those of the (lowest level) primal sketch as informative or useful except as an intermediate stage in visual information processing? If we consider 'lower' animals, the answer is clearly 'yes'. For example, descriptions whose 'vocabulary' is along the lines of Marr's primal sketch seem suited to known purposes of frogs, flies and bees.

Frog 'fly recognition' relies on a description that appears to differ markedly from our own. The frog's feeding activity uses 'net convexity detectors', which underlie a description that makes explicit only that a blob of a certain size is moving across the visual field (Lettvin, Maturana, McCulloch and Pitts, 1959). In the context of the frog's feeding program, this suffices for guiding the tongue to the relevant location and type of object. The fly's flight to intercept others of its kind uses a description that makes explicit the direction and angular velocity of a small patch in the visual field (Collett and Land, 1978; Poggio and Reichardt, 1976). For the bee, the extent to which a form is 'broken up' appears to provide an important basis for selection of the best flowers in the course of feeding and pollen collection (von Frisch, 1966).

As far as human infants are concerned, chapter 2 raised the functional possibilities of vision only in so far as something like the 'highest' level 3-D model description was considered a plausible component of mechanisms underlying imitation behaviour involving body parts such as faces and hands. But other, 'lower', levels of Marr's low-level vision scheme also have properties that might make them suitable for supporting infant activities.

Along primal sketch lines, 18- to 36-week-old human infants' *predictive reaching* towards and interception of moving objects (von Hofsten, 1980) may be best supported by a description of the viewer-centered, two-dimensional properties of the object's movement, such as *directional selectivity* information from the primal sketch. In some respects, the requirements are like those of Sloman's continuous-control example. Of particular interest is the question of how the hand might be directed to a location *ahead* of the object's position for successful interception. von Hofsten (1985) suggests that the infant's reaching strategy can be explained quite simply. If the infant's hand is moved *with* the object, matching direction and velocity, at the same time as it moves *towards* it, the resulting hand movement will intercept the object. Reaches need to be made in a coordinate system fixed to the moving object, not to the static, viewer-independent background; there is no need to calculate future object positions or to know how long it takes to reach out.

Reliance on visual processing at a similar 'low' level could explain how infants may attempt to avoid approaching objects without recourse to shape information. Marr (1982) suggested that the raw outputs of directionally selective units could underlie a rapidly operating *'looming detector'* based on comparison of the movements at corresponding points on the two retinas. These points often, though not always, correspond to nearby points on a single moving object, and Regan, Beverley and Cynader (1979) have clarified the pertinent movement–object relations. Incompatible movement directions mean that the point is moving in depth relative to the observer: approaching if it is moving away from the nose, retreating if it is moving towards the nose. Motion to the right on both retinas corresponds to a miss-path to the observer's left, and motion to the left a miss-path to the right. Marr notes that a system combining motion and contour analysis is preferable for seeing the shape of moving surfaces. However, this more complex analysis is inessential to providing the (viewer-centered) approach information central to the looming phenomenon.

From this viewpoint, there is nothing strange about Ball's otherwise paradoxical demonstration that infants show avoidance behaviours to film of a continuously expanding object whose form changes, from a solid square to a Y, during its apparent approach (cited by Ball and Vurpillot, 1981, p. 125). Ball and Vurpillot's (1981) tentatively expressed conclusion is essentially correct, despite being perplexing from a Piagetian perspective: for the infant, perception of an event such as movement in depth may not imply perception of a constant object. However, their Piagetian notion of 'constant object' requires considerable clarification. In particular, the properties of the type of description to which this term is to be restricted need to be spelt out. (There is obviously a complex issue here about what having a 'real' object concept would amount to, and chapter 4 will look at some ways in which concept attribution might – for good or ill – be related to computational–visual mechanisms of the kind that are being considered here.)

A range of findings suggest a connection between infant prehension and object descriptions with the properties of Marr's 2½-D sketch.[3] For example, two main kinds of information appear necessary to support reaching.

First is information about three-dimensionality, especially depth, that is provided by 'real' objects but not by two-dimensional displays such as photographs. The latter fail to elicit reaching during either the neonatal/prereaching period or the early stages of reaching proper at around 5 to 6 months (Bower, 1972; Yonas, Granrud and Pettersen, 1985). In fact, one of the main concerns with reaching has been its potential as a functional index of depth perception (Bower, 1972; Yonas and Granrud, 1985).

Second is information about discontinuities in surfaces, which appears to describe the boundaries of target objects for the infant. Infants have been found to cease reaching for an object if it is placed *on* a larger object, as when a matchbox is placed on a book (Bower, 1979; Piaget, 1955; Wishart and Bower, 1985). Bower (e.g. 1974, 1982) attempts to explain this performance in terms of a 'conceptual rule': an object is a bounded volume of space, in a place or on a path of movement. For the infant, he argues, there is no longer a potentially graspable object if it loses one of its boundaries. However, considering the infant's reaching to be reliant on a description based on the 2½-D sketch format may offer a more precise characterization. The way the 2½-D sketch describes objects in terms of depth and orientation of patches of visible surface and their discontinuities would result in the same 'errors' that are produced by infants if it were incorporated into a mechanism for prehension. Placing one object directly on another would eliminate the discontinuity in depth which, in the 2½-D sketch, could support the separation of object surfaces from their background.

A study by von Hofsten and Spelke (1985), which used reaching as an index of how infants parse scenes, supports this interpretation. As we have seen, motion appears to be a central and powerful determinant of 2½-D sketch construction in 'prereaching' infants of about 16 weeks. When 20- to 23-week-olds who could reach competently towards displays were presented with two rectangular objects before a background, their reaching was directed most frequently towards the object whose boundary would have been delivered by motion. For example, if the smaller, nearer object moved relative to the larger, more distant one, through motion of either the small object itself or the larger object and its background, infants reached more for the small object. If, instead, the small object, adjacent larger object and background were all stationary, infants reached significantly more often to the more distant larger object. The type of boundary segmentation that would be yielded by the 2½-D sketch appears more important than relative size or distance in determining prehension.

Motion is not the only source of segmentation at this age; stereopsis, for example, may also be operational. If both objects and the background are stationary, infants will reach more to the smaller object if it is spatially separated from the larger one behind it. However, motion does appear to override other processes. Even with spatial separation, infants reach more to the larger object if smaller and larger objects move together relative to the background.

Apparently contrary to von Hofsten and Spelke's findings is the claim that the size of this three-dimensional surface area discontinuous with its background may, in fact, also be significant. Before infants become capable of mature reaching, they exhibit characteristic hand movement patterns that might be thought relevant to manipulation and which occur more frequently with an object of *'graspable' size*, as opposed to a larger object (Bruner and Koslowski, 1972). This might be compared with a favourite example in the ecological psychology literature: the praying mantis; this insect strikes at potential prey objects whose diameter is optimally suited to the angle of opening of its pincher (Shaw and Turvey, 1980). There are, however, a number of problems with Bruner and Koslowski's study, most notably that only very small and enormous (10-inch diameter) objects were used. Distinct size-related behaviours may arise because infants could see only part of the boundary of the larger object, not because their visual-motor system is adapted to respond to any particularly fine-tuned, body-scaled description of 'graspable' size (Rutkowska, 1992b).

This style of AI–psychology mapping clarifies a task for interdisciplinary work in AI: to explore the extent to which an input description suited to a particular functional application of vision can be generated from the 'vocabulary' (i.e. primitive elements, properties and relations) of any representational format that is proposed as a component of human vision. A possibility is that processes for building particular types of description within a format may be dedicated, automatic and modular, in a manner similar to the visual processes that mediate between different formats or levels of processing.

Additional candidates for which such an explanation could be sought include what is sometimes called the infant's 'naive theory' of physical and social objects. Visual processes relevant to 'belief in object permanence' (Baillargeon, 1986) have already been noted earlier (page 54). Along similar lines, modular visual processes operating on motion information have been suggested, appropriately in my view, as a possible basis for 6-month-olds' appreciation of certain kinds of causal relations between objects (Leslie and Keeble, 1987). The data to be explained show that infants are sensitive to the spatial relation of contact between a hand and an object that is picked up (Leslie, 1984); and they are more attentive to reversal of an apparently causal event, such as 'direct launching' (where one object moves towards and contacts another which then moves off), than to identical spatio-temporal changes due to reversal of an apparently non-causal event (Leslie and Keeble, 1987).

As far as infants' 'social' knowledge is concerned, it has also been suggested that descriptions based on motion primitives could serve as a hard-wired basis for infants' discrimination between causality and intention, by making explicit, for example, the way in which the intervention of other objects is or is not necessary to alter the motion of non self-propelled and self-propelled objects respectively (Premack, 1990). This computationally compatible approach provides an intermediate alternative to Neisser's (1988, p. 44) insistence on a direct perception style of explanation of the kind(s) of infant interpersonal exchanges that appear to merit an attribution of intersubjectivity. The information used by special-purpose processes applied to the outputs of low-level vision is not 'directly available in the optic array' to be picked up by 'anybody with the right kind of perceptual system'. But neither would the kind of processing involved in supporting functional, appropriate behaviour require the infant to exhibit 'inference' in the sense of attributing thoughts or feelings to the other participant(s) in interpersonal exchanges.

Three important general points follow from the perspective on functional perception outlined here. First, to be *usable* for particular purposes, descriptions need not themselves make explicit what those purposes are. The significance of an object for action does not have to be represented explicitly; it will be *implicit* in the processes that mediate appropriate behaviour towards it. Second, the *salience* of a particular type of information is a function of what the subject is trying to do, not of its level of processing. And third, the *usefulness (efficiency)* of a representation does not lie in how exhaustively it models an object or event but in how successfully it limits and allows for possibilities for action.

Coordination and control: action programs

Once information has been made explicit by visual processing, it can be treated as a unit and referred to, manipulated, or used by other processes within the system. Its most important 'use' is going to be by processes that coordinate perceptual and behavioural aspects of action. *Action programs* offer the best way of understanding the computational mechanisms that underlie this functional coordination in classical computational terms. Introducing this notion is controversial from the viewpoint of ecological psychology, and of recent computational work aiming at alternatives to classical centralized, knowledge- and plan-based approaches to ability. So, invoking it may seem at odds with my here endorsing a more distributed treatment of mind. However, I shall suggest that the action program concept is not incompatible with this direction. Indeed, it may be the best possible route for getting at not only the kinds of infant abilities discussed above, but also at developmental issues concerning different levels of ability in a domain.

It is action programs – not perception directly – which are assumed to control behaviour, a departure that will prove essential to explaining functioning and development. This section concentrates on a general analysis of how this notion contributes to a sufficient explanation of action (with examples of how action programs might be structured and work being developed in chapter 5). In general, an action program will be viewed as a *virtual* structure that provides a *control* level for functional activity, generating and coordinating processes that underlie preadapted links between perception and behaviour. It serves two vital functions, which will be considered in this section. First, it underwrites the 'selective' in 'selective use of perceptually provided (potentially usable) information', by controlling processes for exploiting the descriptions that are constructed by the system's perceptual component. Second, it controls processes for generating (overt) behavioural processes, such as eye movements, reaching, grasping and so forth, and in doing so it underwrites the 'purposive' in 'purposive behaviour'.

The 'how' of selectivity

Part of each action program will have access to the descriptions that are generated by autonomous perceptual processes. It will consist of processes that are activated by, and/or in some cases are 'looking for', a specific description or set of descriptions. For instance, in the (adult) 'sitting' example discussed above, the action program involved might activate processes that 'looked for' a constellation of descriptions relating to surfaces, orientation and material properties.

The essentially pragmatic root of this proposal is sympathetic to the notion of what ecological psychologists have called 'seeing in the affordance mode'. Here, it implies that a range of attributes of a single object type might be used in different ways and in different contexts without an explicit representation of these attributes belonging to a single entity/object. But there are non-trivial differences in the way the two approaches tackle the functional side of perception.

What affordances aim to do is certainly compatible. In the case of sitting, for example, Gibson (1979) also suggests that perception does not need to deliver information about a 'thing' such as a chair to support this activity. It needs to deliver a unique combination of properties of objects and layouts – rigid, level, flat, extended and about knee-height – which specify that there is a place or a thing that is 'sit-onable'. Or, to take one of Turvey, Shaw, Reed and Mace's (1981) examples, a 'climb-upable' thing has to have certain properties of rigidity, area, height and textural quality to support (a snail's) climbing activity. Infants' visual cliff 'avoidance' has usefully been viewed as absence of information for a ('crawl-onable') rigid, flat, opaque surface that affords support (Rader, Bausano and Richards, 1980), as opposed to the unsatisfactory notion of a negative affordance for activity (a 'falling-off place' (E. J. Gibson and Spelke, 1983; Gibson, 1979)).

However, *how* affordances are deemed to achieve this kind of selectivity is unconvincing. Since they are considered to be another kind of higher-order invariant (invariant combinations of 'properties', 'qualities' or 'variables') specifying the activities of a particular animal rather than objects or events, the notion resurrects all of the problems associated with the purported directness of 'direct perception'. Gibson (1979) proposes that infants should begin by perceiving objects in terms of such invariant combinations of properties, only later learning about the individual properties and about what combinations of properties specify objects themselves, which seems fair enough. But to assert that these invariant combinations are easy to perceive directly *because* they are 'meaningful' whereas the individual properties are not is at best a *non sequitur*.

Action programs interacting with the outputs of visual processing support a *generative* mechanism, in the (linguistics) sense of a finite system of 'elements' and 'rules' that can generate (or be used to interpret) an indefinitely large number of structures. Despite restricted perceptual and behavioural capacities, a wide range of activities become possible. The same descriptions can be used (or parsed) in different combinations by different action programs, whose operation constitutes the 'rules' of the system, instead of having to be recomputed each time they are part of a useful combination. This does not preclude the possibility that, at the action program level, through phylogeny and/or ontogeny, the relevant description combination may well be compiled into an 'invariant', template-like form (although even this would allow for a good deal of flexibility, as we shall see in chapter 5).

Introducing action programs, then, points towards a mechanism-level explanation for why apparently available and potentially usable information can be exploited for some activities and not for others. It is the structure of the subject's action programs that determines whether or not information made explicit by preattentive visual processing is also explicit at the level needed to generate behaviour and support activity. An explanation also becomes feasible for how the same information can be used in different ways. For example, the same movement information may be useful for, and used by, action programs controlling the avoidance of looming, visual following and reaching (in addition to its strictly visual use in the processing of surface information).

For this kind of computational approach, control issues of 'when' and 'how' information is selectively used, traditionally grouped under the label of 'attention', feature the structure of programs and their relationship with internal and external body (machine) processes. The relationship between action programs and the system's perceptual component can be viewed as bi-directional. Depending on their specific structure, action programs can be 'activated by' or 'respond to' the presence of a description constructed by perceptual processes; or they can generate processes that 'search for' the presence of a particular description, either through (internal) monitoring of perceptually provided information or through (overt) behaviour capable of influencing the data on which subsequent data-driven processing will operate.

This distinction parallels that between an *interrupt* and a *test* in computer science (Pylyshyn, 1984). An interrupt is a form of data-driven control; some internal or external event causes a state change to occur or a new process to be activated. A test, on the other hand, implies process-driven control, with a currently active process testing for some computational state. It may be that the earliest infant activities owe rather more to the former category, though concurrent operation of both is ultimately necessary. Looking at selectivity in terms of the changing structure of action programs offers a locus for investigating the possibility and nature of such development, though more detail on action programs and their operation is going to be needed before this useful simile can be converted into a fully fledged developmental argument.

The way ecological psychology handles selectivity highlights the potential strengths of a computational treatment of control issues. Even allowing for supposedly direct perception of affordances, as opposed to the kind of psychological decomposition of vision that characterizes the computational account, 'direct perception' metaphors for the perceptual process end up needing, but lacking, an effective means of tackling the selective use of information. The *radio receiver*, as Michaels and Carello (1981) note, cannot capture the link with behaviour that is vital to a systematic explanation of perception's role in action. Plus, it is passive in so far as it requires a 'tuner' if it is to do more than resonate to a single frequency. The same problem arises for the analogy with *smart machines* like the polar planimeter (Runeson, 1977), which are specialized for a single function and whose operation cannot be understood in terms of interconnections between more basic components.[4] As Runeson acknowledges, the non-perceptual aspects of the system he describes remain mysterious and are, in some sense, the (homuncular) 'user' or 'person with planimeter'.

Gibson's (1979) solution was to introduce the concept of *attention* to account for different subjects detecting different information in a single object, or one subject detecting different information in a single object on different occasions. 'Attending' is defined as 'perceiving in relation to a task or goal, internally or externally motivated' (E. J. Gibson and Rader, 1979), and it consists of producing behaviour that will facilitate the detection of appropriate information (looking, rubbing, listening, etc.). This is a good description in so far as it highlights the fact that selective attention is an emergent property of the operation of an action system. But 'attention' is too passive, molar and redescriptive a construct to work effectively as a primitive in a psychological explanation. It remains sensible and necessary to ask questions like: 'What controls attention? What is being "educated" in the "education of attention" that purportedly accounts for perceptual development? How are attention and behaviour coordinated?'

The relation between 'attention' and 'behaviour' is especially significant because, despite ecological psychology's behaviouristic definition of 'attending', the overt side of behaviour alone is insufficient to carry the explanatory force of this construct. Attending is no different from other activities in having an external and an internal aspect. Michaels and Carello's (1981, p. 71) discussion of attention illustrates the kind of problem that arises when it is operationalized solely in terms of the functional properties of overt behaviours. A person looking for a set of car keys, say, might behave just like someone looking for a book of matches. How, then, can they be said to be attending differently? They conclude:

> it can be supposed that it is a brain configuration that distinguishes the would-be driver from the would-be smoker. It would be desirable to be specific about how such brain configurations might be described; unfortunately, at this time, it can only be suggested that the term 'brain configurations' means whatever it is about the brain that serves the same function as the interleaved metal plates (variable condensers) in old fashioned radios (Michaels and Carello, 1981, p. 71).

At best this suggests a solution at the biological level – at worst it is scientistic 'hand waving' – when what should be sought here is a *psychological* configuration. This 'slippage' to a biological grain of analysis is particularly unsatisfactory given ecological psychology's assumption that psychological and biological levels of description cannot be causally related. It suggests that the construct is not only too molar but also of the wrong kind to explain the phenomena.

By way of contrast, for the computational account, action programs supply the missing (internal) difference. There is no such 'thing' as attention; it is an emergent property of the operation of action programs, one of many vantage points on to the processes they control. Since behaviours – eye and head movements (or 'looking') for instance – are simultaneously program-governed processes and overt bodily processes, both internal and overt aspects of selective attention can be handled, making the 'how' as well as the 'what' of seeking action-relevant information amenable to a psychological analysis in computational terms.

Getting the 'purpose' into purposive behaviour

The infant's ability selectively to use information made explicit by perceptual processing is an essential prerequisite for behaving appropriately in the world. But what exactly does this mean – what is behaviour, and what are the implications of conceptualizing it as a program-governed process? A working definition is that behaviour is a transformational process to which we attribute the achievement of some new state of the world: an end, goal or result. I believe that behaviour – not just the end, outcome or goal that it achieves, but the actual *process* of behaviour – is vitally important to our understanding of infant ability and knowledge. Central to this assumption is the dual nature of behaviour. The relationship between action programs (the virtual machine) and the body (the physical machine) provides the requisite dual perspective for exploring this effectively in computational terms. And it offers an invaluable way of conceptualizing how the mind can actually get things done in the world. On this view, psychological phenomena are not necessarily incompatible with causal, mechanistic ones; they emerge from the operation of a certain kind of machine, one which is capable of supporting the right kind of (computational) virtual machine.

As an example, consider a 'successful' manual search task. We cannot describe performance fully either in terms of an outcome, such as 'the infant is holding object A', or by painstakingly delineating the hand and arm movements involved. We intuitively apply superordinate structural descriptions which view behaviour in transformational and teleological terms. Even a simple description like 'the infant *found* object A' links the infant's possession of the object at time-t_2 with our hiding of the object at time-t_1, and with mediating processes contributed *by the subject.* Such descriptions are not inevitable, and it would be easy enough to generate alternative accounts involving mysterious forces taking over the infant and producing the observed movements. With such accounts, however, we would no longer be talking of 'behaviour', because they would lack the element of subjective mediation and *agency* that is implicit in the intuitive account. The infant's arm and hand could be said to have *undergone* a process, but the infant did not *generate* that process, despite providing the physical basis for its implementation.

At one level, behaviour can be analyzed *morphologically*, in terms of the spatio-temporal characteristics of physical movements involved. When we discuss behaviour in this way, we are viewing it in *causal* terms that are appropriate by virtue of its implementation in body processes. The fact that behaviour is implemented in body movements or motor processes, which are spatially and temporally embedded in the environment, makes it possible to capture this causal aspect. However, it is inappropriate to equate behaviour with overt motor response (as many developmentalists who are resistant to the explanatory power of action have tended to do). This reductionist view reifies behaviour by seeing it solely as an entity that is produced, rather than as a transformational component of a broader system.

Of equal importance, action programs offer a level of description that makes it possible to get at the essential element of subjective mediation by capturing the *teleological* or *purposive* aspect of behaviour. Although action programs (the virtual machine) are implemented in a particular type of causal machine (the brain-body), their structure is not restricted by the same, essentially sequential spatio-temporal constraints as the motor processes and movements they govern. Program processes can, for example, be 'caused' to run to generate the conditions for others to be run in future, not just because some process has previously occurred. When we talk about behaviour in terms of 'acts' (or 'effectivities') – or slip into equating it with 'action'– we are stressing this (intentional and) purposive nature, which behaviour attains by virtue of being generated and controlled by action programs. It is the distinction between a program-governed process and 'mere movement' that makes eye movements 'looking' or certain arm movements 'reaching'.

This view fits with certain philosophers' discussions of behaviour. Dretske (1988), for example, distinguishes between 'movement' and 'behaviour' on the grounds that only the latter can be said to be *produced by* an appropriate internal cause; that is, to stand in the right causal relation to an internal process. From the infancy viewpoint, it is worth stressing, 'internal causes' and the movements they produce are of equal interest to an overall model of ability.

In this context, it is important to emphasize that the 'brain' in the program–computer to mind–brain analogy is really shorthand for 'central nervous system'. In the case of the infant human subject, the physical machine with which we are dealing extends beyond the brain to encompass the movable body. This supports the view that arm, leg or eye movements can have the same theoretical status as the unobservable (mental) processes that are usually equated with information processing. Behaviour must be taken seriously as part of the process of knowing; it is not simply a 'product' or 'output' offering evidence of some static internal belief or knowledge.

The style of functionalism that predominates in cognitive science and mainstream AI has shown relatively little interest in how the physical as opposed to the virtual human machine is to be conceptualized. But when the aim is to understand the infant human system, the significance of details of the physical machine – including not only neurophysiology but also body form – is highlighted. *Reciprocal constraints* exist between cognitive and physical-motor mechanisms; they are not just alternative levels of description of the same system. The transformational processes that the subject can execute are constrained by the range of motor activities that are candidates for control by the program level. In turn, those motor activities are constrained by aspects of body form, such as the presence of arms and hands.

In other respects, however, the notion of behaviour as a program-governed process exploits the same mainstream functionalist assumptions that are employed with respect to the relationship between computational (mental) processes and their physical (brain) basis. Once physical-motor constraints are clarified (which is by no means a trivial task), the processes that they execute can be conceptualized independently of their physical basis. Reaching, for instance, can be viewed as the same behavioural process, under program control, whether it is implemented in a human or a robot arm (or perhaps, in the case of a subject with cerebral palsy, in a human leg). This type of functionalist account can be thought of as largely implementation-independent without implying a disembodied or a disembedded subject. It means that action programs and the processes they control can provide a self-contained, consistent level of analysis for behaviour, but also a principled basis for discussing the role of physical-motor mechanisms when necessary.

Further clarification can again be achieved by contrasting this account with ecological psychology's treatment of behaviour. Here, it is interesting to note that Shaw and McIntyre (1974) took the central term *effectivity* from the digital computer's inventor, von Neumann, who used it to mean purposive activities. An effectivity can also be defined roughly as a goal-directed act (Turvey, Shaw, Reed and Mace, 1981). For ecological psychology, however, the aim of describing properties of the environment (affordances) and properties of the subject (effectivities) in this functional manner (i.e. in terms of their roles in activity) is to establish that there are ecological *laws* relating them. This would extend the theory of direct perception to a theory of direct action.[5]

Philosophical analysis of the 'laws' issue is provided in Fodor and Pylyshyn's (1981) critique of the ecological approach, and in Turvey, Shaw, Reed and Mace's (1981) reply. Here, I shall restrict discussion to what an effectivity might look like, with regard to its role in the intimate match between environment and behaviour that is implied by the quest for ecological laws, and to how the notion compares with action programs as far as the psychological explanation of behaviour is concerned.

Affordances and effectivities are to be seen as dispositional properties of environments and organisms – they refer to what can happen when properties of the environment and of the organism fit together in a certain way (Turvey, Shaw, Reed and Mace, 1981). For an organism to effect an activity, certain properties of the organism and of the environment must 'dually complement' one another. Thus, affordances are said to include 'grabbable things' or 'climb-upable things'; effectivities include 'grabber things' or 'climber things'; and the activities they 'complicate' include 'grabbing' and 'climbing'. The content of an effectivity, and the notion's psychological relevance, is made clearer by the example of the effectivity structure relating to the activity of grasping. This is assumed to consist of: intention to grasp; anatomical configuration permitting grasping; and appropriately sized hand (Michaels and Carello, 1981, p. 177). Taken together, these constitute a potential for action that can be realized, given an environmental object with suitable affordances (in this case, an object whose size is graspable by the human hand).

One problem with this perspective, as far as explaining behaviour is concerned, is the way the effectivity notion conflates physical and anatomical properties of the subject's body with the psychological grain of analysis. The ecological framework can legitimately feature such physical-motor properties in explanations at its biological grain of analysis, where the size and shape of 'grabber things' and so forth would seem to belong (alongside cells and organs such as the brain), but not at the psychological grain where effectivities are based. Ecological psychology appears implicitly to acknowledge behaviour's dual nature, but it lacks the more differentiated and explicit explanation of how the subject's purposes and body properties fit together that is possible for the three-way computational notion of a *virtual machine* implementing a *computational process* in a *physical machine*.

Also problematic from a psychological viewpoint, especially a developmental one, is the way in which *intentions* and *goals* are introduced. Ecological psychology treats 'intentions' as primitive properties of the subject at the psychological grain of analysis, where they correspond with 'goal objects' of the environment. Both of these assumptions are problematic.

'Intention' is scarcely suited to being a primitive in a psychological explanation. What is needed is a more differentiated treatment of what intentions might be, and of the role they might play in the subject's activities. Computational mechanisms begin to help here by offering a view of intention as an emergent property of an action system. Boden (1981) is able to show that a range of philosophical problems can be addressed by viewing any intention as having three aspects: *motivation*, which need not be explicitly represented but arises from the broad structure of plans and goals of the system; *procedural schemata*, which can be equated with plan structures at the level of programs; and *bodily operations*, which can be purely 'intellectual' or involve overt motor activity. This offers a route to operationalizing intention in terms of action programs.

As far as our explanations of the infant are concerned, what is powerful about this approach is that it offers a scale of analysis that lets us get at the *structure* of intentions. Differently structured action programs can make behaviour purposive in different ways. One form may well fit with ecological psychology's view of goals as properties of the environment that complement the intentions of the subject. For example, Boden (1983) discusses the way in which animals may often produce complex chains of behaviour without the program plan that controls this behaviour including a goal; that is, an 'explicit representation' of the outcome that plays a role in the execution, sequencing and termination of the activities. Care is needed in deciding what is to be identified as 'purposive' behaviour; individual subunits may be purposive but not the overall concatenation. Ring-doves, for example, may perform a sequence of activities such as nest building, food collection, cleaning debris from the nest and feeding the young, without having an overarching 'goal in mind' corresponding to healthy, grown up chicks. The infant's earliest purposive behaviour may well be of this type (as we shall see in chapter 6).

In contrast with this, ecological psychology's 'intentions' and 'goal objects' can provide only a molar (and possibly misleading) analysis. They lack the potential for analyzing development of structure or, as we shall see at the end of this chapter, of changes in the relative contributions of subject and environment to the infant's functional activities that occur during the course of this development.

Action as an organizing construct

Where, then, is 'action' in the preceding analysis? The model shown in figure 3.2 summarizes the component mechanisms of my functional decomposition of the infant viewed as an *action system*.

Even more than notions like 'attention' and 'intention', action should not be reified by attempting to treat it as a molar conceptual primitive in explanations of ability. It certainly cannot be equated with behaviour, any more than behaviour is reducible to motor movements. I believe, however, that action is an appropriate and valuable *organizing construct*. Interaction between all of the components of this model are necessary to offer a sufficient account of the purposive transformation of the world that characterizes an action system such as the infant: perceptual processing exploiting environmental regularities to make available potentially usable information; action programs underlying both the selective use of information and the generation of behaviour; mechanisms at virtual and physical levels; behaviour which spans subject and environment; and the environment whose structures and processes interact with the operation of these components in the attainment of functional outcomes. These are all *aspects* of action.

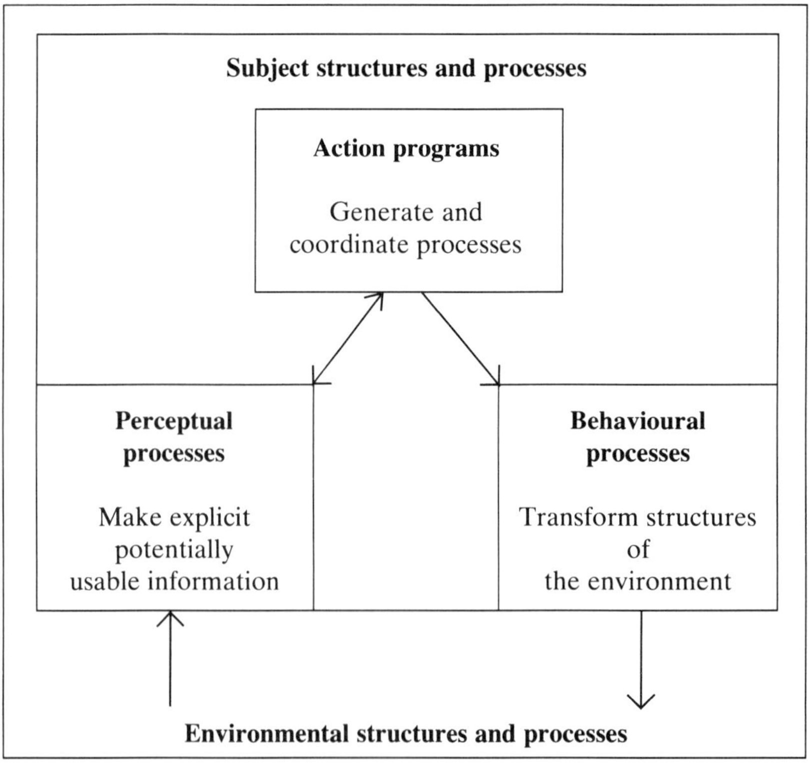

Figure 3.2 Functional decomposition of the infant action system.

A preliminary characterization of infant action could view it, at the most general level, in terms of types of action. Various classifications have been suggested, and their relative merits need not be considered here. Bruner's (1973) is representative: feeding; perceiving and attending; manipulating the world; locomotion; and interacting with others. Each of these types of action can be thought of as based on a repertoire of action programs.

Central to this analysis of infant action is the implication that cognitivist explanations must move beyond the domain of purely intra-subject symbol structures and processes if they are to explain the abilities of naturally intelligent, real-world computational systems such as the infant. An *extended systematic explanation* of ability is called for. This computational view of infant action lacks the rigid demarcation between internal and external, and between subject and environment, that is often considered typical of cognitivist explanations. It is, however, able to draw the distinction between subjective and objective when called upon to do so, without losing sight of the subject's localization in the world.

Attempting to understand the infant mind by moving towards a more *distributed* view of ability is consonant with a number of recent trends in cognitive science. For example, an important direction in the semantics of natural languages depends on viewing the subject as a *situated agent* (Barwise and Perry, 1983, 1984), proposing *situated inference* based on the way the subject is embedded in the world as an alternative to the classical equation of inference with formal symbol manipulation (Barwise, 1987). Related work on *situated action* is characterized by viewing action as a more collaborative, improvised process than is typical of its classical computational treatment in terms of plan construction and execution (Suchman, 1987). In a similar vein, Varela's (1988) *enaction* framework for cognitive science attempts to exploit biologically informed ideas about adaptation, in order to displace task-specific problem solving based on representation in favour of creative *problem definition* and *effective action*.

On the explicitly computational, AI front, a relevant line of connectionist thinking emphasizes that a significant feature of complex cognitive abilities is our reliance on *external support* (Rumelhart, Smolensky, McClelland and Hinton, 1986). That AI is eventually starting to move in the direction of Dennett's (1978b) quest for 'whole' creatures whose abilities extend to exhibiting adaptive sensory-motor behaviours within their environment is signalled by novel labels for computational inquiries concerned with the subject's behaviour. For instance, there is growing interest in how *animats* (simulated animals or autonomous robots) might be empowered with adaptive behaviour (Meyer and Guillot, 1991). And *computational neuro-ethology* assumes that connectionist models of neural systems (for example, so-called 'neural networks') will be meaningless unless they are embedded in a sensory-motor system interacting with an external environment (Cliff, 1990).

Finally, from the viewpoint of theoretical work in computer science, Smith

(1985) argues cogently that *embeddedness* rather than formality is the defining characteristic of computation, and determines its significant properties.

An extended systematic account of infant action is therefore equally relevant to understanding the nature of infant abilities; to constraining the kinds of infant–environment interaction that could be important in development; and to how we conceptualize computation – hence to the general issue of defining the nature of computational psychology.

Redrawing the boundaries of computation and cognition

This computational view of action proposes an infant who is active in both the cognitive and the behavioural sense. It is based on four main principles: *externalization*; *embedding*; *transaction*; and *scaffolding*.

The abilities of naturally intelligent systems such as the infant rely extensively on an *externalization* of processing that exploits and is supported by physical (and social) environmental structure. Action programs, as we have seen, govern both internal and external transformational processes. Covert 'intellectual' processes manipulate internal symbol structures. But overt behavioural processes have a privileged status because they can manipulate environmental structures external to the subject.

The relation between programs and the body means that the symbol structures and program-governed processes of the infant system are actively *embedded* in an environment that can also be conceptualized in terms of structures and processes. In general terms, both the subject and the environment can be conceptualized in terms of a 'state-space' or 'problem-space' analogy. Each consists of a set of states together with processes or operators that link one state to another. The notion of embedding implies that we can represent infant and environment in terms of a 'shared' state-space. As Newell's (1980) account stresses, a physical symbol system – in this case the infant – and its environment form a closed system because the outputs of the symbol system either become, or affect, later input objects. (This is, of course, the viewpoint of an extra-system observer; it is not assumed that infants have a complete representation of the environment or of themselves.)

Action programs, as introduced above, incorporate two types of behavioural operators: those that transform environmental structure directly, such as reaching and grasping; and those that transform the relationship between subject and environment, such as locomotion and eye movements. Both types of transformation change the subject–environment relationship and provide new potential inputs to the action system. This is flagged in figure 3.2 by the unidirectional arrow indicating the relation between action programs and behaviour, with behavioural processes viewed as exerting their main effect on action programs through the mediation of the perceptual component.

In fact, the link between behaviour – specifically, eye movements – and perception may be even more intimate than this environmental loop implies.

Recent work in what calls itself the *animate vision* paradigm looks at how embedding vision in a sensory-motor repertoire, including eye movements such as gaze control, determines what kinds of visual computation will be effective, and which visual computations the system will perform (see Ballard, 1989 for a review). For example, behaviour supplies an important constraint on interpretation, as when fixation of a stationary point permits optical flow to be interpreted as a depth map (an interpretation that becomes invalid if the eyes are pursuing a moving target).

This view of action equally redefines the boundaries of computation and of infant cognition by setting the stage for *transaction* as the basic mode of organization of activity. And it can help to clarify how the adaptive properties of infant (and adult) action systems are partly attributable to the *scaffolding* provided by regularities in environmental structures and processes. Bruner originally used the notion of 'scaffolding' to characterize the way adults enable a child to perform a task that s/he cannot yet do alone (see, for example, Wood, Bruner and Ross, 1976). For example, an infant may be supported by the arms and propelled forward, enabling them to 'walk' before they are capable of controlling the entire complex of movements involved. In my view, however, scaffolding operates in infants' (and adults') 'natural' transactions with the physical world, as well as in socially mediated exchanges. Its key property is that (social or physical) environmental structure reduces the number of degrees of freedom in a 'task' with which the subject's cognitive resources might otherwise be too inadequate to cope.

The way in which these principles work can be illustrated from a range of perspectives. From psychology, work on *everyday cognition,* which focusses on the structural organization provided by the subject's goals and purposes in everyday contexts, and by the physical and social environment in which they are pursued (e.g. Neisser, 1982; Rogoff and Lave, 1984), shows the value for any computational system of restricted capacity of getting its cognition 'into the environment'.

One investigation, of remembering among 8- to 10-year-old children engaged in cake baking and seed planting activities, shows how the *physical structure* of the environment is repeatedly used as a *memory aid* (Cole, Hood and McDermot, 1978). Locating the many ingredients that a recipe calls for is facilitated by the physical organization of the 'kitchen', which keeps them (and groups of other functionally related items, such as utensils) in proximity to one another in a constant location. When melted shortening is needed, it is sought on the shelves where ingredients are stored. As the authors note: 'The fact that many everyday environments (e.g., kitchens) are physically structured in a manner that lessens memory load is so ubiquitous that it is difficult to see (Cole, Hood and McDermot, 1978 p. 368).

On the classical–PSSH computational front, Simon's (1969) exploration of the idea that task complexity might be shared by organism and environment was incorporated into early developments in Newell and Simon's computational models of problem solving activity.

Their initial General Problem Solver (GPS) characterized task solution as a totally internal process (Newell and Simon, 1963). GPS's structure was basically that of a state-space, consisting of objects (in the form of data structures) and operators. This space was explicitly represented 'inside' GPS, which awaited an input in the form of an initial object and a desired object. Then, GPS's problem solving program selected operators to reduce the difference between the initial object and the explicitly represented goal object with which it had been provided.

Later models, however, highlighted ongoing exchange between a problem solving system and its environment through a Production System (PS) program structure (Newell and Simon, 1972). Consisting of a number of independent condition-action rules, which activate operators when the condition associated with them appears in the input or in the system's short-term memory, this permits a much more active interplay between the system and its environment. The changing states of the environment, rather than some totally internal goal, can control the system's actions. (This is the type of program that Boden (1983) finds promising as a potential explanation of the ring-dove behaviour mentioned previously.) Additionally, Newell and Simon noted the computational possibilities of using the environment as a form of *external memory* to augment the system's limited capacity internal working memory.

Similar notions are featured in some connectionist thinking about how sequential thought might be implemented in PDP models (Rumelhart, Smolensky, McClelland and Hinton, 1986). We may solve very complex problems in mathematics, reasoning and so forth by constructing external representations that allow us to break the problem into a series of intermediate steps. Then, we can rely on our relatively automatic and powerful visual abilities for pattern matching, which can be thought of as implemented in PDP mechanisms, to 'see' the answers to the simpler subproblems. In this way, the system is driven from state to state by its output changing environmental 'stimulus' conditions, which in turn function as inputs that change the system's state and hence its outputs.

A different but related approach to subdividing a task and relying on environmental support also characterizes attempts in 'animat research' to devise computational architectures that can provide functional alternatives to central, knowledge-based planning. The overarching idea is of the *emergent functionality* that can arise from a strongly situated system composed of many components that interact with the environment in which it is embedded independently of one another (Steels, 1990). A key example is provided by Brooks's (1991a, 1991b) *subsumption architecture*. Instead of simply subdividing a complex task into sequentially executed steps, this decomposes the overall system into a number of (ideally) independent, data-driven subsystems, each designed to produce a simpler *task-achieving behaviour*. *Layered control* is attained by building the system's lowest level of behaviour first, then, once that is operational, adding a further layer on to that, and so on. Although lower levels continue to operate in parallel with higher ones, their roles can be subsumed when their outputs are suppressed by the higher ones.

The approach is illustrated by implementation of a series of *creatures*, whose sensors and motors are organized to support 'exploration' of the real world (an AI laboratory) in real time. In one version, the task of the lowest layer autonomous system of the creature is avoiding contact with objects, while the second layer aims to try to visit distant places specified by its sensors.[6] When these operate together, the higher layer can command the lower layer's motor component to move towards reachable objects, updating motor commands as it proceeds, while the lower layer leads the creature to move away from initially unforeseen obstacles. Thus, the second layer achieves its 'goal' of exploration without being explicitly aware that there are/were any obstacles. There is, here, no central control system where an explicit representation of the world layout (as understood in most mainstream, classical AI) is built and manipulated in order to choose goals and activities, or to determine their order of execution.

As a final example, a straightforward way in which physical properties of the subject might constrain action, and mesh with the embedding environment to undercut any necessity for central symbol structures providing an exhaustive and/or explicit representation of the 'task' involved, is illustrated by Winograd's (1987) hypothetical pizza-delivery robot, Ramona. Orders take the form of letters making up words in a standard vocabulary, and Ramona enters them into horizontal rows in a grid in its computational field. The lowest row's dropping into an output buffer signals to the chefs which pizza to prepare next, and when they are all ready the order can be delivered.

Since bigger pizzas take longer to cook, preparation sequence is clearly vital here, and Ramona appears to 'control' the sequence appropriately. Two things working together make this possible. First, in the environmental domain of pizza names in which the robot operates, bigger pizzas have longer names; so, they involve a greater, hence heavier, load of letters; for example, an order for 'GIGANTIC:ANCHOVY&GARLIC' weighs more than that for 'MEDIUM: PEPPER'. Second, the height to which the robot arm carrying pizza names to the grid can rise depends on the weight it is carrying, and heavier names end up nearer the bottom of the grid; hence their preparation starts sooner than that of smaller pizzas. Pizza size ends up related to cooking time without recourse to symbol structures representing notions about larger pizzas needing to be requested first or taking longer to cook.

Implications for explanations

Identifying infant abilities with action, together with the principles by which action systems are assumed to operate, has implications for the kinds of explanations of functioning and development that might be worth pursuing.

Most basically, there may seem to be something of a conceptual 'culture clash' in endorsing a decentralized view of infant abilities while simultaneously invoking action programs in their explanation. Ought opting for a distributed view of how cognition, computation and the mind 'work' necessitate abandoning explanations in terms of classical, PSSH-style notions of goal-governed behaviour, where symbol structures representing what the subject knows about how the world works, predetermined goals and plans are of the essence? Their replacement is, after all, a key aim of work such as that of Brooks or Winograd. In the crudely explicit sense in which such notions have often featured in mainstream AI – Winograd and Flores's (1986) view of how infant feeding might be addressed in this light, noted in chapter 1, provides an apposite example – I believe the answer is 'Yes'. However, to the extent that such notions can be translated into types of action program structure, I believe the developmental answer has to be 'No'.

'Goal' is evidently a complex notion, and it often needs to be used more discriminatingly. Some state or event that constitutes a 'goal' from an observer's vantage-point need not do so from the perspective of the subject, even when it is a state that is brought about by the subject's activity. The action-based abilities of naturally intelligent systems such as the infant are characterized not only by externalization, transaction and scaffolding, but by the further principle of *restructuring*. Organizational changes that routinely occur in even 'basic actions' like reaching show there can be *developmentally* different ways of arriving at what seems to be the selfsame 'end'. Infants may not start out with much, if anything, by way of mechanisms for formulating 'goals' or 'plans', but their behaviour will change in ways suggesting that *in some form* they acquire such mechanisms, and *at some level* it then becomes appropriate to attribute 'goals' and 'plans' to them in place of more implicit forms of motivation.

Ontogenetically as well as phylogenetically, we need a way of conceptualizing mechanisms that can handle *both* styles of organization. Distinct forms of control, with a shifting balance between the subject's and the environment's contribution to outcome attainment, are implicated.

Clark's (1989) discussion of my (1984, 1986) view of how computation and cognition include and exploit the system's embedding in the world notes that it recognizes principles similar to some developments in biology.[7] The general '007 principle' that Clark proposes to characterize this mode of functioning captures a key dimension of scaffolding:

> In general, evolved creatures will neither store nor process information in costly ways when they can use the structure of the environment and their operations upon it as a convenient stand-in for the information-processing operations concerned. That is, know only as much as you need to know to get the job done. (Clark, 1989, p. 64)

There is, however, an aspect of scaffolding that is at least equally significant as far as development is concerned, and which may best be thought of in terms of *serendipity*. This involves the subject's not only getting things done with environmental support that they could not do alone, but getting things done that they did not 'know' how to do and were not even 'trying' to get done.

'Serendipity' marks the activity of an embedded system within the scaffolding environment producing results for which the behaviours involved were not designed. In the case of Winograd's (1987) Ramona, for instance, its behaviour is not designed for control of the pizza preparation sequence; this 'comes for free' as a side-effect of what is, from the perspective of the robot, the genuinely 'task-achieving behaviour' of putting names into its grid. The behavioural process under the control of Ramona's action program could be described as something like 'raise your arm until it gets to the grid'. The added dimension of how high up the grid it reaches is determined by extra-program properties of the subject–environment system (physical properties of pizza names and of the arm in which the behaviour is implemented). Constrained in this way, this component of the movement is more akin to Ramona's having its arm moved than it is to Ramona's moving its arm. This aspect of the robot's activity cannot be said to 'exploit' or 'work with' environmental regularities *in order to* control sequencing. A purely *physical–practical contingency* is established: whenever Ramona's arm is raised by a distance proportionate to the size of a pizza name, bigger pizzas get ordered, hence cooked, first.

Developmentally, the important question is: could such serendipitously successful movements come under action program control to become purposive, task-oriented behaviours? Could the robot, for example, learn to raise its arm in particular ways in order to affect sequencing? In Ramona's case, almost certainly not. Even if its sensory capabilities enabled it to 'notice' the contingency, its arm is 'floppy', period; there is no range of variation in arm rigidity that might be so controlled, since variation appears to be constrained solely by environmental factors (the relative lengths of pizza names).

In the case of human infants, however, the general pattern of restructuring in basic activities suggests that such *anticipation* does routinely emerge, when a range of 'serendipitous' consequences are mediated more equally by physical properties of the environment and physical-motor properties of the infant. Staying with arm and hand movements, there are types of visually provided description that, unlike those discussed earlier (pp. 64–6), initially fail to influence prehension but eventually come to do so. Infants are, for example, well able to discriminate visually between *convexity* and *concavity*, but detailed observations of their reaching for flat, convex and concave objects of equal diameters reveal identical hand shaping (whether with the hand loose and fingers lightly folded, or the hand extended and flat with fingers spread wide) before the age of 7 months (Pieraut-Le Bonniec, 1985). Adjustment to the particular manipulatory potential of the objects occurs only after contact has been made with them. This has changed by 10 months of age, when all infants exhibit differentiated hand shaping during the approach phase of reaching (for example, fingers are extended; bent in a claw shape; bent back), in 'anticipation' of contacting the particular object they are aiming at. At comparable ages, a similar pattern is found in infants' acquisition of an action-based understanding of *size–weight* covariation. Their lifting behaviour comes to anticipate the (invisible) weight of graspable objects on the basis of their (visible) size (Mounoud and Hauert, 1982a, 1982b).

The locus of this ability for permanent, foresightful adaptation to environmental circumstances can be viewed as residing in action programs, whose restructuring may provide a basis for the observed reorganizations of infant action. For instance, if some aspect of the environment is to play the role of 'external memory', or support anticipatory activity generally, the perceptual component of the infant system must be able to construct a description of it, and that description must be(come) accessible to the action program level that also controls behaviour. Of course, an explanation in this vein needs to do more than describe two kinds of program structure and then simply attribute them to different developmental levels; a transformational mediating mechanism must also be spelt out. In principle, however, this is a possibility meriting further exploration.

This potential contrasts with the ideas of ecological psychology that were discussed briefly earlier on. Ecological psychology's treatment of 'goal objects' offers an observer's view of what is 'in' the environment for the subject; not only does it fail to make the distinction between purposively and serendipitously attained outcomes, it is not clear that it would be able to do so. Its notions of 'intentions', 'goal objects', 'affordances' and 'effectivities' lack flexibility because they appear to refer to a single style of organization that the subject simply either has or has not 'got' (or else ecological laws may abstract over outcome-equivalent yet distinct mechanisms, which raises the same difficulties for developmental analysis). As far as developmental mechanisms are concerned, however, it probably makes little or no sense to explain the prehension of, say, both a neonate and a 9-month-old in terms of identical effectivities and goals, even if we can observe them both attaining what are, from our viewpoint, identical outcomes. More analysis of psychological mechanisms is essential, whether as an addition to or as an alternative to the quest for 'more physics of the kind appropriate to living things and their environments' and 'physical principles at the scale of ecology' on which Turvey, Shaw, Reed and Mace (1981, p. 254) insist.

However flawed the 'scheme' mechanism of Piagetian cognitivism may be, there were valid reasons for its introduction. One of these was the requirement for a modifiable structure. The presence of underlying schemes was meant, by Piaget, to distinguish between *voluntary behaviours* such as looking (or 'directing the glance'), which will be open to development, and *reflexes* such as the pupillary reaction to light, which will not (Piaget, 1953). The way the action program construct handles behaviour in the environment holds out at least the possibility of exploring such essential development.

As far as 'PSSH versus PDP' conflicts go, this chapter's preceding analysis offers little support for the idea that connectionist models, if taken outside the context of their relation to PSSH concepts, have superior explanatory status.

The last section showed that compatible and similar examples of an action system's transactional functioning are available from classical and connectionist attempts to conceptualize problem solving. For the PDP sketch of how sequential activities might be organized, there are outstanding questions about how the system arrives at the subproblems into which a complex task is decomposed, and whether their ordering and execution needs some means of controlling overall organization in addition to the state–state transition mechanism.

Rumelhart *et al.*'s discussion of such issues does not conflict with Newell and Simon's (1972) classical analysis:

> Goals can be explicitly represented as patterns of activation, and can thus provide one source of input to the thinking process. . . . a PDP network could learn to establish specific subgoal patterns in response to particular superordinate goals and inputs. (Rumelhart, Smolensky, McClelland and Hinton, 1986, p. 46)

In so far as such classical notions are used, connectionism cannot be considered a self-contained alternative to PSSH (Rumelhart *et al.*'s treatment of some forms of thinking even includes an internal 'model of the world' with connections to the system's input and output units, which is considerably more traditional than the action model being developed here). As Minsky and Papert (1988) put it: even if the brain is composed of distributed systems, it may not *be* a distributed system. Classical concepts may still play an important role in characterizing architectures and their control structures, certainly so far as phenomena grounded in action systems are concerned.

An area where the view of action developed here may seem directly at odds with current thinking, but in my view turns out not to be, is Brooks's work on subsumption architecture. This, Brooks (1991) claims, is a radical departure from other methodologies and motivated primarily by engineering considerations. It is not classical AI, since symbolic representations, production rules and so forth are not supposed to be involved; nor is it connectionism, since its layers of dedicated simple machines are supposedly incompatible with uniform networks, particularly with hopes that they will give rise to explicit distributed representations.

In fact, the functional architecture of Brooks's creatures has a great deal in common with the view of infant action that invokes the classical (symbolic) computational notion of an action program. A set of action programs that can exploit the same 'vocabulary' of perceptual descriptions and motor processes fits with Brooks's rejection of the traditional clear distinction between serially ordered 'perception', 'central' and 'action' subsystems, particularly in so far as there is no homogeneous central representation of the world. Perceptual processing that constructs multiple descriptions of the environment, each accessible to different action programs, rather than an integrated object description, fits Brooks's claim that there is no single (central) place where *a* description of the world is delivered. It is also compatible with his assumption that a number of algorithms (from different layers) may be run on the same sensor outputs. (Cf. Brooks's notion of higher layers exploiting the motor processes of lower ones.) And the program-based action system described here compares with Brooks's creatures in lacking, at least initially, any explicit representation of its intentions.

With the aim of understanding the infant in mind, rather than sheer engineering success, classical program concepts are likely to be useful in accounting for the generativity that is central to action, and to underwrite scaling-up to more complex tasks and forms of organization. This concurs with some of the sentiments of Kirsh's (1991) appropriately sceptical critique 'Today the earwig, tomorrow man?', which doubts the possibility of an architecture like layered control of task-achieving behaviours handling the *full* range of ability.

The way in which an extended systematic explanation of action handles ability also has implications for cognitivist proposals that forefront formal explanation in one guise or another. It makes it difficult, if not impossible, to carve up the tasks performed by the infant in a way that is amenable to an abstract, universal analysis independent of processing considerations. The more mutual approach to subject and environment reveals that transaction comes into play in even apparently simple tasks – or in single experimental trials – making it difficult to apportion responsibility for aspects of performance between subject and environment. And formal task analysis is straightforward only where performance results in a clear 'product': some static, isolable entity (such as the 'sentence', which provides the unit of analysis for much contemporary linguistics) whose structure can be described independently of whatever process(es) arrived at it. For this reason, Keil's (1981b) analysis of formal constraints is clearest and most convincing when dealing with syntax or the organizational structure of ontological categories. In action, however, attempting to isolate such a product is generally an artificial exercise since process is central: we are invariably dealing with one state of the subject–environment system being transformed into another, generally with intervening states.

Similar difficulties affect Marr's (1977, 1982) prescriptions for the priority of a top-level computational theory of the task the information processing system is tackling. In Marr's vision model, the construction of descriptions in each of the three representational formats constitutes a task in its own right, with its own computational theory (and possible algorithms); and those outcome descriptions have a product-like status, analogous to linguistics' sentences. But Marr's entire vision model is no more than a fragment of the systematic model of action that was sketched above. At the action level, the computational transformations underlying even a basic performance will generally subsume numerous information processing tasks at Marr's scale of analysis. For example, several information transformations are involved in the standard manual search task where an infant uncovers and retrieves an object that s/he has previously watched being hidden. For such (transactional) tasks, there may be no 'top-level' computational theory in Marr's sense. Even if a formal explanation proves useful for understanding the properties of isolated components of a task or internal mechanism, it will lose the whole that is action.

Marr's method, it is worth noting, is predicated on being able to break down a larger-scale 'task' into its component computations. But the method itself does not provide a way of achieving this preliminary functional decomposition (cf. the similar problem, noted above, with regard to connectionist models of sequential abilities). Marr himself relied on eclectic insights from a range of disciplines to arrive at his suggested structure for low-level vision. Conceptual and empirical analyses of infant abilities should, therefore, contribute to our understanding of the functional architecture of action systems, prior to (or even in the absence of) subsequent formal (sub)task analyses. To the extent that this analysis leaves us with a process model without a computational theory, explaining 'action' is likely to be an uncongenial problem from Marr's perspective on explanation.

Marr's (1977) ideas about the kinds of research problems that we should, ideally, be tackling mesh with Fodor's (1983) speculations on explanation in cognitive science. He suggests that not all tasks will decompose into a 'clean' computational theory and an algorithmic account of the processes employed (in some cases, many processes may operate simultaneously to solve a problem, and their interaction will be its own simplest description). However, achieving 'deep' scientific understanding means focussing on isolating and explaining tasks that do decompose in this way. Marr (1977, p. 45) considers the best source for such problems as: 'operations that we perform well, fluently (and hence unconsciously) since it is difficult to see how reliability could be achieved if there was no sound underlying method'. Tasks like problem solving will be excluded from this category (as the kinds of action-based abilities of concern here are likely to be); Marr suggests that they are based on a vast amount of knowledge and expertise, and that we perform them poorly, even if we understand them well intellectually.

This perspective leads to conclusions very like those of Fodor's (1987) attempt to specify the implications for explanation of his distinction between *input* and *central* systems in cognition. Hard-wired, modular, data-driven and 'informationally encapsulated' input systems should be at least possible, if not easy, to explain because the processes that they execute, and the information on which those processes operate, are restricted and can be precisely specified. (Visual and speech processing are, predictably, his key examples.) By way of contrast, 'higher' systems' greater flexibility should make them difficult to explain, because their processes are assumed to function across knowledge domains, to be under voluntary rather than automatic control, and to have unrestricted access to the subject's beliefs and knowledge. For Fodor, too little is known about them to say even which these 'true higher cognitive faculties' are, but thought and problem solving are suggested (cf. Marr, 1977).

Such pessimism seems to me to be unwarranted. Adopting 'action' as the scale of analysis for naturally intelligent abilities cuts across Fodor's modular–central distinction in several interesting ways. Once 'output' (behaviour) is taken seriously, and action programs replace supposedly mysterious 'central' processes, the differential properties (and explanatory possibilities) that Marr and Fodor attribute to input and central processes appear questionable.

First, despite the fact that the kinds of action-based abilities that are foremost with regard to infancy incorporate (presumably non-modular) central processes, they are, most certainly, things that all normal humans come to do very well indeed. Second, at different points in development, the same action can appear relatively automatic *or* voluntary, with restructuring marking a common transition to increasingly flexible patterns of performance. Finally, even if central, coordinating processes are domain-neutral, they need not prove to be mysterious or unrestricted in the way that Fodor's thesis leads us to expect. Where activity in the sense of action is concerned, as opposed to the purely central, cognitive activity of Fodor's model, there are interesting proposals concerning common processes in different problem solving situations and representational systems (e.g. Bruner, 1970; Karmiloff-Smith, 1991; Mounoud, 1986). A computational analysis of infant action – in the process rather than the formal sense – adds further possibilities.

Studying central processes in the functional context of action, not misleadingly as ends in their own right, may offer a valuable developmental route to understanding what exactly they are. The general aim is, for example, to understand the use of vision in guiding purposive behaviour, or of speech processing in communicative interchanges. If formal analysis is unable to handle such phenomena, then it is probably better to redefine our definition of a useful 'scientific' explanation than to succumb to the view that 'the limits of modularity are also likely to be the limits of what we are going to be able to understand about the mind' (Fodor, 1983, p.126).

Notes

1. Drawing or discriminating between drawings could be everyday uses, although they were not for these student subjects.
2. Prototypical examples are said to be at the center of most categories. They are the best or 'focal' examples, providing a basis for concept and language acquisition because they correspond to the way the world is actually structured (Mervis and Rosch, 1981).
3. At a related level of processing, Fisher (1986) suggests that his 'surface clusters' may prove useful for navigation, and work in robotics forefronts the role of 2½-D sketch information from stereopsis for this purpose (Brady, 1985).
4. The polar planimeter is an arrangement of rods and wheels that measures the area of planar surfaces, without measuring lower-order variables such as length of side and multiplying them together.
5. The relationship that Gibson saw between perception of the environment and the

subject's behaviour is captured in his use of the causal term 'induce', which prefigured the notion of affordance (e.g. Gibson, 1950, p. 199).

6. To date, Brooks (1991b) reports real-world implementation of at most three layers, although more have been simulated.
7. For example, it has become clear that the ability of sponges to filter water for feeding purposes exploits ambient water currents to restrict the amount of pumping they have to do (Vogel, 1981, cited in Clark, 1989, p. 63).

Chapter 4

Information and meaning

In the background to the infancy–computation mappings of the preceding chapters are a number of large-scale theoretical debates whose proper place is generally deemed to be with philosophy. Most widely acknowledged in the infancy literature is the prevailing tension between *realist* and *constructivist* methodologies for understanding 'mind'. While this book does not aim to pursue a philosophical analysis of the issues, the present chapter takes time out from developing computational details of the model of the infant to clarify to what kind of perspective my approach commits itself by endorsing computation, action and their extended systematic explanation. The intention is to sketch the kinds of conclusions that I would expect to arrive at in view of the theoretical choices about infant abilities that I have so far made.

My starting point is with an ubiquitous notion that often slides past explicit analysis: *information*. Researchers of most persuasions into infancy have found it reassuring to conclude that infants have access to information from the outset, although cognitively oriented theorists have lacked the consensual definition put forward by direct perception theory (or, indeed, often any definition at all). None the less, they have been liberal in their use of the term, from early on in the field's development concluding that even newborns 'store and process simple visual information' (Friedman, 1972), and that very young infants 'screen input with reference to information from everyday experience' (Carpenter, 1974) and are able to construct 'a relatively persistent organized classification of information, a model which the organism uses in arranging information' (Lewis, 1969, p.84).

But why is this? What *is* information? Dretske (1981) appositely observes that numerous book titles include the term, but few of their indexes list a definition; Marr's (1982) (most) comprehensive computational framework for explaining the 'processing of visual information' is certainly no exception. Is it the foundational concept it seems to be or merely, as Varela (1988) contends, the phlogiston of cognitive science, invoking as it does pre-existing order to explain regularities in the way we know the world? Similar qualms are expressed in Oyama's analysis of the concept of 'information' in biological explanations (1985, p. 66). She notes how information, defined in its traditional, quantitative, information-theoretic sense of reduction of uncertainty, quickly changes from being a measure of unpredictability to being invoked as an explanation of predictability, from where it is soon put in the role of causal control of development and adaptation. The problematic nature of information has also been acknowledged widely in philosophy (e.g. Churchland, 1980; Dennett, 1983; Dretske, 1981, 1983). As Churchland stresses:

> The question of how to characterize information processed by the mind-brain and how to characterize the processing of information is a question for *theory*, and it cannot yet be said that there is an anointed theory. (Churchland, 1980, p.194)

The simple reason why information, opaque as the notion may often be, takes on such significance in explaining infancy is that possessing and/or having access to information has been taken to imply a foundation for *order* in the infant's experience – *contra* Piaget, an ability, however rudimentary, to make sense of the world. Hamlyn (1981) points out that the folk psychology notion of 'information' can often be equated with 'knowledge'; for example, when characterizing the content of memory. Yet the notion of 'information' crops up in all kinds of places, including explanations of purely physical systems; and the quantitative treatment that might be invoked in explaining, say, the workings of a telephone system is unlikely to be what excites this enthusiasm.

What is (often implicitly) being acknowledged as interesting is some idea of *semantic information*. Broadly speaking, this singles out the information we invoke in discussing things that are about something or that are meaningful as far as our subjects are concerned: the information we intuitively think of as playing a role in perception; acted on the basis of; thought about; and transmitted in communication. This suggests an intricate connection between *information*, *mind* and *meaning*, and the philosophically central notion of *intentionality*; and a more than usually complex field of discussion, for questions of meaning present appalling complexities and are invariably described as 'thorny', 'slippery', 'tricky' and so forth. Following the discussion of chapter 3, this issue is discussed from two viewpoints. First, I consider a traditional, centralized view of meaning, which sees perception as a relation between the input to a sensory system and central processes, and poses the question: how can the mind make sense of that? Then, I follow up the idea of changing focus by increasing the scale of analysis to a more systematic account, and consider the implications for meaning of perception's nesting in action, and action's embedding in the environment.

Reservations about information 'recovery'

Faced with ecological psychology's often strident metatheoretical claims about what cognitivism *cannot* do (for example, equip the subject with veridical information about the world) there is a tendency to be led into describing how it *can*, rather than questioning whether these are things that our theory really needs to endorse. At one level, Marr's framework for visual information processing can be used in this way, and it has been viewed as establishing a degree of *rapprochement* between (realist) ecological and (constructivist) cognitivist perspectives by showing the compatibility of Gibson's focus on the rich (external) information available for perception and Piaget's assumptions about the need for complex (internal) processes (Boden, 1985; Rutkowska, 1988). Certainly, its novel, AI-based suggestions for the kinds of computational cognitivist mechanisms that could be preadapted clarify that Piaget is very much the straw man of the developmental side of the direct versus mediated perception debate. However, highlighting isolated, seemingly desirable features of the two perspectives is by no means the same as effecting an epistemological synthesis of the concepts and conflicting styles of explanation that they offer. The extent to which any full synthesis of the roles of perception and action in the structuring of understanding can be achieved depends closely on precisely how perception is being viewed. Perhaps enthusiasm for the problem-eliminating notion of 'information' has led computational cognitivism to present itself in a more conciliatory mode than is appropriate or desirable?

Marr's sympathy with Gibson's perspective, and his own commitment to realism, is emphasized in his talk of information 'recovery' as the goal of visual processing. This kind of processing, Ullman (1980) is at pains to assert, does not create information, but extracts it, integrates it, and makes it explicit and usable. 'Recovery' is typical of the metaphors adopted in both PSSH- *and* PDP-inspired AI work on information processing (cf. 'reconstruction', 'modelling', 'coding' or 'discovery') – but is it an appropriate metaphor? Although I have previously been free with it myself (Rutkowska, 1988, 1992a), I have a number of misgivings about it; hence my persistence, in the preceding chapters' discussion of vision, with what I consider to be the more neutral characterization of processing's role as one of 'making information explicit'.

In a number of non-trivial respects, the kind of contemporary AI vision model that is typified by Marr's work fails to conform to the direct realist assumption of an unambiguous causal connection between things in the world and the invariant structures that are equated with the subject's 'information' about them. Most obvious is the fact that the *locus of invariance* is in (internal) structural descriptions produced by processing in the visual system rather than in the (external) optic array. For example, both direct perception (e.g. Gibson, 1979; Michaels and Carello, 1981) and computational (e.g. the constraint analysis style of computation described below) approaches to the perception of reflectance (roughly how light or dark a surface looks under varying circumstances) would view its constancy as emerging from relations such as that between the amount or intensity of light at a surface, the intensity of illumination and the surface's inclination to the illumination source. Their dispute is over whether this relation is available to the subject in preexisting, mind-independent information or is 'imposed' by visual processing.

Even more important, to my mind, is the significance of the *imaging process* in defining the boundaries of the *perceived domain*: the physical constraints that are assumed to make visual information processing possible include not only interactions between physical structures in the world but also their interaction with the image; hence the centrality of the optics of natural-image formation to a top-level computational theory of vision. Strictly speaking, then, *task structure* cannot be defined independently of *subject structure*.

Furthermore, the way that such AI vision models work departs from the idea of a completely unequivocal relation between things in the environment and the informative structures that might be assumed uniquely to specify them. Consider briefly the process of *constraint analysis*, which for some time has been central in AI vision. This is based on the assumption that entities that can be described in several ways while considered in isolation can be given a unique 'interpretation' when relationships between them are taken into account.

The 'line labelling' approach to scene-analysis, whose input is a 2-D line-drawing and whose output is a description of a 3-D scene, can serve as a simplified microworld example of the kind of processing involved (Clowes, 1971; Huffman, 1971). Any individual picture line can be viewed as a candidate for a number of potential labels or 3-D 'meanings' in the scene domain, which reveals the difficulty of arriving at a consistent labelling for the whole picture. Even in a rarified microworld of trihedral bodies without cracks, laminae or shadows, the Clowes–Huffman scheme allows it to be 'interpreted' in four ways (as a concave or convex edge; or as two kinds of occluding edge, depending on which object surface is visible to the camera/eye). So, each junction type can have many possible labellings (4 to the power of the number of lines making up the junction), and the possibilities to be considered in the picture as a whole become vast as the number of junctions increases.

The style of parallel computation that copes effectively in such contexts does so by establishing relations between picture lines on the basis of constraints in the trihedral scene domain. For example, the potentially infinite number of pictures depicting trihedral objects can contain only 4 types of junction; and only 16 labellings of these junction types correspond to 'legal' physical entities such as corners in a trihedral world, in contrast with the 208 types that would be derived by considering all possible edge-labels at all 4 junction types.

One of the first computational algorithms to implement such constraints was Waltz filtering (Waltz, 1975). It eliminated all impossible labellings for individual lines through a pairwise comparison procedure and extension or 'propagation' of the results to neighbouring junctions. This usually made it unnecessary to employ a traditional linear search procedure: using constraint analysis, the number of possible labellings for each junction was typically reduced to one. At a junction, then, the interpretation of a particular line was constrained, rather like the value of an algebraic variable being constrained by a set of simultaneous equations.

One of the most impressive strengths of the PDP systems that are the descendents of Waltz's work has turned out to be their capacity for implementing simultaneous satisfaction of multiple constraints such as those involved in vision (for example, see Ballard, Hinton and Sejnowski's (1983) discussion along these lines of the parallel stereopsis algorithms that Marr proposed for 2½-D sketch construction).[1] The style of computation in such systems can be predominantly characterized in terms of *relaxation*, where a network gradually settles into a configuration or 'solution' by attempting to satisfy a large number of 'soft' or 'weak' constraints that have implications only if considered collectively, rather than the 'strong', 'hard' and consequently 'brittle' constraints exploited in the predominantly sequential rule following of classical systems. Stable states are arrived at through *statistical* rather than *syntactic/logical* inference (and are not obviously compatible with ideas of either pick-up or recovery of unambiguous information). Thus, Rumelhart and McClelland view connections 'as representing constraints on the co-occurrence of pairs of units' (1986, p. 135), not as analogous to wires that carry current in an electric circuit.

This process-based approach suggests a systematic explanation of perception and information that cannot be reduced to its individual components. Thus, it does not fall easily under traditional classifications. It is not just the *transduction* that is sought by naturalistic accounts of the information for perceiving, since more is going on than changing the input from one physical form into another. Nor is it *constructivist* or *representationalist* in the sense of involving interpretation of 'cues' through top-down 'hypotheses' derived from knowledge outside the perceptual process itself. Even ecological psychology's evaluation of 'seeing machines' excludes Waltz-style computation from its criticism of cognitivist vision systems' reliance on 'abductive inference' or inference to the best hypothesis (Turvey, Shaw, Reed and Mace, 1981).

Shepard (1984) suggests 'mirroring' and 'resonance' as metaphors for the biological origin and synchronic functioning of constraint-based perception. However, it seems to me that it is not obvious that such processes in natural vision *must* have arisen, as he suggests, from the evolutionary *internalization* of constraints in a domain of the external world (although both mirroring and a clear-cut external–internal split are implied by characterizing the subject's processes as *exploiting* physical constraints). Rather, the processes that comprise constraint analysis might better be viewed as a further component of the subject's contribution to *domain definition* (which would remain consistent with the notion that these processes *implicitly embody* assumptions about the domain in question). Shepard himself comes closer to this view when he stresses the 'functional mesh' that must hold between perception and the external objects and events that he believes it represents.

The key notion to which I want to draw attention here is that the world and our subjects' descriptions of it are probably, in Varela's (1988) sense, '*co-relative*': 'external' structures and 'internal' structures are mutually defining and operate together to determine a known domain. 'Chicken and egg' questions like 'Which came first, the world or constraint analysis?' seem to be sensible, and to have a straightforward answer only in so far as constraint analysis in low-level vision is thought of as operating in a bottom-up, data-driven mode. But the causal, external → internal scenario that this implies may be mistaken.

The Mach–Eden illusion provides an interesting challenge to the assumption of such inevitable *directionality* in low-level visual information processing. The basic version (Attneave, 1971; Mach, 1985/1959) involves 'reversing' one's perception of the surface orientations of a card that is folded into one or more V-shapes and laid on a table with the fold(s) uppermost. On reversal, two effects are striking: initially lighter surfaces now appear fluorescent, and initially darker surfaces very dark indeed; and, if you move your head, the card appears to be moving with you. A 3-D cube with a segment cut out gives an even better effect: it appears to be rotating about its main axis in the same direction as your head movement and about twice as fast.

The standard cognitivist observation about these phenomena (e.g. Ullman, 1980), is that top-down knowledge about object shape is overpowering input information, a seemingly clear-cut interpretation when Gregory's (1970) hollow face-mask, which exhibits the same phenomena as the Mach–Eden illusion during its generally spontaneous 'reversal' to a normal face orientation, is concerned.[2] But there is a more important, and possibly more valid, point: what we are seeing is the kinds of processes that are thought to mediate 2½-D sketch construction operating *backwards*. Rather than 'shape from shading' (Horn, 1975), or 'shape from optical flow' (Clocksin, 1980),[3] the visual system is getting shading and optical flow from shape. It seems that there are constraints *between* shading and shape, and *between* optical flow and shape, but shape is not simply or inevitably being 'recovered' *from* shading or optical flow. In the general case, image structure is more approriately viewed as providing a constraint between properties X and Y (cf. Draper, 1980), rather than information about X from which information about Y must be derived.

Varela (1988) phrases a similar argument in terms of colour vision. This can be viewed much like constraint analysis, as 'largely independent of incoming wavelength' but resulting from 'cooperative comparison between multiple neuronal ensembles in the brain, which specifies the colour of an object according to the global state it reaches given both a retinal image and a certain inner expectation of what the object should be' (1988, p.57). (Here 'inner expectation' is akin to the 'implicit assumptions' of Marr's evolutionarily determined visual processes.) He notes that supposedly illusory phenomena like *colour induction*, where, say, a piece of grey paper laid over red paper looks greenish, actually reveal the core of normal colour vision. The *relative activity* of the three classes of cells that we traditionally call 'colour receptors' *defines* what colour is for us. Contemporary attempts to restate colour vision in terms of surfaces' reflective properties instead of wavelength do not salvage the idea of 'recovery'. They simply introduce the question of how surfaces are defined in the first place (and, as we have just seen in considering the 2½-D sketch and the Mach–Eden illusion, this appears to be due to the very kind of computational process that gives rise to colour vision).

The circularity of action (read in the broad sense of subject processes or activities) and interpretation that is typical of constraint analysis is what leads Varela to argue that the appropriate metaphor for knowing the world is an epigenetic one of 'enaction' or 'bringing forth', as opposed to 'representation' (in the realist sense involving internal modelling or substitution). Only a pre-given world that is definable independently of the subject's activity can be represented in that traditional sense, but that does not seem to apply to the input to visual information processing: 'Information must be seen not as an intrinsic order but as *emerging* from the cognitive activities themselves' (Varela, 1988, p. 3). Varela (cf. Maturana, 1970; Maturana and Varela, 1980) aims to ground a replacement to 'recovery' and 'substitution' metaphors for information and representation in the biologically informed notion of *structural coupling* between a self-organizing system and its surrounding environment:

> the view of cognition is not that of solving problems through representations, but as a creative bringing forth of a world where the only required condition is that it is effective action: it permits the continued integrity of the system involved.
> (Varela, 1988, pp. 59–60)

Koenderink (1980) draws a similar conclusion with regard to Ullman's (1980) comparison of direct-perception and Marrian-recovery treatments of the relation between motion and surface perception:

> 'Solid shape' is not present in nature but is a mutual property of perceiver and environment . . . you do not 'extract' what is already there: what is there depends on *me*. In this sense, I do not become attuned to things: the things are what they are because I am what I am. (Koenderink, 1980, p. 391)

This view of the roles of subject and environment in domain definition fits with earlier chapters' doubts that infant or adult vision is in the business of constructing structures whose purpose is to underwrite veridical perception in the sense of standing in an unambiguous one–one relation to objects and events in the environment. In so far as notions of 'structural coupling' and mutual 'domain definition' appear more apposite than 'recovery', the structures in the world that contribute to this coupling will not necessarily carve up in the same way, or be categorizable in the same 'vocabulary', as the subject structures that support functional correspondence with them (*contra* ecological psychology's realist assumption of an environmental vocabulary of goal objects, threats and so forth, to parallel psychological accounts of the subject at the psychological grain of analysis). The idea is that the way 'things' are viewed depends on mind and environment, but not on the environment's being ready populated with the categories of objects and events that are constructed. The potential of connectionist ideas about computation to clarify this position seems to me to be their most attractive property, though there is a considerable way to go for it to be realized.

The possibilities come from such systems beginning to show how computational symbol structures – hence the subject's domain knowledge and information about the world – could be an *emergent property* of the parallel operation of multiple subsymbolic units of the brain-mind; and its rules an emergent property of the kind of parallel computation operating with multiple 'soft' constraints that was noted earlier. (Marr's vision model begins to show some of the requisite properties, since it is a hybrid that relies on modular computations using primarily cooperative, parallel modes of processing to output symbolic structural descriptions in the classical style.)

As Clark and Lutz (1992) emphasize, the most philosophically interesting of these systems offer what they call 'dimension-shifted subsymbolic representations'. This means that they employ a particular kind of *distributed representation*. Their units never code whole, linguistically and semantically meaningful entities such as 'cat', unlike individual symbols of the molar, mainstream AI models that were criticized in chapter 1. Instead, such 'macro-concepts' are coded by activation patterns over many units, which may correspond to subsymbols or microfeatures that can neither be linguistically labelled in this way nor be equated with semantically transparent 'macro-subsymbols' such as properties like 'furry', 'whiskers' or 'tail'.

Identification of subsymbolic processes with a 'subsemantic' level could play a very significant role in psychological theory, although a clear basis for making a subsemantic–semantic distinction still seems lacking. The supposedly subsymbolic elements of many connectionist networks may simply turn out to be the fine-grained symbols that were initially promised by PSSH. In particular, clearer criteria are needed for determining when the microfeatures of the dimension-shifted variety of PDP model are shifted enough to count as subsymbols (the lines that are input to the scene-analysis program noted above certainly are not). Systems (or parts of a single system) that support the emergence of higher-level symbols from lower-level ones need to be distinguished from those where symbolic functioning emerges from a genuinely non-symbolic level.

Finding a way of ensuring that a distinction between symbolic and purportedly subsymbolic levels reflects more than relative ease of attaching verbally meaningful labels to them is needed if it is to hold true, as Smolensky (1987, p.101) argues, that:

> The higher semantic level is not necessarily capable of being formalized precisely, and the lower level is not 'merely implementation' of a complete higher level formalism. Both levels are essential: the lower level is essential for defining what the system *is* (in terms of activation passing) and the higher level is essential for understanding what the system *means* (in terms of the problem domain).

Some of the pertinent issues here relate to the problem of characterizing the level of explanation at which the subemergence level of such connectionist networks is assumed to operate. In the human case, neurons provide the subsymbolic structures from whose physical coupling with the environment our symbolic computational abilities emerge. But PDP can be viewed as exploring the kinds of algorithms that collections of connectionist units can support, and may best be located at a level of explanation intermediate between the levels of physical implementation in neural wetware and of representations and algorithms at the symbolic computational level (Rumelhart and McClelland, 1986; Smolensky, 1987, 1988). As such, they cannot provide a complete account of the brain's structural coupling with its environment.

Connectionists use neurons as a productive metaphor for conceptualizing computational processes (which sometimes suggests a seductive, albeit potentially spurious, biological validity for their models). However, PDP units are not neurons or groups of neurons; their properties are a selective abstraction over what are believed to be some relevant properties of neural organization and functioning, and detailed evidence of how their functioning relates to neuroanatomical evidence of 'real' neurons and of how the brain works is yet outstanding (Crick, 1989; Crick and Asanuma, 1986). As far as drawing and understanding a hoped for subsymbolic–symbolic boundary goes, especially close scrutiny is merited by ways in which we might view the retinal image, the content of whose analog 'representation' of the environment can be couched in the language of physics, and its relation with the discrete symbol structures to which its processing gives rise on the basis of connectionist-like constraint analysis. (We cannot lose sight, however, of the fact that the 'retinal image' is itself an abstraction from neural functioning, not an objective given from which to start our inquiries.)

Despite such unresolved issues, connectionist systems undoubtedly open up a lot of intriguing possibilities. It is therefore disappointing that their prevailing theoretical direction appears to be relatively unquestioning perseverance with some version of realism. For example, PDP models are regularly proposed as a strong basis for *rapprochement* between direct perception theory and computational explanations of perception (e.g. Humphreys and Riddoch, 1986), whereas reasons like those sketched above suggest that a valid merger along these lines is unlikely. In particular, the notion of 'emergence' is significantly different from ecological psychology's assumption of physical and psychological grains of analysis as alternative descriptions of a single system. A wider-ranging theoretical exploration by connectionists might play a key role in fostering a genuinely epigenetic alternative to traditional realist and constructivist perspectives.

The 'what', 'where' and 'how' of information

Where do reservations about the prevailing computational view that vision 'recovers' information leave the notion of 'information' in relation to the supposedly key function of the brain-mind: information processing? 'Information' in cognitive science brings to mind van Brakel's (1991) discussion of 'sweet potatoes' in the Dani culture of New Guinea. Sweet potatoes are implicitly or explicitly featured in over half of Dani conversations, and over 70 terms are used to refer to them; yet these terms appear to be used without precision, with the Dani rejecting the idea that 'different susceptibilities' are to be associated with different kinds of sweet potato. Conventional use of vocabulary appears to function adequately without fixed meaning based on a one–one word–world mapping, and to have as much to do with establishing and maintaining social relations and group bonds as with transmission of thought about an objective world. The fact that 'information' is a single term that we use in many contexts does not make its reference to structures and mechanisms in the world any more direct.

The role that information plays in explanations makes it an *abstract* concept. Information is not a thing that can be *reified* or *localized*, and it makes no real sense to talk of information quantity or content as a property of any structure in isolation. It is a way in which we try to characterize certain aspects of the way that structures in the world (including those within the subject) interact with and affect one another; and to attribute information content to those structures in terms of their place in the interaction. With regard to vision, for example, the notion that 'it' can and must be localized 'in' the optic array or 'in' the subject sets up a misleading dichotomy; the 'information' on which perceiving is based requires a systematic explanation in terms of interdependent structures and processes within the subject and between subject and environment. Either cognitivist metaphors in which information is 'transmitted' and 'stored', or ecological psychology's view that it remains in one place (in invariants), involve reification and consequent oversimplification.

Information, then, is not a primitive concept that is going to explain how or why the subject–environment system 'works' as it does (cf. arguments due to Varela and to Oyama, noted above); it is an abstraction that depends on that very understanding. From this perspective, information talk is misleading unless it acknowledges the notion's inherently systematic *relational* nature. Across its many uses, it has two major dimensions: *correspondence* and *control*, which fit closely linked answers to the question '"What" and "where" is information?' If certain types of relationship hold between structures, then access to one can influence your behaviour (more precisely and generally, the behaviour of some process) with respect to the other. Thus, it is common to talk about the content of information in terms of both what it is about (what the informative structure is assumed to correspond with) and what it is for (what the informative structure influences or controls).

In Gibson's (1979) ecological approach to perception, these dimensions map onto the notions of 'invariant' and 'affordance'. Described in terms of invariants, the structure of the ambient optic array is assumed to correspond to some object or event about which it carries information; described in terms of affordances, the structure of the array is assumed to control some behaviour for which it provides information. But even at a very basic level this raises problems. Gibson needs affordances to be objectively real – 'in' the environment independently of any subject – yet simultaneously to capture the subjective aspect of perception in action. His assertion that affordances are *neither* objective nor subjective is unconvincing. It is surely less paradoxical to agree that the phenomenon of functionally meaningful perception is a *joint product* of environmental and subject structure.

The computational approach to information processing can accommodate the same dimensions of correspondence and control, albeit in a different way. Newell's (1980) definition of representation in terms of a physical symbol system's fundamental capacity for *designation* implies them both. Designation has two aspects. There is 'action at a distance': when X designates Y, a process that has X as an input behaves as if it has access to Y. This captures the notion that the symbol structure X is informative in corresponding to something. Of equal importance, designation is always 'relative to a process', and its implications depend on the nature of that process. This captures the notion that X is informative in terms of the part it plays in influencing or controlling a process P.

This fundamental relationship of designation (or symbolization) is *transitive* (Newell, 1980). So, at each level of processing, the subject's structures continue to designate the entity responsible for the input. This captures the idea that something is staying the same – on the correspondence dimension – even though information processing involves transformation between symbol structures. Emphasis on *explicit naming* (Marr, 1976, 1982) captures the idea that there is also a sense in which something appears to change – on the control dimension – when information is processed. Every symbol structure makes explicit different information, usable by different processes. Both of these aspects are important: without transitivity it would not make sense to talk of one symbol structure implicitly containing information that can be made explicit by further processing.

The idea that the information in a structure can be either *implicit* or *explicit*, which was central to chapter 2's view of what the processing of an AI vision system is doing, and to chapter 3's account of vision's role in action, shows how difficult it is to talk of the information content of any structure outside the context of the system in which it is embedded. Palmer's (1978) metatheoretical analysis of 'representation' argues that the information content of a structure can be defined fully *only* by including the processes that can be applied to it.[4] This clarifies the apparently paradoxical idea that implicit information in a structure 'is not there at all' unless the system has processes to derive it, but if the system has such processes it 'is there' despite appearing not to be. It also supports the important point that 'it is possible to have a representation that seems nonsensical by intuitive notions but is appropriate and sensible given the processes that operate on it' (Palmer, 1978, p. 265).

As far as the question of *how* subjects gain access to information goes, the key conclusion of the preceding discussion must be that it does not mean acting as a 'sieve' for that information. Nor does it mean that nothing more than active selection is involved, a view that would remain compatible with information being an objective 'commodity'. Non-optional processes of constraint analysis make a non-trivial difference, with their 'implicit assumptions' and unique perspective on how links between environmental and subject structure could be related to emergent symbolic functioning.

Some interesting recent attempts to define information and how subjects gain access to it also forefront ideas congenial to a systematic account of information, even though they start from a realist perspective sympathetic to direct perception theory. For example, Dretske's (1981) analysis of the 'flow of information', which aims to view information as an objective, mind-independent commodity that can provide a *naturalistic* foundation for psychological phenomena, ends up with a definition of the information content of a signal as relative to what the receiver already knows about circumstances at the signal's source. Say you know that an object is in one of a chest's three drawers, a signal carrying the information that it is not in the second drawer (for example, that drawer's emptiness) carries the information that it is in the third drawer only if you already know the first is empty.[5]

From the language-based perspective of situation semantics, Israel (1988, p. 125) views information as 'an irreducibly extrinsic, relational phenomenon'. He argues that 'one situation (that is, what is going on at one spatiotemporal location) carries information about another situation relative to a law or constraint relating types of situations' (Israel, 1988, p. 124), noting that 'such constraints are valid only under certain kinds of background conditions or in certain circumstances'. The world is said to be rich in information, on the terms of this definition, which is said to provide a mind-independent basis for intelligent action.

It is, however, important to note that this approach sees a lot more to the subject's gaining access to the information carried by a situation than just gaining access to the situation itself: it is much more subject-dependent than it may superficially seem. The kind of information carrying situations that are generally featured are intra-subject situations such as what is going on in the retina. For example, if red spots carry the information that someone has measles, and your retinal pattern includes red spots, do you have the information that your child has measles? If not, what more is needed? 'World', then, is taken to cover structural relations within and between subject *and* environment, not just what is going on in the subject's environment. And an important part of the answer to the preceding and similar questions is that information in the world 'informs only those attuned to the relations that allow its flow' (Barwise, 1983, p. 65). That is, in some sense, the subject must be 'attuned' to the relevant constraints (of which more will be said later in this chapter).

Such ideas are certainly uncongenial to ecological psychologists (for example, see the discussion between Turvey and Carello and Barwise and Perry (Barwise and Perry, 1984)). They may be more compatible with the idea that subject and environment play mutual roles in domain definition, as sketched in this chapter's preceding section, than with hard-line direct realism.

Trying to 'see' versus 'see as'

Ecological psychology's realism is able to simplify the issue of how infants make sense of the world, by equating semantic content with information content: meaning does not have to be constructed or contributed to by the mind; it is pre-given in environmental information. (As noted in chapter 3, Gibson (1979) asserts that infants should directly see the meaning of at least some things *ab initio* by picking up their affordances.) Starting to look at information in systematic terms, and acknowledging the subject's role in gaining access to it or making it explicit, means passing over the chance of such oversimplification. The way that traditional philosophical accounts of the mind and its meanings have looked at the pertinent issues also sees them as far from straightforward.

The intentionality that philosophers consider characteristic of the mind is said to involve more than straightforward correspondence or correlation: an 'arrow of reference' from some things to others. This is illustrated when we talk of someone thinking X, believing Y or knowing Z and so forth. We cannot understand or explain what they think, believe or know on the basis of purely objective criteria concerning X, Y or Z. The content of someone's thought, belief or knowledge depends not only on the reference of the terms X, Y or Z, but on their *meaning*.

The validity of such account(s) always depends on facts about the *subject*. For example, their validity will not be maintained if other descriptions are substituted for X, Y and Z, even if these new descriptions are, objectively speaking, referring to the identical object or state of affairs. Someone may believe that the cat is a Siamese, without its being true to say that they believe that the cat is a cat, or that they believe that the cat is a mammal. (This resistance to substitution is generally called *referential opacity*.) In contrast with this, if we talk of someone stroking a cat, it is thought legitimate to substitute any (co-designative) term that describes the object (such as 'Siamese' or 'mammal') for 'cat' without upsetting the validity of our account. Stroking is not considered an intentional notion.

An example of this treatment of intentionality that is particularly relevant to understanding phenomena that come under the general rubric of how infant's perceive the world involves the distinction between *seeing* and *seeing as*.

'Seeing' – often termed 'just', 'simple' or 'merely' seeing – has typically been treated by philosophers as falling into the non-intentional category. As far as the validity of the account goes, describing someone as seeing the cat, the Siamese or the mammal are thought to mean the same thing. In contrast with this, 'seeing as' demands close links with the subject's *conceptual* abilities and is thought to be intentional, along the lines of 'recognizing as', 'conceiving as' or 'thinking of as'. Seeing a cat *as a Siamese* (or *seeing that* it is a Siamese, to make clear this version's propositional attitude lineage) has a meaning quite different from seeing a cat that just happens to be a Siamese.

Three things about this traditional way of looking at meaning are fairly clear. One is the enormous extent to which its account of being a psychological subject is dependent on properties of language, and on conceptualizing the mind in terms of mental states whose structure is assumed to be usefully characterized in sentential–propositional terms – a far more molar level than that aimed at in the computational structures being explored in this book. Second, its centralized view of the mind gives rise to the notion that meaning is localized in the content of mental states, and that inquiry should focus on relations between the inputs to a system and those mental states. Finally, it has been considered – both implicitly and explicitly – to hold out serious theoretical possibilities for infant development. It suggests a way of simultaneously addressing two key issues: the meaning of things for the infant and the appropriateness or otherwise of attributing concepts to them.

Rather than assuming that this dichotomy from the philosophical literature can be employed as an analytical device, and that our data and theories must somehow be fitted to it, I believe that it is more worthwhile to ask: is this dichotomy useful? If it is to fulfil its explanatory promise in relation to the infant, it should be possible to make the distinction clearly by mapping the two classes of seeing onto plausible mechanism distinctions. Two approaches directly relevant to the view of infancy developed so far will be considered here.

Fodor and Pylyshyn (1981) discuss 'seeing as' in so far as they propose failure to account for it as one of a range of philosophical objections to the direct perception thesis. Their argument need not be detailed here, but it does provide an important focus for Bruce and Green's (1985) developmentally relevant attempt to reconcile direct perception and cognitivist approaches, including Marr's vision model.

Bruce and Green argue that the theory of direct perception offers a theory of 'seeing' (whose physiological underpinnings may well be understood in terms of Marr's work on information recovery) but a notion of 'seeing as' is needed for most human perception:

> Most human activity takes place within a cultural environment, and we see no alternative to the assumption that people see objects and events *as* what they are in terms of a culturally given conceptual representation of the world. (Bruce and Green, 1985, p. 333)

This implies that infants must simply 'see' prior to the impact of cultural experience. Bruce and Green's distinction places the explanatory burden on the notion of 'conceptual representation'. How might this be distinguished from the mechanisms with which the infant starts out, or from the kinds of mechanisms possessed by animals capable only of 'seeing'?

Two senses of representation are distinguished by Bruce and Green. First, they take examples like fly vision, which was briefly considered in chapter 3 in relation to Marr's (1982) ideas on primal sketch construction; chapter 3's discussion of infant vision, looming and the primal sketch, and of prehension and the 2½-D sketch, would offer comparable human examples. The fly's vision system works to provide the information needed to 'modulate' its moment-by-moment activities; for example, flight or landing. Analysis of its information processing at Marr's algorithmic level is said inevitably to involve 'representations'. But this is not to say that the fly builds up a conceptual representation of the world which a 'little fly in the head' consults to decide on a course of action. This is certainly a valid point (cf. chapter 3's action program alternative to a central homunculus), but Bruce and Green go on to argue that this sense of 'representation' is compatible with direct perception because the fly's representations are simply *momentary patterns of activity in a neural network*. In contrast with this, conceptual representation is said to necessitate *memory*.

Separating 'seeing' from 'seeing as' at the mechanism level cannot succeed on the basis of this distinction. On the one hand, the plea to a 'neural' basis for 'seeing' does not follow through. The fly's temporary patterns of neural activity are, after all, determined by permanent mechanisms based on evolutionary as opposed to individual experience. 'Seeing as' must likewise depend on underlying features of the neural network, whose patterns of activity are just as transitory while they are in operation. On the other hand, different functions are attributed to Marrian and conceptual representations, but the mechanisms underlying the latter are elucidated primarily in terms of what they are *not* (the former being the descriptions in representational formats that featured in chapter 2). The question is: what additional and/or different mechanisms would the fly/infant need in order to be building and/or consulting a conceptual representation?

What Bruce and Green's distinction really comes down to is the familiar dichotomy between representation as 'selective correspondence' to the world and representation as 'substitution' for it. Viewed computationally, however, both are based on symbol structures. Marr's theory of how descriptions are constructed in vision, even if they are seen as an emergent property of connectionist-style computation, are not reducible to the status of a purely physiological account of information recovery.[6] It is notable, for example, that some properties of intentionality are associated with isolated early/low-level vision. What counts as information for vision, and how, depends on subjective factors about the processes that will operate on it, and on their role in constructing particular kinds of description, not solely on its objective, physical characteristics. This has led some to conclude that all seeing must be intentional 'seeing as' (e.g. Clowes, 1971). As we shall see below, they are probably wrong so far as strict philosophical treatments of intentionality go, but this serves only to reinforce the difficulty of clarifying and exploiting the dichotomy.

Dretske's (1981, 1983) attempt to provide a theoretical bridge between the often very different ways in which questions about perception, knowledge, belief and so forth are discussed in philosophy and in cognitive science provides a more liberal approach to clarifying and applying the notions of 'seeing' and 'seeing as', so far as the prospects for non-human species and human infants go. He associates the two forms of seeing with a computational distinction between an initial *analog* coding of information and a subsequent *digital* coding. His ideas thus prove readily comparable with Marr's work, which provides a typical AI model along these lines, even though Dretske does not draw on it explicitly.

'Seeing' is to be equated with perception. Our rich experience of the world is assumed to be based on a sensory representation that codes information about the world in an analog, continuous form (cf. Marr's retinal image). 'Seeing as' is due to cognitive or conceptual processes, which abstract and select from this information, resulting in digital representations, which can be equated with the subject's concepts (cf. the digital-symbolic nature of all post-image descriptions in Marr's model). Meaning is thus located and identified with the content of structures at the symbolic level. Referring back to the earlier feline example, the analog image would be said to contain information that there is a cat, a Siamese, a mammal and lots more besides, all of which the subject would 'see'; but discovering what things are being seen or recognized 'as' requires understanding the types of symbolic descriptions that the subject can construct (cf. Marr's representational formats).

Following through this approach, equating symbolic descriptions with infant *concepts* could allow us to see them as the content(s) for infantile *beliefs* that go to make up the kind of *theory* of the world that increasingly is being attributed to the infant. The existence of innate concepts is seen by Dretske (1981) as an empirical issue, and he draws attention to the well-known visual cliff phenomenon to suggest an affirmative answer:

> Some information is automatically digitalized for the determination of important responses (avoidance, flight, pursuit, concealment, etc.). In virtue of this inherited alignment between functionally significant structures and their informational sensitivity, such structures have a meaning or content *prior* to any actual exposure to signals bearing the relevant information. What we are talking about, of course, is *innate concepts*: functionally significant internal structures that, prior to learning, prior to exposure to signals bearing the information that *s* is *C*, constitute the system's way of digitally representing the fact that something is *C*. (Dretske, 1981, p. 231)

It is impossible here to miss Dretske's stress on *functional significance*. Looking from the vantage-point of information in the world, however, Dretske is at pains to insist that the meaning of such innate structures is firmly located in the system's input. The meaning or content of either innate or acquired conceptual structures is to be derived solely from their '*informational origin*', not from their output or from their successes or failures in producing appropriate behaviour. This view develops (Dretske, 1988), as we shall see in this chapter's final section, to become more compatible with seeing action as significant to our understanding intentionality.

The way the analog–digital distinction handles concept attribution relies on evidence of selective input–output relations, and fits the way that infancy researchers have seen infant performance on discrimination tasks as evidence for more than categorization of input. It squares not only with the sense in which Spelke's theory is attributing to the infant a precocious 'object concept', which 'arises from an intuitive theory of the physical world and its behaviour'

(Spelke, 1988, p. 230), but also with key properties of the mechanism distinction that it is trying to make. For Spelke, perception is said to transform physical energy into 'knowledge of a *continuous* layout of surfaces', while central thought systems then 'package the world into "*units*" or "*things*"' (Spelke, 1988, p. 229; italics added). While somewhat out of line with details of the computational vision model – which would see surface descriptions as digital-symbolic, hence as 'packaging the world' into entities of *some* kind – this view is grounded in a type of analog–digital distinction that attempts to distinguish perceiving from thinking on the basis of their different inputs and modes of organizing experience.

Trying to understanding infants' concepts in terms of an explanatory parallel between the dichotomies of 'analog/continuous versus digital' mechanisms and 'seeing versus seeing as' raises two main kinds of difficulty.

One kind involves a fairly introverted look at how Marr's model fits Dretske's requirements (Rutkowska, 1988). For example, not only the analog image but also subsequent digital symbolic descriptions can be said to contain implicit information that has the potential to be made explicit, rendering problematic the equation of conceptual content with information content that this version of Dretske's position requires. This implicit–explicit distinction, especially as it relates to vision in action, may get closer than analog–digital to capturing the fact that our representations – not only 'sensory' but also 'conceptual' in Dretske's sense – contain far more information than we are normally able to use.

A broader problem with this way of trying to do things is that it turns out to be a fairly blunt instrument so far as developmental issues are concerned. The nature of the intentional level of functioning was opaque in Bruce and Green's analysis. Dretske's liberal treatment, which would allow low-level vision to be viewed in conceptual 'seeing as' terms, could underwrite concept attribution to virtually any species and at any developmental level – from flies to the youngest infants. Often, however, we find that talk of young infant's supposed concepts, theories or thoughts is a prefix-rich field, riddled with *proto*concepts (e.g. Woodfield, 1988) and *naive* or *intuitive* theories (e.g. Carey, 1985; Keil, 1990; Spelke, 1988). Sometimes, this is due to the notions being used in a metaphorical sense. More usefully, though, the implication may be that initial mechanisms are *precursors* of some mature form, not simply in the trivial sense of preceding them chronologically, but in the more interesting sense of bearing a causal relation to some mature form, with which they already share some properties and into which they will be transformed through development.

Prematurely equating such early mechanisms with 'conceptual' and 'theoretical' abilities might mask the absence of important properties that are – or possibly are not – being picked out, albeit opaquely so far as the details of computational mechanisms might be concerned, in many philosophically motivated accounts. As a way of carving up and understanding the world, concepts and their associated theories are often assumed to involve 'more' than input–output regularities that an observer may interpret as the subject obeying principles.

A significant dimension that is reputed to characterize a 'properly conceptual level' (Woodfield, 1988) or concepts in the 'full-blooded sense' (Kirsh, 1991) emphasizes the *flexibility* and *interrelatedness* associated with this style of functioning. For example, it is central to Carey's (1985) attribution of a naive physics to the infant to claim that Spelke's perspective on infant object understanding establishes that 'the infant's concept of *object* cannot be specified apart from related concepts such as *space*, *number*, and *physical causality* . . . that articulate a set of physical beliefs' (Carey, 1985, p. 397). Keil (1990, p. 152), on the other hand, questions whether infants seemingly carving up the world in terms of 'principles' is sufficient to endow them with a theory – even spiders and suchlike qualify as object theoreticians on this criterion. Just behaving in a way suggesting that only certain inputs are 'meaningful' is surely not enough?

At stake here are the *kinds* of relations that underlie the infant's understanding. These should go beyond 'recognitional capacity' (Woodfield, 1988) in the sense of an ability to pick out similarities between things; for example, responding one way to things that are 'X' and another way to things that are 'not X'. Seeing or recognizing at a truly conceptual level that 'the cat is a pet', say, requires the subject to have both the concepts 'cat' and 'pet', and as part of a conceptual system whose component concepts work together in specific ways. It must be possible to believe or think the thing is other than 'X' (for example, to entertain the idea of the cat being wild, a mammal or a Siamese) and that other things, including arbitrary things, can be 'X' (for example, to entertain the idea of a dog, a snake or a rock being a pet). This goes alongside the view that the application of a theory furnishes subjective *reasons* and *justifications* for the subject's beliefs and/or actions, and these are distinct from the objective data to which they are applied. Such properties, it is worth noting, propose criteria for 'conceptual representation' that do not necessitate the temporal delay between input and output that was a feature of Bruce and Green's (1985) account, though they are closely allied with a view of how representation involving substitution might work.

Kirsh (1991) follows this line to give a computational definition of the conceptual that is more restricted in application than Dretske's. He argues that symbolic representations are essential for concepts, but they must be of the right *syntactic* kind, which can be combined and manipulated to support *predication*. This essential feature of full-blooded concepts is defined as the ability 'to identify the common property which two or more objects share and to entertain the possibility that other objects also possess that property' (Kirsh, 1991, p. 163).

In a similar vein, Davies (1992) draws on Evans's (1982) 'generality constraint', which refers to the kind of flexibility of concept combinations that was noted above, to identify propositional attitude states with cognitive states whose content is conceptualized in this sense. However, he disputes the view that semantic content must be restricted to this traditional arena; and that there is an essential dichotomy between non-scientific 'folk psychology', with its explanations in terms of attitudes and reasons, and cognitive science.

Davies contrasts two ways in which cognitive science can challenge this position. The first, unsuccessful route would be to try extending the domain of reasons and justification to encompass phenomena like linguistic understanding (cf. visual understanding). This introduces the notion that it depends on justificatory reasoning; then, that this must be operating with tacit or unconscious rules, since non-linguists' scant conscious knowledge of linguistic rules is well known. The idea of '*unconscious justification*' that is implied here is, however, 'totally mysterious'. There comes a point, as Wittgenstein argued, where 'doubting', hence justification, gives out; and the kinds of processes at issue here are most likely it (cf. Hobson, 1991). More profitable and coherent, to Davies's mind, is the strategy of extending the domain of semantic content beyond the realm of reasons and justification. This assumes that there is a subclass of cognitive states, referred to by Stich (1978) as *subdoxastic* states, which have semantic content without being intentional states of the propositional attitude variety. They may, however, play a causal role in explanations of the subject's construction of these *doxastic* states; for example, the key beliefs, desires and intentions. Doxastic states *must* meet the generality constraint, but there is no necessity for subdoxastic states, such as tacit knowledge of linguistic rules or the registration of retinal disparity by stereopsis in the visual system, to do so.

To the extent that infants' 'object' discriminations are interpreted, as they were in chapter 2, in terms of visual processing and the kinds of descriptions that it constructs, their object understanding does not obey the generality constraint. There is no need for it to do so, since this interpretation locates their understanding squarely at the subdoxastic level.

In visual processing through constraint analysis, as noted earlier in this chapter (pp. 93–4), the input elements to such processes can be interpreted in different ways, depending on their context. This could be said to involve a kind of generativity: a limited number of individual elements potentially have multiple 'uses' and can feature in an indefinitely large number of different combinations. But this kind of processing (or 'input') system is also characterized by *rigidity* with regard to combinatorial dimensions where the generality constraint demands flexibility; although generative, it is not syntactic in the relevant sense. Being able to make explicit some aspect of the physical world – for example, a surface – over many situations does not entail any ability to represent it as a *property* that is common to that range of situations, let alone potentially applicable to others. Visual processing does not deliver a description of anything *as,* or with the property of being, a surface; it simply delivers or fails to deliver a description of a surface. Nor does it justify such descriptions; characterizing its processes as embodying implicit assumptions implies something more akin to blind prejudice than to reasoning.

To the extent, then, that infants' scene parsing abilities are grounded in visual processing, they need not involve 'concept' or 'theory' in the technical sense being discussed here. The visually-based knowledge of surface-segmentation and motion that underlies the infant's organized behaviours towards objects is not available in a 'conceptual' form that could be used by a central thought system to define properties of a physical object (for example, that *it* is cohesive; or that *it* is bounded), or to entertain the idea that they might be properties of anything else. The infant's appreciation of 'objects' and of 'space' is so intertwined that they cannot be considered to be independent concepts. The 'object concept' that we see in such studies implies no unitary conceptual content, but emerges from the operation of multiple process-embedded constraints whose role in 2½-D sketch construction sometimes appears to have the goal of making explicit 'unitary objects'.

Empirical evidence that gets closer to supporting a conceptual level of knowledge organization during infancy is due to work by Mandler (1992, in press). Infants were 'familiarized' with a series of small models of either animals (for example, dogs, fish, rabbits) or vehicles (for example, aeroplanes, cars, motorcycles), then presented with novel objects from the same and the contrasting category. From 7 to 11 months, their examination time for the new model revealed discrimination between the global-level categories. Even models of aeroplanes and of birds with outstretched wings were discriminated by 9-month-old infants, despite their apparent perceptual similarity.

In the current context, the most important feature of these data is that the models of animals and vehicles do not move. Infants' different treatment of them cannot therefore be considered to depend on a purely perceptual analysis of the forms of motion that are believed to be at the core of this global-level category distinction (for example, animals move by themselves, in rhythmic but unpredictable ways). If, however, perceptual information is being transformed into 'simple concepts of animate and inanimate things', then the models could be *recognized as* representing or standing in for categories of objects whose properties have already been learned, and discriminated on the basis of those conceptual categories.

Unlike the scene parsing studies, this work is, of course, featuring infants in the second half of their first year. However, Legerstee (1992) has proposed a very similar 'conceptual' interpretation for the differential behaviour of much younger infants towards people and things. This conclusion seems far from compelling. All of the data reviewed provide evidence of different behaviours towards people and objects, or of different 'search' behaviours when people or objects move out of view, and many current or recent between-condition perceptual differences are or might be implicated.

For example, in a key study, infants were presented with people and objects that alternated between being passive and acting contingently to the infants' responses. By 2 months of age, infants showed more social responses to a person than to a doll. Their affect was also sensitive to the person's behaviour (positive when the person responded contingently, negative when they were passive, cf. Trevarthen, 1974), but it did not vary in this way with the contingently moving and passive objects. The questionable argument is along the lines that a passive person and passive doll are perceptually similar; hence, the fact that infants do not confuse these 'exemplars of the social and physical domains' is strong evidence that 'performance on the person–object task is based on conceptual understanding' (Legerstee, 1992, p. 63).

Probably, there are many perceptual differences between a person and a doll besides the motion patterns involved in their contingent activities. Leaving this aside, however, it is more important to the issue of concept attribution that a doll, which is a person-model, and a person are treated *differently*. By analogy with Mandler's data and argument, behaving towards the doll *as if* it were a person, in the *absence* of perceptual similarity based on person or object category-defining properties, would support the conclusion that conceptually represented properties are being attributed to the doll. In this study, however, it seems that infant mechanisms are responding to properties of people and of objects, not treating either as 'exemplars' of anything at all.

This general conclusion is compatible with Hobson's (1991) cogently argued critique of increasingly influential attempts to explain infants' and young children's social understanding by attributing to them a 'Theory of Mind'. One of Hobson's fundamental arguments is that infants' social knowledge does not involve 'theoretical' concepts, and it fits closely the conclusions about infants' physical knowledge that have been reached here. If infants have and use a theory, it should be possible to distinguish between their theory and the data or evidence that they draw on to support it. But neither object-related nor person-related experiences have this data-like quality. For young infants, their experience of other minds, like that of physical objects, is a 'fact' not a 'conjecture'. It is more appropriate, he concludes, to understand the infant's social understanding in 'pre-Theoretical' terms of 'innate capacities for perceptually anchored personal relatedness' (Hobson, 1991, p. 45), and generally in terms of 'innately constituted propensities to perceive the world in such a way as to engage their own tendencies to action and feeling' (1991, p. 36).

The developmental problem is clearly whether these two ways of understanding have to be fitted together, and, if so, how? It is illustrated in Premack's (1990) wish to attribute to the infant a Theory of Mind, while contrasting the infant's understanding of intention in terms of a 'hard-wired perception based . . . . on appropriate stimulation' (Premack, 1990, p. 2) with the 'common sense' of folk psychology for which intention is said to be 'an inferred state of mind based on evidence for *desire*, *belief* and *planning*' (1990, p. 12). It is not just that the infant has a 'theory' whose content differs from that of the adult. As Premack notes, their 'subject matter' is at odds too: the adult's common-sense theory is applied to identifiable objects, but the infant's is based only on the motion in which an(y) object engages. There must, he suggests, be a (human-specific) developmental transition from 'perception interpretation' to 'conception' with relative freedom from physical parameters.

For the time being, I shall leave open the issue of whether the properties thought to define 'real' conceptual functioning can provide a useful way of characterizing the abilities of *older* infants. In later chapters, 7 and 8, when development is considered, this issue will feature in two closely related questions: does it pick out a useful mechanism–ability demarcation? As we saw in this section, attempts to answer this in the affirmative for the example of the 'seeing versus seeing as' dichotomy were far from straightforward. And what happens to action-based 'precursor' mechanisms?

Meaning in action: the infant as a causally embedded system

We have seen how semantic content need not be restricted to the traditionally defined realm of intentionality; and a systematic account of infant action matches a coherent alternative to the centralized 'mental state' approach to meaning. That the explicit information in visually constructed descriptions cannot become available to the infant in 'conceptual' form, does not mean that it is unavailable to other kinds of process in the system. This was illustrated in chapter 3's outline of a mechanism-based distinction between information being explicit at the level of vision, and at the level of 'central' action programs (pp. 67–70). As far as 'meaning in action' is concerned, this distinction is central to understanding the nature and implications of the claim that infants make sense of the world not conceptually but *pragmatically*.

Some views support a systematic explanation that explicitly considers the defining characteristic of *semantic information* to be its role in the generation of behaviour. For example, Koenderick's (1980) mind-related notion of information leads him to the view that: 'In nature there is *structure* (information in Shannon's sense), but no *meaning*. . . . Only if structure is able to change the state of the perceiver, that is, influence his future behaviour, can you speak of information in the sense of meaning' (Koenderick, 1980, p. 391). In a similar vein is Israel's (1988) answer to the puzzle, posed earlier, of what is needed to 'have' as opposed to 'carry' the information that a child has measles; that is:

> states of some agent, say of a doctor, that both carry the information about the spots and also *lead to behaviour appropriate to the tale they are telling* about the infection – appropriate, that is, given the goal of a healthy, happy child.
>
> (Israel, 1988, p. 124, italics added)

Such states are termed 'information and control states', and they suggest a view of what it is to be 'attuned to constraints' that bears a close relation to the computational notion of information being explicit at the level of action.

Viewing semantic information as the information used by action programs, which underlie selective correspondences between infant and environment, sets up the possibility of exploring a *causal semantics*. This sees the meaning of a symbol structure as dependent upon its causal links with other structures and processes; and it seems to me a perfectly 'respectable' and useful way for infancy theorists to think about 'semantics', even though it is less common than the 'denotational semantics' whose focus is on the mapping between a structure and what in the world (or in some possible world) it 'denotes' (that is, 'refers to' or 'stands for').

The general idea behind this view can be illustrated through the kind of solution that it suggests to a key question that is relevant to clarifying how the infant understands things: 'under what conditions can a machine be said to refer to the world or attribute meaning to the symbols it manipulates?' This is to argue that the machine would need to have sensory and effector apparatus and be *causally embedded* in the world to which its symbol structures refer. McDermott's (1983) broad outline of a causal embedding solution to the meaning problem is relevant here because it is principled in the sense that he insists we can and must address the issue within the confines of the physical symbol system hypothesis.

McDermott argues that meaning is just one of many mental phenomena in which there is a discrepancy between what is happening and the way we think it happens. In particular, we are misguided in our intuition that a symbol has meaning for us because we *interpret* it; that is, have some way of connecting the symbol to concrete external reality which tells us *what* it refers to. There is no need to assume that we ever refer to the world in a manner different from a robot which can manipulate symbols without knowing what they 'mean' in this way. The robot simply performs elementary operations on uninterpreted symbol structures.[7]

The everyday misguided intuition about how adult knowledge refers to the world would seem to be what makes Piaget's traditional 'conceptual-representational' view of knowledge and meaning so plausible. The essential error is the attempt to drive a wedge between internal and external by considering perception to relate to the province of external reality, and concepts or representations to an internal domain that refers to the external reality experienced directly through perception. Within the analysis here, however, both are mediated by uninterpreted symbol structures. Adults do not need more, and infants do not begin with less.

Nor is there any need to move outside the domain of uninterpreted computational symbols to explain how adults can use what they normally think of as symbols. If I believe, say, that a word has meaning for me because I know what object it refers to, McDermott suggests that three types of computational symbol structures are involved: my knowledge of the word is one symbol structure – including information about its sound, etc.; my knowledge of the object is another symbol structure; and my belief that one refers to the other is a third symbol structure. Knowing what the word refers to does not involve directly apprehending the object in the external world.

A causally embedded system works when it has an interpretation in which the assignment of 'real-world' properties to its symbol structures satisfies three conditions in our particular world. This *standard sound interpretation*, under which the functioning of the symbol system makes sense to the person trying to understand it, is a logical tool to explain how the system works; it is not something that the system itself has access to above and beyond its symbol structures.

A standard sound interpretation requires three things, all of which are achieved by examples of early infant action. Infant avoidance of approaching objects can serve as an example. The infant system must exhibit *input soundness*: symbol structures constructed by the sensory systems need to be 'true' in the sense of corresponding to actual states of affairs in the world. Under conditions when 'OBJECT APPROACHING' is appropriate, for example, they should not indicate 'OBJECT FLYING PAST' or some other irrelevant alternative. There must also be *inferential soundness*: the system should be constructed so that it tends to infer 'true' symbol structures from others. The action program that exploits visually provided information to make explicit 'looming' should 'infer' that behaviours such as 'MOVE HEAD BACKWARDS' and 'RAISE ARMS' should be generated as opposed to 'KEEP LOOKING HARD' or 'REACH'. The final requirement is *output soundness*: the effectors should be constructed so that they tend to perform the behaviours that the interpretation assigns to the symbol structures connected with them. If the symbol structures 'MOVE HEAD BACKWARDS' or 'RAISE ARMS' are produced, the effectors should actually produce the behavioural processes of backwards head movement and arm raising.

This kind of system's achievement of functional correspondence with its environment is heavily context dependent, relying on the way system and environment operate *as a whole*. As stressed in chapter 3's analysis, the usefulness of a description lies in how successfully it constrains possibilities for effective action, not in objectively and exhaustively 'modelling' the environment. There is reliance on what Dretske (1981) terms 'channel conditions'. These are conditions that could vary but are usually fixed, stable and enduring, hence generating no relevant alternatives between which the subject is forced to choose. They can be taken to include things like the floppiness of the pizza-delivery robot Ramona's arm, or the natural environment of evolutionary adaptedness of infants, frogs, flies and other pragmatically knowing creatures.

Chapter 3's examples of the role of low-level, visually provided information in functional activity suggest that *viability* (that is, subject–environment structural correspondences that are reliable enough to 'work') can be achieved without the onus being on the 'input' system taken *alone* always to get things 'right' by 'recovering' or uniquely specifying external 'things' to which behaviour should be directed. For example, we saw how a primal sketch level description might support avoidance of approaching objects, by exploiting implicit assumptions that we would normally define in terms of object boundaries and their movements. To be more accurate, however, the kind of primal sketch primitive whose shorthand label might be 'edge segment' is actually describing something like 'an intensity change in the image that corresponds to *some* physical phenomenon, which is *likely* to be a change in reflectance, illumination, depth or surface orientation'. Nevertheless, such structures can play a useful functional role when placed within the context of the rest of the subject's processing potential and behaviour in the world.

A distinction relevant to appreciating the broad aims of adaptive action comes from Simon's (1957) early work. A prevalent view of the overall aim of psychological organization is that it should be striving to get the best possible solution to every task or problem that the subject faces. This *maximizing* strategy (maximize gains – minimize losses) may be applicable in competitive situations, and effective with finite problems that have a circumscribed, known set of conditions, outcomes and feasible paths of activity within the problem-space. But it does not work so well when conditions are changing over time, there are time constraints on solution and so forth. Then, it may be more advantageous to adopt a *satisficing* strategy – that is, finding a path of activity that is 'good enough' to keep the individual concerned going. Many action-based solutions to getting on in the environment may exhibit satisficing rather than maximizing.

The example of 'looming' highlights the way in which the viewer-centered nature of low-level vision's descriptions can be positively useful to the infant. In this case, the neurophysiological mechanisms concerned, together with the computations that they support, do not result in an objective description of movement in the environment. But they make explicit information whose content can normally be taken to be something like 'direction of something's movement with respect to my nose'. Such *subjective* solutions to the description of object location and movement are *locally* successful in spatio-temporal terms. The directionally selective elements on which this computation operates do not inevitably or unambiguously pick out object edges or boundaries, but behaving as if they do is a viable strategy in the *particular* environment in which we evolved, and for which the infant displays evolutionary adaptedness.

In the non-human world, the utility of frog vision's apparently ambiguous, impoverished and subjective description of a 'bug' relies on the fit between the frog's behaviour and its environment. Even if its 'bug detectors' are neither objective nor foolproof, aiming its tongue towards the location they specify in the visual field supports feeding behaviour that is viable in the frog's usual environment; but a hungry frog will starve while surrounded by stationary insects (Oatley, 1972), and a frog with an immobilized tongue would fare no better. The male hoverfly's visual description of a 'female hoverfly' could map onto all sorts of things, such as a rapidly moving elephant a long way off; and they have, in fact, been observed to chase distant birds, pebbles and midges, in addition to female hoverflies (Collett and Land, 1978). If its description is ill founded in this respect, the hoverfly's flight behaviour will normally make the discrimination on its behalf by failing to get it to the target, or there will be no life-threatening consequences should it get there. As far as such action-based pragmatic understanding is concerned, 'truth' may be 'what works' reliably, *enough* of the time.

The implications of viewing the meaning of computational symbol structures in causal embedding terms counter worries about the mechanism implications of PSSH accounts of the young infant, such as Rumelhart and McClelland's (1986) connectionist concern that 'interpretation' of 'elaborate symbol systems' would be an inescapable yet implausible corollary. What they mean by 'elaborate' is opaque, but it is clear that the symbols of a causally embedded system should not require any 'interpretation' over and above that due to the 'inferential' mediation of action programs.

Drawing on both connectionist and classical computational concepts may, in fact, be essential to a full account of how causal embedding is able to handle the 'symbol grounding problem'. Harnad (1992) takes up both in his account of how a machine's symbols could mean something for the machine and not just for the person who designed it. Since symbols cannot be grounded in a pure symbol system, they must be grounded non-symbolically. The right place to look, Harnad contends, is the system's robotic capacity. In particular, the *transducer/effector* and other *analog* structures and processes that constitute its interface with its environment are not just 'peripherals' whose implementation is no more than a matter of convenience:

> The hybrid approach to grounding . . . is non-modular, in that it cannot be decomposed into autonomous symbolic and non-symbolic components, and it is not homuncular, in that you have to *be* the transducer/effector and analogue structures and processes (not just the symbolic ones) in order to be a mind: It is not that the mind *receives* the transducer/effector or analogue activity (or, for that matter, the symbolic activity) as data. If the mind is grounded this way then it just *is* the activity of those structures and processes. (Harnad, 1992, p. 80)

The key to understanding symbol grounding, as far as Harnad is concerned, is a *sensory category learning mechanism*, because our robotic capacity goes beyond straightforward visual discrimination to the sorting and labelling of objects in terms of *categorical representations*. Connectionist models are a clear contender for this role, and Harnad reports his own success in training a simple network to split a continuum of 8 lines into the categories 'long' and 'short'.

Harnad argues that the network is not itself the grounding – this must include the analog structures and processes on which the network operates. His view that current connectionist systems are probably not 'brain-like in the right way' is well founded. However, his avowed agnosticism about connectionism's symbolic/non-symbolic status might be better replaced with further analysis of the properties of connectionist systems that are able to operate close to the transducer level, rather than taking high-level features as input as did his own line-length categorization example. This might clarify whether a more evolved version of the emergence view of connectionist-symbolic functioning would have the potential to provide a reasonably close fit to his hybrid analog/transducer plus symbol structure model. A further issue for Harnad's model, which is taken up in the next section, is the omission of any analysis of the 'effector' side of action, and of how being located in a transducer/effector *system* is relevant to our understanding of meaning through causal embedding.

Representation and misrepresentation

So far, discussion has included three senses of 'representation': the traditional Piagetian view that ties it to concepts, evocation and substitution; a more technical sense of describing the primitive elements, properties and relations of a structure that refers to something (for example, the descriptions central to AI vision); and the broader idea of a system that mediates selective correspondences between subject and environment, and can thus be thought of as their overall 'interpretation' of things. On the computational side of cognitive science, however, the term 'representation' at times seems to be used promiscuously and trivialized by becoming synonymous with (virtually any) symbol structure.

From the viewpoint of a systematic explanation of infants' action-based abilities, and from that of the basic PSSH notion of designation, no isolated symbol structure 'means' very much at all; certainly, anything that it might be taken to 'mean' is opaque. For example, in chapter 3's account of vision's place in action, and in the preceding introduction to the idea of causal embedding, the symbol structures constructed by vision are nearer the beginning than the end of the story of how the infant comes to terms with 'perceiving' the environment. It is, therefore, risky to call them representations, since it could be mistaken to imply that the infant represents the world and then acts on it (hence my persistence with the less familiar, hence more association-free, 'descriptions').

Mechanism-wise, however, I believe that action (more precisely, the computational mechanism that underlies it) *is* the infant's *representation*. And I intend a stronger sense of representation than that implied by a purely definitional equation such as 'action mechanisms' = 'computational symbol structures' = 'representation'.

Dretske's (1988) distinction between three types of representational system proves useful for clarifying this position. Central to his analysis is the relation between the way a system's elements can be said to *indicate* things, and what it is the *function* of those elements to indicate within the system. As far as the system's meaning is concerned: 'what a system *represents* is *not* what its (expressive) elements indicate or mean. It is what these elements have the *function* of indicating or meaning' (Dretske, 1988, p. 59). For example, an animal's distended stomach will indicate how much it ate recently, and may indicate how much food is around in its environment at a particular time, but it cannot be said to represent these things since nothing has assigned it the function of indicating them.

A Type I system of representation is one whose elements have no capacity to indicate anything except through the function assigned to them by some external source. Maps, musical notation and some features of natural language are such purely *conventional* systems. By way of contrast, the elements of Type II systems have an intrinsic capacity to indicate due to 'the way they are objectively related to the conditions they signify' (Dretske, 1988, p. 54), whether this arises from natural laws or human design. Examples include 'natural signs', such as birdsong or fingerprints, and the readings of human-designed gauges, such as bathroom scales or voltmeters. Central to a Type II system of representation is an external source using these *unconventional* elements in a conventional way. An individual element may indicate many things, but the function that it is assigned within the system will normally exploit only one of special interest. For example, a petfood manufacturer could assign the function of indicating the palatability of a new petfood formula to the extent to which an animal's stomach becomes distended after feeding.

The computational account of infant action meshes with the requirements for a Type III system of representation, because neither what its elements indicate nor their function are externally determined. Such systems:

> have *their own* intrinsic indicator functions, functions that derive from the way the indicators are developed and used *by the system of which they are part*. . . . These functions are not assigned. They do not depend on the *way others* may use or regard the indicator elements. (Dretske, 1988, p. 62)

These *totally unconventional* systems are what Dretske calls *natural systems of representation*, the functions of whose elements we have to discover rather than to assign.

This perspective supports the view that the visual descriptions constructed by the perceptual component of infant action are not equivalent to the infant's representation, and also helps to clarify that the infant can be said to have a representation and what kind of representation it is. In particular, Dretske considers natural systems of representation (such as infant action) to exhibit two key properties of intentionality as it is traditionally characterized by philosophers.

If we first consider the descriptions constructed by the visual system, there is a clear distinction between what they indicate (their status as 'natural signs') and what they are assumed, supposed or taken to mean by the action program level of the system. For example, as we have seen, the directionally selective elements of a primal sketch level description indicate something extremely general, basically the existence of a discontinuity in something in the physical domain. But the action program that selectively uses them in 'looming detection' assigns them the function of indicating points on the surface of an approaching object. This function assignment can thus be said to determine a *content* or meaning for the elements of the infant's representation, in this case of 'something to be avoided'. On this kind of analysis, our familiar frog and fly examples illustrate how elements with similar indicator functions (primal sketch level, moving-blob descriptions) can be assigned quite distinct indicator functions by their respective action programs: in the one case, something to eat; in the other case, a potential mate.

A second important feature of this type of representation is the ability to *misrepresent* the way that things are. The heavy reliance of causally embedded systems like the infant on a context-dependent style of functioning means that we should not expect any standard sound interpretation to hold up under unnatural or unusual conditions (cf. Dretske's (1988) notion of 'local validity'). For example, changing a creature's natural environment by taking it into the laboratory will change its channel conditions and prevent its visual descriptions from being able to indicate what they are supposed to indicate within the system. The consequence of this kind of breakdown in input soundness will be the system representing things in a manner that no longer corresponds appropriately to the way things are in the environment – an ability (to represent things as being a certain way even though they are not) that is shared by words, beliefs, thoughts, etc.

Facing an infant with an expanding shadow presented on a shadow-caster (Ball and Vurpillot, 1981; Bower, Broughton and Moore, 1970a), or moving the walls of the room in which s/he is sitting or standing (Butterworth and Cicchetti, 1978; Butterworth and Hicks, 1977), produces just such consequences in a system that evolved in an environment free of experimental psychologists and laboratories. The infant's visual indicators misrepresent that something is approaching, or that s/he is losing balance. Likewise, the design of the male hoverfly means that it cannot but misrepresent small cylindrical objects projected from 'pea-shooters' by experimenters in the laboratory as potential mates (Collett and Land, 1978).

Misrepresentation depends on two things: the *condition* of the world that is represented; and *how* it is represented. To determine the latter needs a principled way of working out what a system's indicator function is, and Dretske argues that this cannot be done unequivocally for any indicator element taken in isolation. Say a visual description that is normally constructed in the presence of an approaching object is produced in the presence of an expanding shadow, it could simply be assumed that its function is to indicate expanding shadows. This means that our infants' visual descriptions lack the ability of the system *taken as a whole* to misrepresent things, a conclusion that fits the systematic account of action. It is also suited by Dretske's (1988, p. 70) conclusion that working out the way things are being represented must depend on indicator elements being involved in the control of behaviour: 'only by *using* an indicator in the production of movements whose successful outcome depends on *what is being indicated* can this functional indeterminacy be overcome'.

This conclusion converges with the feeling of many infancy researchers that discrimination measures alone cannot establish that infants perceive 'distal' environmental properties such as depth rather than 'proximal' sensory stimulation that has the potential, with appropriate processing, to specify such properties. Evidence of functional overt behaviours that are thought to be appropriate to perception of such properties is argued to provide a more valid measure (e.g. Bower, 1972; Bremner, 1988).

The kind of inferential soundness with which the infant's action-based knowledge is being credited owes more to what Barwise (1987) calls *situated inference* than to the type of *formal inference* that characterizes the central processes of Fodor's (1980) view of computation.[8]

For Fodor, the computation that underlies psychological processes involves the execution of operations over formal properties of mental representations, which are generally thought of as being sentences in a 'language of thought'. Hence, the purported flexibility of central as opposed to input systems is associated directly with a type of representation that has the combinatorial/ syntactic properties that were discussed earlier (pp. 107–8) in relation to the nature of concepts. 'Formal' is generally taken to imply that the syntactic form of these representations alone determines the processes applied to them, not properties such as their reference or meaning. An inference in such a system is therefore formal in the sense that its validity depends solely on the appropriate rules having been executed.

By way of contrast, a situated agent – like the infant – can be envisaged as relying on mechanisms that are computational but whose inferences are not formal in this sense. When the content of a representation depends on its embedding circumstances, as it does for causally embedded action, then valid inference will also depend on those circumstances. The simplest kind of mechanism is one exploiting constant environmental features. For example, the validity of the inference 'if corresponding points in my two retinas are moving away from each other, then I should take avoidance measures' depends on the reliability of the representation involved, which depends on the continuation of natural environmental conditions. Under altered, unnatural circumstances, such as those associated with misrepresentation, this situated inference ceases to be valid, even though identical computational steps may have been executed.

In fact, Fodor's (1980) elevation of formality to computation's defining characteristic is explicitly challenged by Smith (1985). In (that intuitively most formal of formal cognitive science disciplines) computer science, it proves impossible precisely to define 'formality' in the way that a 'form versus reference' dichotomy suggestion requires. These properties are not orthogonal to one another; the disconnection between a computational system and what it represents is always a question of *degree*, with different degrees of disconnection being possible rather than a connected/disconnected dichotomy. In any case, 'formality' is inessential to such a definition: everything *substantive* about computation is captured by saying that it is symbol manipulation – not formal symbol manipulation; by acknowledging that many properties of computation (for example, temporality and locality) emerge from constraints due to *physical embodiment*; and that this embodiment gives rise to the *embedding* which is at the heart of computation's semantic properties (that is, what it is *about*).

This account of infant action as *natural representation* bears some similarity to both ecological and Piagetian accounts of action, and it also diverges significantly from them.

It is sympathetic to ecological psychology's view that, in relation to perceiving, understanding involves *pragmatic knowledge*, not the construction of traditional central beliefs whose content can be expressed in propositions about the world that are 'true' or 'false' (Shaw and Turvey, 1980), although its focus on computational-representational mechanisms leads to a very different view of action. If all knowledge involved justification of the kind focussed on in philosophy, then we would never be able to talk of 'innate knowledge'. From some perspectives, this is indeed the case. For example, Winograd and Flores (1986) consider knowledge of what exists as synonymous with consciousness and constructed through shared language in the context of activity; this means that infants cannot really know anything. Hamlyn (1978, 1981) argues that being able to get something right (as opposed to wrong), in the way that 'knowledge' implies, requires some concept of truth, the understanding of a norm. This, he suggests, cannot be acquired by an individual acting alone, it originates in social relationships; 'whatever is cognitive is social too' (Hamlyn, 1981, p. 17; cf. Russell, 1984).

In the absence of a detailed consideration of how early action works, however, such a clear-cut distinction appears arbitrary and unnecessarily restrictive. As we shall see further in later chapters, infants' interactions with the physical environment do appear to exhibit a developmental appreciation of rules that govern the success of their behaviour. To say that this is not knowledge because their 'rules' are not 'norms' and 'success' is not 'truth' means little unless the differences can be shown to correspond with distinct processes with distinct properties. It seems more appropriate to say that infants – or frogs or flies – may not know very much about objects, than to argue that they do not *really* know anything.

My position also meshes to an extent with Piaget's (1969) ideas of *practical knowledge* and activity-derived subjective categorization. When, for example, the infant shakes his or her hand upon seeing a rattle, or makes sucking movements upon seeing a feeding bottle, the action scheme that is assumed to be involved is described as a 'practical concept . . . as if it were here a question of "definitions by usage" characterized by the words "it is for . . .", "it is to . . ."' (Piaget, 1963, p. 124). But the Piagetian view is very motor- as opposed to action-based, and my view of action's development will part company with Piaget's view that the infant must go 'beyond' action to a 'real' form of intentionality that supersedes reliance on environmental constraints.

In fact, Dretske's (1981, 1983, 1988) work potentially provides a way of using the familiar terminology of folk psychology and the propositional attitudes to make a whole set of intentional attributions to the young infant. To begin with, whatever the infant's representational system enables him/her reliably to identify in the (normal) environment might be equated with the infant's *concepts*. And it is interesting to note that when Dretske (1988) considers things from the viewpoint of explaining behaviour, his analysis comes up with a more plausible mechanism candidate for a 'seeing versus seeing as' distinction than it did when information was the focus (that is, the distinction between a structure's indicator function and its assigned indicator function, as opposed to the problematic analog–digital demarcation) (Dretske, 1981, 1983). Even a basic action like reaching towards something, which on the face of it fits the category of non-intentional activities such as stroking or eating, appears to support something akin to the intentional notion of 'seeing as', once we distinguish the 2½-D sketch structure that describes the situation from the function of indicating a 'graspable object' that is assigned to it by the action program involved.

These concepts can furnish the content for infant *beliefs*. When the way things are being represented is related to the normal environmental source of information, we might consider this a true belief, and count this 'information-caused' belief as *knowledge*. When changed environmental conditions give rise to misrepresentation, we might consider this as the construction of a false belief.

There are, however, a number of reasons to resist overreliance on such terminology. First, there is the danger, raised already in relation to infants' concepts and theories, of premature attribution passing over what may be developmentally significant dimensions of the traditional picture. The infant's first pragmatic concepts would certainly diverge from the traditional 'objects of thought' as far as their (syntactic) combinatorial powers go; they are not 'informationally unencapsulated' in the way that Fodor's (1983) central processes purport to be. For example, the kind of misrepresentation that the current analysis would attribute to Ball and Vurpillot's (1981) infants, who 'avoided' a shadow that expanded while changing shape, needs no recourse to any mechanism suggesting independent, recombinable concepts of object, approach and/or shape. A purely pragmatic interpretation of infant representation and knowledge makes more sense of their findings than would a 'strictly' conceptual interpretation.

As far as beliefs are concerned, Dretske (1988, p. 107) comes to suggest that we might refer to 'these simple and relatively isolated representations as proto-beliefs', to stave off similar objections. His own emphasis is unconcerned about whether or not the language of intentionality is used, stressing rather that the interesting thing is that these structures possess what are reckoned to be important properties of 'genuine' beliefs:

> Even at this level . . . we have internal structures whose relevance to the explanation of behaviour resides in *what* they say (mean, indicate) about the conditions on which the success of behaviour depends. Even at this level . . . we have internal structures that not only mean something but also mean something *to* the organism in which they occur. (Dretske, 1988, p.70)

However, Dretske subsequently introduces further distinctions – for example between these 'implicit' beliefs and 'explicit' ones – implying that straightforward belief attribution is unlikely adequately to capture the distinctive properties of infant mechanisms.

We can best conclude that action-based representation and the pragmatic knowledge that it supports exhibit important properties of intentionality, but relying on the terminology of intentional notions to highlight infant action's important properties may produce a misleading picture, not least because we do not know that 'genuine' intentionality sets up developmental parameters in an appropriate way. What does seem to be the case is that the more we understand about infant action mechanisms in computational terms, the better we are able to understand the properties of intentional phenomena, and how and when they are properties of the infant.

Notes

1. Earlier, Marr (1976) noted how implementing the 'least commitment' principle usually goes alongside a filtering/constraint analysis style of computation.
2. A further implication of such phenomena is that even a low-level, hard-wired computational process such as vision is at least partially *cognitively penetrable* in Pylyshyn's (1980, 1984) sense.
3. In the former case, these processes generally rely on 'assuming' the surface's reflectivity function; in the latter case, that the subject is stationary.
4. This supports his distinction between *direct* and *derived* representations. These can be *informationally equivalent*, even though one may represent a relation as a primitive while the other requires its computation from more primitive relations.
5. Haber (1983) offers an important comment on this from the viewpoint of cognitive psychology's failures to find objective notions of information useful. He argues that Dretske overminimizes the implications of this relative, subjective dimension of information.
6. It is unlikely that direct perception theorists would concur with Bruce and Green's designation of it as a 'physiological' theory (cf. Reed, 1986).
7. The notion of 'uninterpreted symbols' should be taken in McDermott's sense that there is no need to propose an additional 'interpretive' layer of symbol structures over and above those that mediate the subject's causal embedding in the world. In the computational (computer science) sense, these symbols may be 'interpreted'. The notion is closely related to designation, and Newell (1980, p. 158) defines it as: 'The act of accepting as input an expression that designates a process and then performing that process.'
8. Fodor (1987) and Barwise (1987) address their respective definitions of the two forms of inference; the appropriate treatment of inferential legitimacy and validity; and what kind of inferential model we need and want for a psychological account.

Chapter 5

Action as procedural representation

In the psychology of infancy, the mainstream view of action is largely attributable to the impact of Piagetian theory, and its defining characteristic is *deficit*. Many conceptual difficulties stem from the part action is called upon to play within his framework. Presupposing that development involves a transition from biological to psychological knowledge, Piaget allocates action to the biological pole while giving it a central role in the construction of conscious understanding. This means that he works *backwards* in his analysis of infancy, always charting progress towards the mechanisms that he believes will have to exist at the end of the sensorimotor period. He constantly stresses what the infant lacks as opposed to the complexity and sophistication of the mechanisms that are operating. This appears to tie action irrevocably to a deconstructed view of the infant, and it is unsurprising that it has proved hard for cognitively oriented theorists who wish to stress the infant's competence to associate that competence with action.

We cannot begin to tackle the problem of development unless we work *forwards* from an effective conceptualization of the known phenomena of infant ability. This chapter is devoted to fleshing out the skeletal outline of action programs that has been presented so far, and to showing how they provide a view of action as a form of *procedural representation*. This aims to overcome some of the main limitations of the 'scheme' construct upon which Piaget's precomputational cognitivist view of action is founded, in particular to establish a better foundation from which to begin thinking about development.

Sensorimotor schemes – a lost cause?

First, we need briefly to consider what Piaget is asking 'schemes' to do. The most fundamental assumption of his framework is that all 'knowing' is associated with transformational activity in a broad sense. This underlies his insistence that 'any piece of knowledge is connected with an action' (Piaget, 1971, p. 5). Piaget's perspective on perception typifies the way he sees information about states of the world as a misleading and unreliable route to knowledge. As far as the infant is concerned, the essential task of relating discrete states is to be effected by sensorimotor schemes, which are assumed to underlie overt action.

Piaget's work is complicated by spanning four domains of inquiry. There are the usual accounts of *performance* (what subjects can achieve, as indexed by behaviour in context) and of *mechanism* (the psychological structures and processes that could generate this performance, described in terms of internal constructs such as the 'schemes' with which we are concerned here). More unusually, Piaget attempts to address *experience*, by inferring the subject's awareness or phenomenology, and *structure*, by offering a logico-mathematical formalization of properties of the psychological mechanisms that enable the subject to handle transformations in the world. To avoid any future confusion with the broader sense of 'structure' as organization, for example, as in computational symbol structures and the way they are organized, Piaget's formal usage will be capitalized as 'Structure'.

A clear distinction has to be drawn between mechanism, which is the main concern here, and Structure, whose relation to it proves contentious. But this distinction is often blurred in Piagetian theory. Piaget's structuralist preoccupation is with *describing* action schemes treated as a *primitive* unit of analysis. In particular, he seeks the Structure or general form of the system into which he assumes initially independent actions to be progressively coordinated. When schemes are defined, it is along the lines of 'the structure or organization of actions as they are transferred or generalized by repetition in similar and analogous circumstances' (Piaget and Inhelder, 1969, p. 4).

Any attempt to describe the (primitive) structures and processes that constitute a mechanism necessarily involves a degree of abstraction from whatever 'real' system is being described. But Piaget's treatment seems to me generally to be pitched at too rarified a level, and to exhibit a surprising lack of concern with the nature of internal processing mechanisms. He does not share functionalism's concern with understanding behaviour *for its own sake*. The 'structure or organization' of actions should help us to get 'inside' action mechanisms, but this Piaget's scheme construct conspicuously fails to do. As far as individual actions are concerned, they are at best attributed to the coordination of two schemes, one perceptual and one motor, whose inner workings remain opaque. The primitive status of schemes, description of which is tied to their behavioural aspect, means that the various forms of assimilation ('functional' or 'reproductive'; 'generalizing'; and 'recognitory') that are supposed to get the infant to this ultimate perceptual-behavioural 'reciprocal assimilation' are not and can not be spelt out in mechanism terms.

A useful initial-state model from which to approach development depends on an effective way of characterizing structure in the 'lower case' sense, and is more likely to come from a computational treatment of representation. Piaget denies any representational ability to the young infant, on the grounds that they have no 'mental images' (Furth, 1969), but pursuing the link between representation and pragmatic knowledge in computational terms can remedy

Piaget's failure to offer anything approaching a sufficient account of initial context-bound perceptual-behavioural abilities. His focus on the overt behavioural aspect of action fails to clarify how any object or event can selectively activate behaviour – any number of things could coincide temporally and spatially with a given scheme. And his treatment of 'looking' as an action scheme confounds the overt behavioural aspect of attention with the information it potentially makes available. Looking behaviour may play a role in seeking information, but it cannot construct it in the sense that Piaget requires; the possibility of selective attention is predicated on the availability of visual processes for making explicit information.

The scheme notion, then, is too molar and undifferentiated to play a satisfactory explanatory role in relation to action. But Piaget was not on the wrong track altogether, and something to do the theoretical work that the notion of schemes *ought* to be doing is essential for exploring the still important questions about the organization of *psychological* mechanisms that it was meant to address. The basic requirement of being able to clarify what underlies *control* of voluntary behaviours, hence how they might develop, was noted in chapter 3. Additionally, schemes are invoked to explain the *functional equivalence* of behaviour patterns despite differences in the details of the precise movements used or the situations to which they are applied; for example, when a toy, a feeding bottle or a blanket is grasped by an infant, the behaviour involved shares a common pattern despite clear differences on motor movement dimensions such as degree of finger separation. And finally, although Piaget's conclusions concerning the nature and degree of organization of infant perception seem almost entirely to be mistaken, he raises a broad issue that remains important: how can, and should, we conceptualize the relationship between *specific* and *general* aspects of the infant's understanding, and between the *content* and *process* of knowing?

Procedural representation

To spell out in more detail how action programs can offer better explanations of infant abilities – and suggest new questions – I intend to draw on ideas from the so-called *procedural programming languages* of classical AI. All programming languages are formally equivalent but, in practice, each was devised to facilitate programming a computational system that was designed to tackle a certain range of tasks. Procedural languages such as PLANNER and CONNIVER were primarily devised for systems engaged in problem solving (Bobrow and Raphael, 1974; Shapiro, 1973), a domain whose concern with purposive transformation of the environment shares many of the characteristics of action. So, ideas that were thought important about how computation might be organized in that domain may provide us with insights relevant to the conceptualization of action programs.

High-level AI languages are considered to be knowledge representation languages as much as programming languages (Barr and Feigenbaum, 1982). Their symbol structures support two closely interdependent aspects: a *data-base* of assertions, which describes 'objects', events or states of affairs; and a set of *procedures*, which, when activated, generate processes that change, delete or add to assertions in the system. Central to their relevance, as far as action is concerned, is that much of a system's knowledge may be *implicit* in special-purpose procedures; and, as we shall shortly see, that 'data-base' is a complex notion, not necessarily reducible to the 'store' of static facts that it generally brings to mind. An account of the infant's action-based knowledge that draws on these ideas can be thought of as a *procedural representation*. This does not imply commitment to each and every detail of procedural languages. What it does lay stress on is a *process-oriented* approach to the mechanisms underlying ability, from the earliest infant pragmatic knowledge through to adult linguistic and reasoning skills; and viewing development in terms of changes in those underlying processes.

The classical concepts involved can be seen as foundational rather than outmoded, and some of their key properties are closely linked to recent sources of interest in connectionist networks. For example, Churchland's (1987, 1989a, 1989b) views on the merits of PDP systems' highly active 'representations' leads to an explicitly procedural view of memory. If memory comes down to a functionally persistent trained network, then the question arises of what this distributed representation implies for the common view that memory often involves storing 'facts'? The solution is to view 'knowing' as a procedure; for example, knowing a 'fact' such as the location of Paris would be equivalent to having procedures for doing things like answering the question 'Where is Paris?' Along these lines the symbol structures of a procedural system could broadly be thought of as emerging from patterns of activation of a connectionist network that can be 'switched on', 'switched off' or 'altered' by what happens elsewhere in the system, and/or that can 'switch on', 'switch off' or 'alter' other parts of the system.

An essential qualification about the idea of procedural representation is that it is orthogonal to the *procedural–declarative* controversy in mainstream AI, which concerns the relative merits of actually programming a system in a procedural programming language versus one that relies more on data-bases together with relatively few general-purpose procedures. Using computational ideas to specify a theory of psychological mechanisms is an enterprise distinct from programming to test its sufficiency. Once a computational model has been developed, it might be expedient to test its sufficiency by implementing it in a declarative language, even if the mechanism model itself is highly procedural or process oriented. (For the purposes of conceptual analysis, however, there would be less explicit correspondence between bits of the program and elements of the theory.) This is not to dispute that in the case of the infant (and other real-world subjects), where procedural concepts are used to clarify psychological mechanisms rather than as a programming formalism, notable properties of their mechanisms for representing the world may stem from the way they are physically implemented. The possible link, noted above, between the symbol structures of a procedural representation and processes of an underlying connectionist network, is suggestive of this direction – though no more than this since, as noted in chapter 4, connectionist systems operate at an explanatory level somewhere between neural/physical implementation and the representations and algorithms that are of concern here.

It is also important, as Johnson-Laird (1983) appropriately points out, to appreciate that the programming distinction is not equivalent to differentiating between *knowing how* and *knowing that*.

It has been argued, for example, that infant knowledge develops from a procedural form of representation to include generally accessible knowledge stored in a declarative format (e.g. Mandler, 1983, 1984). On this view, infant procedures are limited to sensory and motor routines involving (a primitive) recognition of familiar things or the execution of 'motor programs', and conscious access to them is very restricted. To become aware of and able to recall things, the infant must analyze the outcomes of applying procedures and transfer the results into a new representational format – a declarative knowledge-base of concepts and facts, whose proposed nature is highlighted by the assumption that it can also directly accept verbally coded information.

But changes in the content and organization of working knowledge need not demand changes in the way it is 'stored', or in the *type* of representation involved. A computational theory of what we think of as concepts or 'knowing that' may well be grounded in procedural representational mechanisms, as much as a theory of overt action or 'knowing how' more obviously seems to be, provided we get beyond a misleading identification of procedural knowledge with nothing more than sensory and/or motor routines. (For example, Johnson-Laird's (1983) 'mental models' approach to adult reasoning and language understanding makes extensive use of process-oriented procedural constructs.)

There is long-standing agreement in the infancy literature that subjects' performance changes in ways that suggest decreased reliance on immediate input and increased reliance on memory (e.g. Bower, 1982; Mandler, 1983, 1984; Nelson, 1977). However, what is 'stored' as the result of a program's operation, and how this affects later functioning, are extremely difficult problems for which, I believe, a process-oriented approach should be kept in mind. Considering the consequences of activity as a 'declarative' representation risks driving a misplaced wedge between knowledge and procedures.

As far as knowledge representation is concerned, the suggested distinction is between fact-like *static structures* and *processes* for operating on or using them (Stillings, Feinstein, Garfield, Rissland, Rosenbaum, Weisler and Baker-Ward, 1986). But one of the most significant insights behind the notion of program procedures is their *dual nature*: they can be activated to generate processes; or they can be manipulated or reorganized by other procedures, in which case they are being treated as part of the program's data (Barr and Feigenbaum, 1982, vol. 2; Longuet-Higgins, 1978; Smith, 1985). It is thus inappropriate to think of (either transitory or more permanent) data-base structures as 'what is known' and procedures as 'what can be done' with or to that knowledge. There is an important sense in which 'data-base' and 'procedure' are ways of referring to different functions of symbol structures, and those functions need not be mutually exclusive. When it is suggested, then, that a procedural system has access only to the outcome states produced by the operation of its procedures, or to independent 'conclusions' about how they work, this, strictly speaking, is not true – it *can* have access to the symbol structures that designate its procedures.

So much emphasis on the procedural nature of knowledge may seem to make the notion too general. As Keil (1981b) has argued, claims about the nature of structures underlying knowledge are essentially empty unless those structures place constraints on what can be known. For example, he suggests that the notion of young children and adults encoding knowledge in 'scripts' is trivial since any knowledge whatsoever can be captured by the script formalism. The same argument can be made in relation to Production System models of cognitive development (Thornton, 1987), and, indeed, for procedures (Marr, 1982). Here, however, my claim is not just that infant knowledge could be expressed in a procedural programming formalism – which would be trivially true – but that *action* provides the infant with a procedural representation of the world. There are spatial, temporal and causal *constraints* on what infants can know through action. As we began to see in chapter 4, these lie in the interaction between environmental structure and the cognitive and physical structure of the infant. To see what the nature and limits of action-based knowledge are, we need a more differentiated model of the mechanisms of action than that provided by ecological psychology or Piagetian theory, and it is as a conceptual tool in developing this model that the notion of procedural representation proves useful.

In terms of Marr's three levels of explanation, my use of procedural notions to formulate ideas about the organization of action is most closely allied to the intermediate algorithmic level. Procedural representation of action is intended to draw on two senses of the notion of 'representation': the broad sense of the subject's overall structures and processes for selectively interacting with the world; and the more specific sense of using procedural concepts to help specify the primitive components of the system, their properties and relations. Talking about action in this way can, I believe, expand our view of action as a natural system of representation in Dretske's (1988) sense, although it is not going to fit Marr's (1977, 1982) requirements for a representation. His prescriptions, as we have seen, demand that any account of input and output representations used to solve a problem should be based on a formal computational theory of the information processing problem the representations are designed to tackle. In Marr's (1982) view, procedures are merely implementation: 'the decision about what is to be made explicit and what is not is a decision about the representation itself and is independent of the implementing mechanism' (Marr, 1982, p. 342).

As was noted in chapter 3, however, the scale and transactional nature of the performances that an account of infant mechanisms must tackle may simply debar it from the formal side of Marr's idealized explanatory scheme. We lack the insight into the nature of the 'problem' that would make it possible to start with a formal mathematical analysis of such activities; to clarify the *what*, *why* and *how* of infant action, drawing on classical process-oriented AI, psychology and philosophy too has to be a valid option. As far as the current goal of formulating ideas about mechanisms at the algorithmic level goes, approaching this from a direction that takes account of properties of underlying implementation may, in any case, be no bad thing, as connectionist work is beginning to establish.

Operationalizing action

In this section, procedural ideas are introduced by showing how they might help to clarify four features of action and the action program mechanisms underlying it: functional equivalence of behaviour; control; selective perceptual-behavioural coordination and attention; and perceptual control of the form of behaviour. Some of these ideas map on to commonplace terminology, so the meanings intended here merit spelling out to avoid possible confusion.

Generativity and functional equivalence

Functional equivalence in behaviour is related to the *generative* nature of programs. This introduces a level of symmetry between perceptual and behavioural components of action. Both are general, abstract or generative in so far as programs governing perceptual analysis can handle an indefinitely large number of inputs to generate structural descriptions of them; and programs governing behavioural processes can generate an indefinitely large number of movement patterns.

The way that behavioural processes might operate echoes the constraint analysis style of parallel computation that was discussed in relation to vision; they involve establishing a movement pattern that depends on getting a number of independent muscles to operate together. At a higher, classical-symbolic level of description, we can think of such perceptual or behavioural processing as exhibiting the functioning of *procedures*. In the broadest sense, a constraint can be defined as 'a relation between a set of qualitative or quantitative descriptions' (Steels, 1982, p. 75), and it can be viewed computationally as a set of program procedures that can complete a relation from partial descriptions and maintain it through changes to any of its parts.

This approach means that a behavioural procedure need not involve the kind of computational organization, or attendant implementation difficulties, often associated with a 'motor program' that is viewed as a set of instructions to individual muscles (cf. Hinton, 1984). It is notable that this computational view of behaviour mirrors closely the requirements for ecological psychology's idea of a *coordinative structure* (Kugler, Kelso and Turvey, 1982; Turvey, 1977). This is thought of as a group of muscles that operates as a collective, being 'tuned' or 'modulated' directly by perceptual information. Interestingly, however, a coordinative structure is assumed to obey constraints, which maintain a relationship between its variable parameters while allowing their values to vary (cf. the satisfaction of constraints by low-level visual processes).

A fair amount has already been said about perception, so behaviour will be taken up here to provide an example of how procedures work. Figure 5.1 shows two programs that might underlie behavioural processes for drawing triangles (Winograd, 1973). Each is based on *procedures* that govern the processes or activities of a mechanical drawing device, and can be thought of as a procedural representation of 'triangle knowledge'. They illustrate several relevant features.

A program can be just a fixed series of instructions (strong, inflexible and hence brittle 'rules'). Program A consists of two procedures, which control the processes of moving forwards 100 units and changing direction by turning through 120 degrees. These processes are repeated, with the frequency and order specified, whenever the program is activated. The outcome of operation of this very rigid program will always be an equilateral triangle of side 100 units. To draw a different triangle, a wholly different program is needed.

But programs can be made more complex and flexible by giving their procedures *inputs*, which determine further when and how they should operate. A key method for generating the inputs that such procedures 'require', 'call for', or 'expect' for their activation involves introducing *variables*, which can take different *values*. When a particular value is ascribed to a variable associated with some procedure – the value of the variable is bound – this provides an input to that procedure and it can then generate the process that it governs.

In program B, the procedure FORWARD now requires an input provided by the variable SIDE. This must have a value ascribed to it; in this case, a number of length units. (In other programs, variables can take qualitative

Program A	*Program B*
To Triangle	To Triangle: Side
Forward 100	Forward: Side
Left 120	Left 120
Forward 100	Forward: Side
Left 120	Left 120
Forward 100	Forward: Side
Close	Close

Figure 5.1 Two triangle drawing programs. (After Winograd 1973.)

rather than quantitative values, as we shall see in later examples.) This single program can now produce a potentially infinite number of different equilateral triangles by operating with different values for its variable. Both programs *implicitly* embody a kind of triangle knowledge, severely limited in A and more general in B. (This embodies a larger-scale version of chapter 2's idea of perceptual processes implicitly embodying knowledge of physical constraints.) B's more general knowledge, it is important to stress, is not achieved simply by adding information to a static memory or data-base – it involves changing the program so that its operation can generate a wider range of outcomes.

The nature of behaviour can be captured in these computational terms. By adopting the term 'procedure' instead of Piaget's 'scheme', we can legitimately view behaviour as a program-governed process that is generated whenever the procedure(s) governing it operate. I do not claim that procedures are what Piaget 'really meant' by schemes. If anything, his precomputational concept is something like an undifferentiated notion of program. But this is the type of concept that can do the job we need.

Control

A program procedure is essentially a disposition to perform a certain process under particular conditions. In the case of the human infant, we can consider preadapted movement patterns, which are present at birth or appear during the course of maturation, as providing physical constraints on the causal, process basis for behaviour. Neonatal examples would include reaching movements of the hand and arm associated with prehension (von Hofsten, 1984); stepping or kicking leg-movement patterns associated with walking (Thelen, 1984); and hand-to-mouth movements that may play a role in later feeding behaviour (Butterworth and Hopkins, 1988). Contrary to some views, neonatal 'reflex' movements do not disappear during development; motor skill research has established that even the movement pattern characteristic of the apparently 'useless' tonic neck reflex appears in adult behaviour (Keele, 1982).

By analogy with early/low-level vision, such elementary behaviours may have their own computational principles of operation – an effector-motor analog of Fodor and Pylyshyn's (1981) 'autonomous computational reflex' interpretation of input-sensory systems. This has to be considered a working assumption, since the details of such principles are currently far less clear than they are for vision. To explain the nature and development of the purposive aspect of such movement components, we can view them as controlled by procedures, whose activation and ordering is determined by the overall organization of the action program of which they are a part.

As far as purposive control is concerned, the triangle programs illustrate how the end or 'goal' of an activity can be *implicitly* embedded in the procedures that will be executed. These procedures will be run off one after the other, but the activation of the program already 'implies' the future state that will, normal circumstances permitting, be achieved. The type of control shown by these simple programs gives a *hierarchical*, integrated organization to what might appear to be simply serially ordered behavioural components. But just what kind of program structure – hence, what form of computational control – is required to explain the infant's early action?

Bruner (1968, 1969, 1973) proposes something like hierarchical control to explain the relationship between the activation and successful execution of early skill. In early prehension, for example, preadapted component movements may be executed in a more variable order than will be seen in the mature ability. From about 10 weeks of age, the mouth may open and close before the reach is completed, as if in 'anticipation' of the consummatory act of sucking the captured object; and a finger placed in the mouth before the object is retrieved will be sucked and the reach terminated, suggesting that 'The object is plainly destined for the mouth' (Bruner, 1968, p.39). Bruner suggests that individual behaviours are controlled, from the outset, by an organizing *intention*. It is as if infants are trying to fit their movements to the requirements of a precociously activated hierarchical program with a predetermined *explicit goal*. On this view, when young infants see a small object and bring their hands together close to their body, where they are clasped, handle clothing or go to the mouth, this is interpreted as implementation of the final 'object manipulation' behaviour of a putative reach–grasp–retrieve sequence (Bruner and Koslowski, 1972).

In a recent study of my own, however, I have been unable to find behavioural evidence of the mouth being the object's 'destination' before or during reaching. Anticipatory mouthing sometimes occurred from about 16 weeks of age, but only after the infant got a good grip on the object concerned, and after experience of getting things that found their way into the hand to the mouth. And the most striking examples of 'proximal midline activity' in young infants (12 to 14 weeks) occurred as part of a pattern of wariness or distress that questions its status as a prereaching behaviour in Trevarthen's (1974) sense (Rutkowska, 1992b).

It may be that sequences like reaching, grasping and retrieval are not initially achieved by three behavioural procedures that are part of, and whose order of activation is thus controlled by, a single action program. An alternative interpretation could be based on notions of emergent functionality (Steels, 1990) such as Brooks's (1991) idea of layered control of parallel task-achieving behaviours. This would suggest that three independent, more simple action programs control the sequence (starting with reaching towards a seen object and ending with retrieval of a grasped object to the mouth); the successful completion of each one would establish the conditions for activation of the next, and none would need explicitly to represent the final 'goal' or any relation to the procedures of other components. While 'more simple', such programs would by no means be trivial for, as we shall shortly see, even an apparently direct perceptual-behavioural coordination actually calls for the interaction of a number of procedures. For the moment, there is no need to choose between the options, but questions of the general form of early action programs will be addressed explicitly in chapters 6 and 7.

Selective perceptual-behavioural coordination and attention

Whether the organizational unit of early action involves 'simple' coordination between perception and individual behaviours, or more 'complex' coordinations involving multiple behavioural procedures, mechanisms capable of mediating the *selective* relation between the action system's perceptual and behavioural components have to be clarified. The procedural analysis of infant ability can be extended to show how action programs offer a route to operationalizing the notion of descriptions of objects and events in the environment playing a role in generating and controlling behaviour.

Two concepts that were originally introduced in procedural languages such as PLANNER prove useful: *pattern matching* and *pattern-directed invocation* of procedures. These enable procedures to be triggered when symbol structure patterns associated with them occur within the system, whether this has been achieved through response to events 'outside' the system or 'internally' as a consequence of the operation of some other procedure.

Prazdny (1980) provides a particularly relevant example of such processes since his work is concerned with modelling infant performance on visual search tasks. It includes pattern-directed invocation in the form of IF-ADDED and IF-REMOVED procedures, which specify processes to be executed when some assertion whose symbolic form matches the pattern associated with them is either added to or removed from the data-base. With the inclusion of variables, a range of different assertions can match any given pattern, enabling this process to underlie *general strategies*.

Figure 5.2 presents an extract from a (hypothetical) piece of program dealing with coordination of procedures to produce successful search for a moving object that is briefly occluded. It provides an account of the first of a cluster of IF-ADDED procedures; taken together, these would enable a system to 'interpolate' the invisible segment of a vanished object's movement path to 'anticipate' its reappearance.

Consider the program line by line and what it would do in a specific context – when a moving toy that is being 'followed', a REDCAR say, disappears from view behind a GREENBOX (for clarity, high-level object descriptions are being used, but I am not assuming that they correspond to what a real infant would be using):

Line 1

The system has a list of patterns that can be added, removed, or looked for. The procedure OCC1 is triggered whenever a two-element pattern is added to the system. The first element of this pattern is fixed (DISAPPEARED); the second element is a variable (OBJECT-X) and can take as its value any suitable object. In our specific context, OCC1 will be triggered when the assertion [DISAPPEARED REDCAR] is added to the system. This provides the variable (OBJECT-X) with a value (REDCAR), which remains constant for this run of the program.

Line 2

OCC1 searches the system to see if a three-element pattern, whose first element is PAROCCLUDES and whose final element is the second element of the previous pattern of Line 1, is already present (cf. kinetic occlusion of the thing being followed). OCC1 should find [PAROCCLUDES GREENBOX REDCAR] and the variable OBJECT-Y gets as its value the occluding object (GREENBOX). If no such pattern can be found, the program requests help (Line 5) by adding the assertion [GRIPE REDCAR] to the data-base.

Line 3

If OCC1 has succeeded in finding an appropriate three-element pattern, this pattern is removed. [PAROCCLUDES GREENBOX REDCAR] will be removed from the data-base.

Line 4

A replacement three-element pattern is added to the system, whose first element is different (OCCLUDES) but whose remaining elements are the same. A new assertion [OCCLUDES GREENBOX REDCAR] will be added to the data-base. This assertion has never been explicitly added to the system from external sources, in contrast with [PAROCCLUDES GREENBOX REDCAR] for example. Rather, it arises from a form of implicit inference within the system itself.

```
Line 1  IF-ADDED OCC1 [DISAPPEARED *OBJECT-X];
Line 2  IF-PRESENT ([PAROCCLUDES *OBJECT-Y ^OBJECT-X])
Line 3     THEN REMOVE   ([PAROCCLUDES ^OBJECT-Y ^OBJECT-X])
Line 4          ASSERT ([OCCLUDES ^OBJECT-Y ^OBJECT-X]);
        . . . . . . . . . . . .
        . . . . . . . . . . . .
Line 5  ELSE ASSERT ([GRIPE^OBJECT-X])
             CLOSE;
             PCEND;
```

Figure 5.2 An example of pattern matching and pattern directed invocation of procedures. (After Prazdny 1980.)

Note: (Where: OCC1 = the procedure name
'*' and '^' = commands to bind the values of the variables following them: '*' means 'substitute any suitable object for the variable following me' and '^' means 'the value that is already known at this point'
PAROCCLUDES = partially occludes)

The following sections of the program – not included in the figure – respond to the 'inference' that the object of interest is occluded by creating an IF-ADDED procedure and adding it to the data-base. This new procedure 'looks for' a description of a partial occlusion situation involving the occluding object (OBJECT-Y in general; GREENBOX in our particular example) in a location determined by the disappeared object's path of movement (this will be the opposite edge of OBJECT-Y or GREENBOX).

Prazdny's system's 'interaction with the environment' relies upon assertions describing states of the world at a particular time ('snapshots') being fed into its data-base. Unlike a human infant, it does not exhibit overt behaviour. In the infant system, effective procedures for monitoring partial occlusion in a new location would need to govern *behavioural* processes involving orientation of head and eyes to that location, besides specifying what should be done (covertly and/or overtly) if the reappearance information is input to the system. For the infant, keeping track of a changing situation involves handling information over *spatial* as well as *temporal* dimensions, both major sources of constraint on action-based knowledge.

This fragment shows how precise and fine-grained one's articulation of assumptions about a process can and needs to be to implement a computational model. In chapter 8, I discuss the specific model Prazdny develops and raise objections to it. Here I am claiming no more than that he presents a *possible* mechanism for trajectory interpolation. For the moment, it is the ideas relevant to how a process model of action might work that are relevant. The fact that these ideas can support working programs suggests they might contribute to a sufficient explanation where even a proliferation of Piagetian action schemes would be too poorly specified to provide the necessary mechanisms.

Prazdny's use of variables, pattern matching, procedures and pattern-directed invocation illustrates that they offer a way of thinking about how a perceptual-behavioural system could represent a *general* class of objects and a *general* event or situation, as well as the general patterns of behaviour to which Piaget's schemes were addressed.

The notions of pattern matching and pattern-directed invocation offer a way of operationalizing the *selective use* of visually provided information by action programs. Additionally, Prazdny's example suggests how, in the absence of appropriate pattern matching processes, information about a situation might become explicitly available within the system's perceptual component yet remain unavailable to some or all procedures concerned with behavioural control.

Taking procedures for monitoring occlusion to be *behavioural procedures*, as suggested above, both internal and overt aspects of attention are dealt with. So is the way behaviour ('looking') could be invoked by the occurrence of visually provided information ('disappearance'), or generated in order to search for such information ('reappearance').

Couched in Piagetian terminology, the internal constructs proposed for each of these phenomena would seem circular, empty or mentalistic (for example, schemes are supposed to function in 'similar and analogous circumstances', as we have seen, yet those circumstances are assumed to become similar and analogous only through the operation of schemes in the first place). But this example illustrates how proposing computational representational processes underlying action does not involve claiming that infants are capable of some kind of self-conscious imagery or evocation, it thus avoids the fear of 'representationalism' that is shared by Piagetian and ecological psychologies. Whether a procedure is activated by addition of a new description to the system, or is activated from within the system to monitor or 'look for' the addition of a new description, its operation is quite constrained by what is happening in the environment.

The *implicit* manner in which a procedural representation can embody knowledge is clarified by the fact that there is little direct correspondence between the real-world event that is 'known' through a program and the constellation of procedures that enable the system to cope with that event or to generate 'appropriate behaviour'. Knowing is not copying, as Piaget was at pains to assert. In fact, we shall see in chapter 8 that Prazdny's example makes explicit more than may be necessary in the infant case.

Perceptual control of the form of behaviour

Preceding chapters noted the intimate correspondence that can exist between perceptually provided information and the *form* of infant behaviour. For example, chapters 1 and 3 noted how a range of object properties, including boundary location, size, movement and shape, may determine from the outset, or come to determine, the form of infant reaching and manipulation. The explanatory power of ecological psychology's idea of 'mutual constraint' between affordances and effectivities, or of regulation being 'a consequence of the fit between the optic array and the activity' (Michaels and Carello, 1981, p. 54), has already been questioned; and Piaget's schemes fare no better.

The preceding program example illustrates how notions of variables and pattern-directed invocation of procedures can address the issue. 'What' disappears (the value for the variable OBJECT-X) and 'where' it disappears (the value for the variable OBJECT-Y) play a role in determining 'what' will be looked for and 'where' when the program's final procedure checks for the vanished object's reappearance. This points to a general principle: inputs to different procedures can have variables in common, and values for those variables can be 'carried through' from one procedure to another.

In the case of action programs, we can envisage common variables between their procedures for checking descriptions constructed by perceptual processes and their procedures for generating behavioural processes. The pattern that an action program 'responds to' or 'looks for' could contain qualitative and quantitative variables relating to several aspects of an object description, or to several descriptions. Successful matching of this pattern could activate behavioural procedures that include variables whose values are 'carried through' from those provided by the perceptual process.

Imitation provides a particularly interesting example since there is not only correspondence between perception and behaviour, but also isomorphism in terms of both the body-parts involved and the parameters of their movement. One suggested explanation has been that a 'body schema' or representation of the infant's body must be included in a complete account of infant ability (Mounoud and Vinter, 1981), although this leaves open the issue of what possessing such a schema might involve. A further, not necessarily incompatible, proposal is for 'a process of active intermodal mapping in which infants use the equivalences between visually and proprioceptively perceived body transformations as a basis for organizing their responses' (Meltzoff and Moore, 1985, p. 153). Meltzoff and Moore attribute this 'intermodal mapping' to use of supramodal information. Trevarthen (1992) highlights the need for a mechanism that matches the infant's motive or feeling of moving his/her own body-part with what they see of the adult's movement.

In chapter 2, I suggested that a represention along the lines of Marr's 3-D model description would be suited to describing an adult or infant hand, say, in a form suitable for imitative matching. One aspect of imitation might be conceptualized in terms of procedures for matching fixed elements of shape descriptions in the perceptual component (the perception of the adult hand) and elements in the behavioural component (a 'body schema' description of the infant's own hand). This would explain the choice of appropriate body part. It would suggest that it is 'motive' in the sense of behavioural procedures implemented in body processes that is matched to 'seeing', rather than the proprioceptive feeling of moving the body directly. A recurrent, albeit misplaced, criticism of symbolic AI models is that they cannot model processes like analog transformations. But Hinton's (1979) work on perception and imagery shows how structural descriptions can easily include real-valued variables. Transformations such as rotation can then be achieved by continuously changing the real-valued labels on the relevant structural description.

As far as infant imitation is concerned, the description of a hand could represent finger positions in terms of real-valued variables – the angles their volume makes with the hand's main axis. Describing a movement such as opening and closing would then involve increasing and decreasing the values of these variables. Isomorphism of adult–infant hand movement could be conceptualized in terms of procedures for matching real-valued variable elements in their respective descriptions. Other intermodal relations, such as turning to look at the location of a sound source, may also involve perceptual inputs specifying the values for real-valued variables involved in behavioural control via appropriate coordinating programs.

Content and process in action-based knowledge

The relationship between the specific content and general form of knowledge is central to Piaget's view of development, and it is featured in his recurrent distinction between *figurative* and *operative* aspects of knowing. The figurative aspect provides 'an imitation of states taken as momentary and static' (Piaget, 1970, p. 14); for example, perception in the case of recognition, or mental imagery in the case of evocation. It is said to be inseparable from, yet always subordinate to, the operative aspect (that is, action schemes in the case of recognition, or representative-representational schemes in the case of evocation). This distinction forefronts Piaget's preoccupation with what he considered to be the deficiencies of perception and it is, like his scheme construct, too undifferentiated. Procedural computational notions offer alternative, clearer ways of considering specific and general aspects of knowledge, and its content and process.

If organization at a particular time is considered, it will be recalled that Piaget views a process of *accommodation* as responsible for synchronic temporary context-specific modifications in functioning (e.g. Piaget, 1978a). It may help to make sense of this notion if we consider the way variables enable program procedures to control general strategies or patterns of behaviour, or to handle general classes of objects.

For example, in the case of the two triangle-drawing programs (Figure 5.1), program A is incapable of accommodation since it governs a totally fixed series of processes. The outcomes of its operation will not be functionally equivalent so much as literally identical. In contrast with this, the outcome of program B's operation is an equilateral triangle with sides x units long, where x is the value ascribed to the variable SIDE and can be varied in terms of the particular context in which the program runs. This pattern of elements is a very specific configuration, but its general structural properties derive from the program that produced it. It will appear similar, though not identical, to other configurations produced by the same program, such as triangles with sides y or z units long.

Here, different output 'content' is produced by the same behavioural procedures. Similarly, the same perceptual process is able to organize different input 'content'. But the intimate relationship between what a system knows about the world, its changing data-base and procedures makes it difficult to tease apart what might be meant by 'static' and 'active' aspects of knowing.

These examples deal with *synchronic* processes, which characterize a system's functioning at a particular time, as did the computational treatment of perceptual processes in chapters 2 and 4. What of more permanent changes that are made to the structure of a system as a consequence of *diachronic* processes operating over developmental time? In Piagetian terminology, the outcomes of organizational change that enable new contents or contexts to be 'known' can be called *schemas* or – more appropriately and actively – *schematizations* (e.g. Piaget, 1968a; Piaget and Inhelder, 1973).[1]

As a result of Piaget's persistent equation of perception and figurative–static knowledge, such organizations are often considered to be a kind of 'mental picture' of objects or events concerned. This meshes with our intuitive tendency to see perception and behaviour in terms of an input–output relationship in which perception is – most decidedly – the front-end of the system, and behaviour a post-central afterthought. But, in considering a more differentiated, procedural approach to action, it is important to emphasize that procedures governing both perceptual and behavioural processes are part of the subject's knowledge. The developmental process of constructing schematizations of novel content domains certainly involves perceptual *and* behavioural components of action.

Demonstrations are provided by work that makes an unusually detailed study of changes in the execution of seemingly simple activities. For example, chapter 3 drew attention to the fact that early prehension gradually becomes sensitive to object weight. Recordings of the ('kinematic') details of infant movement reveal that this development exhibits three stages or levels of organization of grasping and lifting behaviour (Mounoud and Bower, 1974; Mounoud and Hauert, 1982a, 1982b).

6- to 8-month-old infants are capable of effective grasping whatever the object's specific weight, through a context-specific adaptation to particular circumstances. This type of grasp is just as effective if an object that has been grasped several times is replaced by a much lighter one of identical size. By way of contrast, the grasping of older infants shows disruption, including sudden upward arm movement, in the face of such trick substitutions. By 14 to 16 months of age, there is disruption in the trick context, but it is then speedily compensated for. Between 9 and 16 months, infants appear to have established a pragmatic understanding of *size–weight covariation*. Reorganization following on the outcomes of perceptual and behavioural functioning in the grasping situation underlies a new ability to 'infer' an object's weight from its perceived size, and to contract the hand appropriately, in an anticipatory way, *prior* to grasping.

In Piaget's terms, these phenomena show the development of a novel schematization of things; from the vantage point of ecological psychology, the infant might be viewed as simultaneously discovering a new affordance and constructing a new effectivity. A computational way of thinking about such changes is in terms of an action program constructing a new variable, which derives its values from the description of object dimensions and is implicated in the control of behavioural procedures governing certain of the movement components of grasping. Three points relevant to the content–process relation follow from this.

First, it must be emphasized that these changes involve the way future grasping actions will be programmed, not just a static copy of what has been seen or done. Second, it makes little sense to ask whether the infant's knowledge of weight lies in perception *or* behaviour: it is distributed through the system. Potential values for variables can be considered as knowledge only to the extent that they provide inputs to procedures (one case of the generalization that computational structures are of little use unless there are processes to address, refer to or use them). Finally, it would be simplistic to view perception as providing the static content for action-based knowledge, and behaviour the active process. A more dynamic conception of the content–process relation is required. The new behavioural form 'selectively models' aspects of the environment no less than perception might.

This final point illustrates a key respect in which the perspective on representation that emerges from work considered in chapter 4 (Dretske, 1988; Harnad, 1992) does not seem to go far enough in clarifying behaviour's contribution to action-based representation. Descriptions produced by the perceptual component of action can be agreed to indicate something about the environment, due to their correspondence(s) with it and to the role they play within the system. But there is a sense in which behavioural procedures (the symbol structures that govern motor processes) do so too – through the way their function is to attempt to generate processes that establish correspondence with spatio-temporal features of the environment.

Comparison with other computational and cognitive approaches to action

There are, of course, a range of perspectives on action that employ cognitive and computational concepts, and some of those most relevant to understanding the kind of perceptual-behavioural action typical of infant ability differ in key respects from the view of action in terms of programs and procedural representation that is being developed here.

Especially significant in the current AI literature is the perspective on adaptive action that underlies Brooks's (19901a, 1991b) subsumption architecture. In chapter 3, I pointed out that there are a number of parallels between the way that Brooks's key notion of layered control appeared to work and my view of the general organization of action programs in relation to the perceptual and behavioural components of action. The type of central system that emerges from a procedural account of action programs does not conflict with that conclusion, which is further strengthened by taking into account chapter 4's distinction between 'conceptual' representation and action as a causally embedded system of natural representation. Action programs can provide a procedural representation without raising the spectre that Brooks most wishes to avoid: an explicit central representation of an objective, external environment (a 'world model'), and of the system's goals and intentions. Earlier in the present chapter, I also suggested layered control as a viable functional alternative to hierarchically organized programs controlling the infant's earliest activities. From Brooks's perspective, however, his fundamental assumptions about computation – more precisely, about its *absence* in any traditional sense – lie uneasily with accepting a procedural account of action mechanisms.

Most contentious in this context is Brooks's claim that layered control is totally free of variables and of pattern matching as a means of determining 'rule' selection. These classical notions, which are central to this chapter's account of procedural representation, are increasingly acquiring the status of unwarranted and surplus-to-requirements hallmarks of old-style centralized and decontextualized computational accounts of action (e.g. Agre and Chapman, 1987). All that a task-achieving behaviour is supposed to need is a hard-wired, special-purpose detector that responds to whatever environmental feature is shared by the situations in which the behaviour's functioning will achieve the required (functional) outcome.

One problem with this idea is that it seems unlikely to produce a system with the kind of generativity – hence, functional equivalence of behaviour – that we are used to seeing and expect from even a low-level naturally intelligent system. As Kirsh (1991) notes of Brooks's plans for a robot that could tour offices to collect and dispose of empty drink cans, it would surely be realistic to expect it to generalize from picking up drink cans to, say, coffee cups, without having to reprogram it? With the kinds of special-purpose detectors that are proposed, however, the robot is unlikely to 'see' the similarity, and hence exhibit the generalization that using variable descriptions could provide.

Over and above this objection, however, is the question of whether Brooks's mobots really are as representation- and variable-free as he claims. Kirsh (1991) points out they are almost certainly 'concept free' (in the technical sense discussed in chapter 4) but their all-purpose representation-freeness is compromised by:

> being riddled with wires that carry messages which covary with equivalence classes of earlier signals (e.g. an edge covaries with an equivalence class of pixel configurations) and which often covary with properties in the environment (e.g. real edges, hand manipulations). (Kirsh, 1991, p. 162)

My own view concurs with this. If we take the example of proposed sensors for a wall following robot (Brooks, 1991a), we find that its movement is governed by 'processing' that looks not unlike the variable binding that underlies pattern matching. Its behaviour is to be characterized in terms of:

> allowable ranges on some functions computed on the sensor readings. . . . instead of characterizing wall following as staying within some distance of a wall, and within some orientation range, we could characterize it as maintaining some range of values on three sonars pointing north-east, east and south-east, relative to the robot's orientation being north. We might even say that only two readings had to be within their ranges at a particular time. (Brooks, 1991a, p. 438)

The wire-rich physical mechanisms of Brooks's creatures may be designed to exhibit little direct isomorphism with traditional forms of computation, but they may nevertheless be implementing a virtual machine that supports the kind of causally embedded procedural system of representation being proposed here. As we shall see in later chapters, there are developmental advantages to conceptualizing emergent functionality models within the classical terminology of basic action programs.

The view of action programs proposed here also faces a radical computational alternative from the way the philosopher Churchland (1986, 1986b, 1989a, 1989b) proposes using connectionist ideas to devise a neurocomputational treatment of sensory-motor, and cognitive, activities. This conflicts with both the current emphasis and Piaget's insistence that the organization of even the earliest behaviours could not be reduced to neurophysiology. Churchland's 'geometric' perspective draws on a way of talking about the pattern(s) into which a connectionist network settles in terms of *vectors*, which can be used to describe the overall configuration of activation levels in a network of connectionist units.[2] In his view, a representation *is* a particular vector coding. We are asked to think of the brain representing aspects of reality by positions in a state-space, the dimensionality of which is determined by how many elements of variation there are in whatever is being represented.

Take as an example a system that can reach for things that it sees before it. If its arm has variable shoulder and elbow angles, representing the arm's location will require a state-space with two coordinate axes; so will representing the spatial location at which its two eyes converge. Computation in such a system *is* a vector-to-vector transformation. All that is needed to explain sensory-motor coordination is a general coordinate transformation between state-spaces (what Churchland thinks of as a 'state-space sandwich'); this will take any point in the system's sensory state-space to a position in its motor state-space.

The overarching aim of Churchland's 'eliminativism' is to get rid of intentional notions like belief, desire and intention altogether by explaining psychological phenomena in terms of neuroscience. From the perspective of infant action, however, my feeling is that describing abilities as the operation of numerous 'state-space sandwiches' is far too opaque to provide a useful explanation at computationally and psychologically relevant levels. Connectionist viewpoints that endorse and use the notion of emergence, and which are not intent on reducing psychology to neurophysiology or physics (e.g. Rumelhart and McClelland, 1986), are more likely effectively to capture what is going on in an action system; and they are more compatible with the kind of procedural representation sketched here. For example, Rumelhart, Smolensky, McClelland and Hinton (1986) outline a way in which schemata, their variables and variable binding can be seen as emerging from the operation of connectionist networks. Connectionist systems are especially good for implementing *continuous variables*, but the important point here (and relevant also to Brooks's position) is the explanatory usefulness of the notion of a 'variable', together with the computational roles that it can play with regard to processes such as pattern matching.

Nor is it clear that this approach can be extended to the broader context of action called for by even basic developmental phenomena. Churchland's claims to the contrary, it seems unlikely to be able to 'solve the problem of sensorimotor coordination, in any of its endless incarnations' (Churchland, 1986b, p.12). The action program notion, I shall suggest in chapter 6, provides a better way of understanding the development of complex activities than seeing them as involving a higher-dimensional route to transforming sensory to motor vectors.

Turning to theories that adopt cognitive views of action in an attempt to clarify explicitly developmental themes, the procedural position is clearly opposed to Piaget's treatment of perception's role in action, and to his narrow preoccupation with behavioural process. His treatment obscures the nature of action by failing to clarify the relationship between his notions of 'schemes', 'behaviour' and 'action'. Indeed, the terms are often used interchangeably. Piaget did come to distinguish between 'presentative', 'operational' and 'procedural' schemes, and to talk of the subject's 'procedures' (Vuyk, 1981). But Piaget's notion of 'procedure' is restricted to the overt behavioural/motor aspect of action. Computational program procedures are more basic, being relevant to conceptualizing the control of perception, behaviour and their coordination. We can best view behaviour as a tip-of-the-iceberg phenomenon: an *observable* process coordinated with many that, of necessity, cannot be observed. Computationally, the control of action lies in programs, and these action programs need to include many procedures in addition to those governing overt behaviour.

As far as psychological methods go, behaviour is our only route to getting at these other procedures. This assumption is central to Mounoud's approach (Mounoud, 1981, 1982, 1986; Mounoud and Hauert, 1982a, 1982b; Mounoud and Vinter, 1981), and my perspective concurs with his stress on the limitations of Piaget's attempt to separate action schemes from the classes of objects and situations to which they are applied.

Mounoud (1981, 1982, 1986) emphasizes the necessity for a 'code' that translates or represents object properties, thus endorsing the importance of representation as a *medium* for mediating selective correspondence between infant (or child or adult) and environment (Mounoud and Vinter, 1981). In the case of the acquisition of size–weight covariation, for example, the ability to 'infer' object weight from size is taken to reveal development from a 'sensorial' representation, which is present at birth, to a new 'perceptual' code, which underlies the infant's construction of 'perceptual representations' and permits successful behavioural anticipation of invisible object properties on the basis of visible ones.

The mechanisms involved in these various 'codes' are not clearly specified in any detail, and occasional reference to new representations as 'models' or 'memories' runs the risk of bringing to mind the copy view of new knowledge that was challenged above. However, Mounoud clearly has in mind a far more dynamic notion of change in the infant's action potential. While he notes that the infant's new representations (Mounoud's internal 'organizations of contents' or domain knowledge) are somewhat like 'schemas', he emphasizes that they should not be seen as static. Underlying these representations are *processes* for 'analyzing or sampling and organizing' things, whether 'aspects of reality and their variations', 'objects, properties and their relations' or 'a person's characteristics and their interrelations' (Mounoud, 1981, p. 112). This view, then, sees the content and process of infant knowledge as indissociable in much the same way as this chapter's interpretation of new knowledge acquisition in terms of a changing procedural representation grounded in changing action programs. For Mounoud (1981, 1986), *contra* Piaget, the subject's knowledge is always tied to some specific content.

Central to my view of action programs is their role in coordinating perceptual and behavioural processes. While this is implied in Mounoud's perspective, his discussion of mechanisms handles behaviour fairly independently. Following Piaget's usage, the term 'action procedure', is reserved for a general type of behaviour. The focus of inquiry is redirected to 'action programs', which are seen as actually determining context-specific motor control (these play a role like behavioural procedures with values specified for their variables). However, how the sensory or perceptual code could have its impact on their functioning is not clarified. (No analog to the suggestions made above (pp. 133–7) is proposed.) 'Action program' and 'motor program' appear to be used interchangeably. For example, Mounoud and Hauert (1982a) view the three levels of grasping behaviour found in the size–weight covariation study as evidence for different types of 'motor program'. In contrast with this, my notion of action program aims to emphasize that both perception and behaviour are part of the infant's procedural representational code, and that representational development involves change in how both of these components are being used.

Schmidt (1982b) has proposed a 'schema theory' which subsumes perception and behaviour, albeit in a fairly undifferentiated way. Its main aim is to explain how simple motor skills, such as throwing an object, might be acquired. He proposes the formation of 'generalized motor programs' to avoid storage of separate motor programs for every movement that might be performed, and to make movements in novel situations feasible. A schema or 'rule' is assumed to be abstracted from the relationship between four types of stored information: initial conditions such as size and weight of an object to be thrown; parameters of the movement such as its speed and force; the movement outcome in the environment or knowledge of results; and sensory consequences – what was felt, seen and heard in producing the movement. Novel performances involve assessing current parameters and inserting them into this 'rule'.

Schema theory differs from the concerns of the action program idea in two main ways. One is that it focusses exclusively on issues such as the effects of practice; that is, it deals with increased success in things that the subject can already do, rather than on how they might learn to do something new, such as attaining a novel goal or outcome. Nor does it address more detailed mechanism issues such as: the form(s) the various types of information might take, how they might be coordinated and 'stored', or how the 'abstraction' process might be conceptualized. Also, it purports to explain how the particular muscles involved in a movement are specified and controlled, whereas this is inessential to the action program level of analysis.

Hinton's (1984) computational analysis of motor control establishes that several problems concerning how it is achieved can be overcome by a modular decomposition into four qualitatively distinct, hierarchical levels: the motor problem (for example, 'reach out to object A'); movements of the body, which can be independent of how the selected movements are actually implemented; torques that could cause a chosen movement; and muscle innervation. My present discussion's concern with a general computational framework for the structure and properties of action does not take it beyond the top two levels.

My present concern with generativity is shared by Schmidt's (1982b) and Bruner's (1968, 1969, 1970, 1973) approaches to action. A strategy or skill, Bruner suggests, is like language since, once you have a 'rule', transfer seems to be 'built in'. He talks also of 'program(me)s' of skilled activity. As far as any precise discussion of mechanisms goes, however, he diverges from the view of infant action being developed here by drawing on Bernstein's (and others') neurophysiologically motivated ideas concerning the minimal components for voluntary control of motor activity. These include signalling of the parameters of a desired movement prior to activity ('feed-forward'); comparison with the actual course of movement ('feedback'); ongoing correction based on the results of this comparison; and 'knowledge of results' once the particular objective of the activity has been attained.

Bruner (1968) notes that other units of analysis, including Miller, Galanter and Pribram's (1960) computationally inspired TOTE (Test-Operate-Test-Exit) unit are comparable, yet he hopes the Bernstein variant will be more adaptable to studying voluntary action and its development. This seems most unlikely to me, especially as far as the control of infant action is concerned. It is not obvious that the neurophysiological model *is* applicable to the young infant, nor how it could handle *psychological* issues concerning the infant's 'intentions' or 'goals'. At best, it might say something about how an isolated behavioural procedure, once invoked, is actually run; that is, how the motor process that implements it works.

Bruner tends to fudge levels of explanation here, equating 'intention' with 'feed-forward' – an 'internal discharge in the nervous system' (Bruner, 1973, p. 2). On the other hand, he proposes that 'intentional action' requires that infants have the opportunity to see the consequences of their acts. For developmental analysis, as we shall see in the next chapter, the computational action program notion has the advantage of supporting exploration of changing control at a psychological level. It also helps to establish a sense in which Bruner is on the right lines in believing that the infant both begins with intention *and* needs experience in order to develop it.

Notes

1. Attempts to make the figurative–operative distinction at the mechanism level often adopt the term 'scheme' for the operative aspect of mechanism and 'schema' for the figurative, considering the latter to be some kind of static copy of past experiences (e.g. Gratch, 1975). When Piaget introduces 'scheme' to specify the operative–transformational aspect, 'schema' disappears altogether to be replaced by 'schematization'.
2. See Jordan (1986). In algebra, a vector is simply a way of describing a pattern of numbers, each of which is a component of the vector. Relevant to describing the organization of a connectionist network (e.g. the pattern of activation of its many processing units, or the set of weights/strengths of connections input to a certain unit) is the fact that vectors with no more than three components can be represented as points or arrows in a 3-D space, with one axis corresponding to each component of the vector. Say a system uses age, height and weight to recognize people, we could envisage it as representing each person it encountered by a unique vector determined by the coordinates they generate in a state-space with three axes.

Chapter 6

The structure of infant intention

Infants' understanding of objects and of their own actions are intimately connected. In chapter 4, I proposed that infants can be said to understand and know something about the world around them. What of their own behaviour? When young infants exhibit appropriate behaviour, do they know what they are doing and, if so, in what sense? For Piaget (1953, 1955), the answer is traditional and negative: they clearly do not. The acquisition of intentions and goals must await the development of 're-presentation' and the 'object concept'; only then will the infant's behaviour assume intentional status.

Otherwise divergent viewpoints accept this general conclusion. For example, despite opposing Piaget's deficit view of early abilities, Mounoud (1982) denies that any intention is involved. Surely, he argues, the infant cannot be said to stretch out a hand 'in order to' close it about an object, or to turn the head 'in order to' get an object felt against the face into the mouth: 'The baby does not behave in these ways intentionally to gain a certain effect but is linked to the environment by a specific organization (i.e., the nervous system at a certain point of development) that directs his or her reactions' (Mounoud, 1982, p. 126).

Similar considerations figure in Winograd and Flores's (1986) claim that an infant who cries when hungry is exhibiting only inflexible reflexes. We may talk of the purpose of these reflexes, but this purpose is not 'in' the infant in the sense of a *plan* based on a *representation* of the situation; if they fail to 'work', there is no *generalized ability* to discover 'other ways to eat.' The kind of programming that they take to task corresponds to the General Problem Solver style, noted in chapter 3, which they appropriately criticize for its centralist tendencies. But, as the analysis of representation in relation to causally embedded action shows, quite different kinds of computational organization are possible, some compatible with the notion of 'structural coupling' that informs Winograd and Flores's view of the infant as part of a biological system that is 'mechanistic but not programmed'.

As far as psychological mechanisms are concerned, the essential question is: what is the infant's contribution to the shared subject–environment state-space in which *we*, as observers, see goals, ends or objectives being attained? Even the earliest infant abilities prove to be unlike reflexes as Winograd and Flores conceptualize them and open to development – though not necessarily in the way that Piaget thought. The infant system may be mechanistic *and* programmed, but at first not necessarily along the lines of traditional program structures. The notion of changing computational control of action offers a route to understanding the infant's changing potential for action, and here I consider how aspects of the restructuring of action programs can help us to understand the nature and development of infant intention.

This task is tackled in three stages. First, I give an overview of the main themes of Piaget's theory, to provide a background against which to locate the developmental issues of this and subsequent chapters. Some of his ideas about mechanism and Structure were briefly considered in chapter 6; performance and experience are also taken up here, to clarify the way he hopes to link conceptual development with conscious awareness of the mechanisms of action. Then, the computational ideas developed so far are set to the task of analyzing a central example: the fabled stage IV *A$\bar{B}$ error*. If an infant (around 9 to 12 months) retrieves an object that is hidden at a location A, s/he will perseverate in searching at A, even if s/he sees it hidden at a new location B (hence the 'A not B' error; Piaget, 1955). This has attracted immense interest, and continues to do so, due to Piaget's interpretation of it in terms of the infant's progression from action to representation. Finally, what the computational alternative has to say to the issue of infant intention is considered. Contrary to ecological psychology's position, intention is not viewed as a primitive property of appropriate activity. Nor is it characterized by a single qualitative discontinuity as Piaget and the traditional philosophical assumptions on which he draws assume. A range of systems turn out to share properties that are pertinent to understanding the purpose behind infant action.

Knowing what you are doing: Piaget's developmental framework

My primary concern is to highlight which of Piaget's intuitions and questions are of most general relevance and remain important for contemporary research. Piaget (1953, 1955) demarcated six substages within the sensorimotor period; his major observations, and his own and competing interpretations of the reasons for the infant's successes and failures, are regularly reviewed. It is, however, all too easy to construct *local alternative theories* which address only a narrow range of observations. These are frequently orthogonal to the concerns of Piaget's genetic epistemology. A more relevant critique can be achieved by taking account of Piaget's later work. There we find a clearer articulation of his attempt to span distinct levels of description of human abilities, and of the concepts and themes that are central to his approach. This prevents us from fixating on selected observations and losing sight of their role within the theory as a whole.

Several major themes recur throughout Piaget's work on the nature of, and developmental relation between, action and representation (Piaget, 1968a, 1968b, 1970, 1974, 1976, 1978a, 1978b). Significantly for our understanding of possible developmental relations between types of representation, they can all be found – explicitly or implicitly – in his discussion of infancy (Piaget, 1953, 1955).

The most basic theme, that all knowing involves action in the sense of transformational activity, was discussed in chapter 5. Mechanisms underlying both action ('schemes') and thought ('operations') are said to permit the subject to know the world, but the former confer purely *practical* knowledge, tied to context, while the latter are associated with 'objective' *logical-conceptual* knowledge. In the six sensorimotor substages, we are supposed to be witnessing a transition from knowing as *overt* action to *internal* actions or thought associated with 'an image of absent objects and their displacements' (Piaget, 1955, p. 4).

For Piaget, this experiential change is closely linked with the Structure of action. Sensorimotor actions and thought operations share the property that they do not exist in isolation but are gradually coordinated into systems with a lawful logico-mathematical Structure. At the sensorimotor level, the developmental target is taken to be a system whose functioning obeys the laws of Poincaré's (e.g. 1905) 'group of displacements'. Piaget considers the logical concept of 'group' to provide a useful characterization of 'every system of operations capable of permitting a return to the point of departure' (Piaget, 1955, p. 101). These laws are such that not everything changes at once when transformations are executed. Something always remains invariant, and this underlies the infant's grasp, in performance and experience, of the continuing *physical identity* of objects – the first *conservation*.

As we have seen, the many infant perceptual-behavioural coordinations that have now been uncovered question Piaget's emphasis on the developmental necessity of 'reciprocal assimilation' of initially independent perceptual and behavioural schemes, but a key question remains: we are still far from understanding how infant abilities become more complex.

Piaget adheres to qualitative distinctions between types of knowledge, even if these sometimes result in identical performance and despite their functional parallels. For example, in the penultimate stage of the sensorimotor period (starting around 12 months of age), the infant can search successfully for a hidden object even if it has undergone several visible spatial displacements *en route* to its hiding place. But, for Piaget, this appropriate behaviour does not mean that s/he is able to 'think' about the object concerned. Action, no matter how sophisticated, remains reliant on perception. Thought necessitates a capacity to re-present objects and events in their absence so that their relations can be manipulated and a problem solution arrived at prior to overt action. The infant may sometimes *succeed* in practice but does not yet *understand* in principle why or how this success was achieved. S/he may know things at a practical level *through* action, but does not know *about* action itself. This deficit is demonstrated in two ways: difficulty in searching for an object whose displacements are not visible; and inability to manipulate object relations in novel situations in a foresightful way (cf. the combinatorial/syntactic flexibility attributed to traditional notions of concepts and thought). Only the stage VI infant is supposed to possess such abilities, and their acquisition marks the end of the sensorimotor period. According to Piaget, they reveal that s/he is now capable of deduction based on representation and not acting merely on the basis of habit in accordance with perception.

Piaget's belief that the terminus of infant development is marked by a *qualitative discontinuity* between two types of knowing is related to a set of assumptions concerning developmental mechanisms, which can, of course, be no clearer than the scheme mechanisms on which they need to operate. Despite their autonomy, action and thought are considered isomorphous in Structure. This, Piaget argues, is evidence that one develops from the other, and it is the role of developmental study to verify this fact. Thought is said to be *interiorized* action since it embodies the general form of the coordination of action, which is abstracted from its particular content through feedback from action itself. This *reflective abstraction* is to be the primary mechanism of *equilibration*. For the sensorimotor period, Piaget argues that the Structure eventually found in the direct coordination of actions in context is *reconstructed* at the level of representation through the interiorization process.

These changes in performance, mechanism and Structure are assumed to be paralleled by changes in the subject's experience. Successful action may involve 'elementary awareness', but true understanding goes beyond this to consciousness or *cognizance*. Cognizance relies on a degree of conceptualization that permits awareness of the structure of the task, of how and why action succeeded: 'The child is fully aware of what happens only when his conception is backed up by an inferential or operatory coordination drawn from the coordination of the actions themselves' (Piaget, 1976, p. 41). The central hypothesis of *The Grasp of Consciousness* (1976) appears in the first of the infancy volumes: 'Consciousness arises from dis-adaptation and thus proceeds from the periphery to the centre' (Piaget, 1953, p. 148).

During the sensorimotor period, we can consider cognizance to be at the 'periphery' of action when the infant reaches stage IV. Then, coordination of reaching and cover removal makes possible successful retrieval of a hidden object, even if the infant is not reaching before the object disappears from view under the cover. For Piaget, this means the infant has moved beyond a transitory awareness of interesting events to cognizance of the end achieved – the *goal object* – as independent of the infant's own experience and action. Cognizance of the 'center' of action is inferred from the performance(s) characteristic of stage VI. To Piaget, these suggest a conscious grasp of the *spatial* and *causal* relations involved in successfully finding and manipulating objects; that is, an awareness of aspects of action that are not observable.

Alternative explanations of the AB̄ error

Despite the many claims that contemporary research has made regarding precocious object knowledge on the infant's part, two Piaget-inspired questions still remain constant features of the field: why do young infants fail to search for hidden objects (Baillargeon, Graber, Devos and Black, 1990); and, when they eventually do search, why do they make the errors they do (Harris, 1989)? For an action-based account of infant knowledge, the first question does not present the paradox that it does for approaches that attribute a precocious object concept, existence constancy and permanence to the infant on the basis of discrimination-style data. Since the infant's object knowledge is pragmatic, it depends on developing procedures directed at objects, including those involved in searching for them. The 4- to 5-month-old infant cannot be said to represent objects independently of this procedural representation; hence is not failing to search for an unseen object despite believing in its continued existence and probably knowing its location too. How the infant develops the ability to search, and what s/he has when s/he has developed it, are the key issues, and search 'errors' may illuminate them.

Piaget's (1955) explanation of his AB̄ error discovery reveals his ambivalent attitude towards action-based knowledge. This first means–end behaviour is supposed to provide evidence of some ability to bear the object 'in mind' during its absence, and of an intention to act as opposed to react to a current situation. On the deficit side, however, Piaget argues that liberation from reliance on perception is incomplete. The cover acts as a 'sign' or 'perceptual index' of the object's continued presence, a kind of residual stimulus. Furthermore, search at the initial hiding place demonstrates that the infant's memory of the object is egocentric, dominated by previously successful actions. For the infant, the cover is not really a place where the object is hidden so much as a 'hallmark of the act of accommodation' by which the object was first found.

Contemporary research challenges Piaget's action-based explanation (for reviews, see Bremner, 1985, 1989; Gratch, 1975; Harris, 1983, 1989). Of the numerous explanations put forward, three types of theory have recurred and are particularly interesting, although they remain far from compelling if considered as *alternatives* to one another or to the Piagetian account.

First and foremost, infant performance is something to do with *memory* to the extent that the error occurs if a delay is enforced before the infant is allowed to search, but not in a zero delay condition. From this, it has been suggested that memory failure is responsible rather than action-based object understanding: the infant simply forgets the new hiding place (Gratch, Appel, Evans, LeCompte and Wright, 1974).

Of course, this simple forgetting hypothesis fails to explain the rather obvious fact that the infant does not 'forget' when the object is initially hidden at A. Interestingly, Gratch *et al.* observed that some of their younger subjects avoided making the error if they kept looking to the new hiding place throughout the delay period (a factor that continues to prove important in modifications of the task, as we shall see below); they suggest that successful search in such cases is based on a set involving some type of 'motor memory'. Perhaps this underlies infants' success on the original A hiding trials? If it does, it is notable that this strategy cannot be explained solely in terms of a *motor set* towards a location. If it were, the perseverating infant should be unable to reinstate it after having broken the set to watch subsequent hidings at B.

A study by Diamond (1985) makes the role of memory in infant performance more opaque. As infants progressed through the second half of their first year, they had to be prevented from searching for an increasingly long time before they would fail to search accurately on B trials; by 12 months of age, a delay of around 10 seconds was necessary for infants to revert to (now errorful) search at A. However, any idea that search is simply determined by a memory trace of the most recent hiding, and that this gets longer/stronger with age, can be ruled out. It would obviously not explain the infant's return to A, but equally noteworthy are behaviours during the search period that involve *both* A and B. Sometimes, the infants were observed to keep fixating B (cf. Gratch *et al.*, 1974, above) but to reach towards A; sometimes they uncovered the object at A, but paid no heed to it before proceeding directly to B. Diamond's own explanation of the data is along the lines of a conflict between the contents of short-term and long(er)-term memory in the control of search behaviour, pitting memory of the object's most recent hiding place, B, against the acquired habit of reaching to A. For errorless search, the infant must not only maintain information about the object's existence at B as a basis for search activity, but also inhibit the acquired habit of reaching to A; and the longer the delay, the less successful does the information about B become at controlling search behaviour. Developmental reduction and elimination of the error may be associated with improved neurological control of behaviour.

It is perhaps easier to accept the idea that familiar responses might be resurrected in a search situation than it is to understand the proposed temporal dimension of the infant's current description of the situation losing its potency for response control as delay increases. A proposal for the strength of both new and old potential responses declining with delay time would seem equally plausible, although it would not, of course, fit the data. Explaining infant performance means moving beyond viewing memory as storage and use of experience over time to consider more precise details of *what* is being 'stored' and *how* it is used to guide activity. In these respects, it is problematic for Diamond's model that infants may be no less likely to exhibit the error when they *observe* a series of hidings at A without being permitted to search for and retrieve the object there prior to B trials (Evans, 1973), there being no 'old response' in the usual motor sense. And the error has been found to *increase* near the start of the infant's manual search career, rather than emerging at the same time and declining thereafter (Harris, 1976). Some change in the organization relating perception, memory and response generation appears to be involved, not just in its degree of efficiency. It appears to be an instance of what Bruner (1966) termed a 'growth error': a case of becoming more mistaken about something in the service of improving understanding.

A further type of theory, which addresses what the infant might be acquiring, is phrased in terms of the infant's 'rules for controlling behaviour'. Infants may revert to A even if the object is now visible at B and A is visibly empty (Butterworth, 1977; Harris, 1974), an embarrassing observation for explanations that rely on a traditional, evocation-substitution view of memory or representation. There is certainly some descriptive sense in which the infant lacks a 'rule' stating that an object can be in only one place at a time (Harris, 1975, 1983), yet it seems unlikely that such a rule ever becomes an explicit or *functional* component of the infant's search system. The absence or presence of such a rule always needs to be clarified as an emergent property of the infant's procedures and the way they are coordinated over time, to eliminate the mysterious quality associated with the notion that infants behave *as if* they 'think' or 'believe' an object can be in two places at once.

From the computational perspective, error in the face of a visible object is not that paradoxical. There is no simple one–one isomorphism between the state of the world and the infant's procedural representation of it. Descriptions constructed on the basis of environmental information play a role in determining behaviour, but the infant's action programs control what that role will be. For example, if an object is visible behind a locked perspex door at B (Harris, 1974), it seems that when the execution of one procedure for retrieving it fails (go-to-B), the next most closely matching the current situation description may be activated (go-to-A). Such alternatives may not always be available to the infant. Le Compte and Gratch (1972) found marked developmental differences in infant reaction when an object was hidden at a single location and surreptitiously replaced with another before the infant recovered it. Although 9-month-old infants appeared surprised, they simply continued to stare at the object. In contrast with this, 18-month-olds exhibited additional search, especially thorough investigations of the trick box in which the hiding had taken place.

A third type of approach to explaining search error, and one particularly relevant to clarifying the nature of the infant's 'rules' in terms of behavioural strategies, concerns the infant's *spatial understanding*. Systematic manipulation of the task environment can reduce or eliminate the infant's tendency to return to the initial hiding place (e.g. Butterworth, 1975, 1981b). This can be achieved by maneuvers such as using distinctively different covers at the two locations; or by positioning them vertically above one another and placing a 'background' surface behind the A location. Butterworth argues that infants search correctly provided adequate support is available from a stable, external spatial frame of reference. Many experimental settings eliminate or reduce the necessary perceptual information – for example, identical covers may be used on a homogeneous ground – and force the infant to adopt a self-referential or 'egocentric' coding of object location. Reliance upon such a code will generate search at the egocentrically defined A locations on B trials or, more generally, 'conflict' between the frameworks will result in search distributed between A and B. From a computational viewpoint, however, we cannot simply conclude that: 'the rule that an object can only be in one place at one time is implicit in perception' (Butterworth, 1981b, p. 162).

First, more needs to be said about the mechanisms underlying how information made explicit by perception affects the infant's understanding of and interaction with the situation. Especially interesting is the fact that the way the infant describes events at A has implications for subsequent decisions concerning the import of new environmental information. Following adoption of a spatial code, the location to which search is directed appears to be updated when the object moves; with an egocentric code, such updating seems not to occur. This phenomenon converges with computational issues concerning data-base management. Chapter 5's example (pp. 133–6) of Prazdny's (1980) work began to illustrate the importance of being explicit about how a system operates to add, remove, or change assertions at any time, to understanding how it maintains a consistent data-base and a viable picture of the world.

Solving this problem is going to be non-trivial for the infant, who is faced with a world in which some but not all changes are relevant to ongoing and future activity. For the sake of efficient functioning, it is essential that action programs are selective about the visually provided information they respond to or look for. But the processing consequences of proposed spatial and egocentric codes are not altogether clear. If the egocentric code is based solely on immediate perception of object location with respect to the self, why is it not deleted and replaced whenever new information, and hence a new potential egocentric coding, becomes available?

Second, we must be cautious of the implication that infants can and will use visual spatial information provided it is available 'in the environment'. The exploitation of such information to guide search behaviour appears to be a *developmental* phenomenon. For example, Acredolo (1978) found that 6-, 11- and 16-month-old infants could learn to anticipate the appearance of an adult at one of two windows, to their right and left, but they were affected differently by the presence or absence of a *landmark* (a large yellow star) at that window. Moving the infants so that their viewpoint on the windows was reversed resulted in most 11-month-olds continuing to turn to their right or left in the no-landmark condition. With the landmark, however, half changed the direction in which they turned so that they were again looking towards the appropriate window – the equiprobable pattern so commonly found in the two-location manual search paradigm. Acredolo (1988) considers this as a 'paired-associate treatment of landmarks', with the 9-month-old linking action at the landmark with the event. By way of contrast, presence or absence of the landmark did not affect the performance of younger or older subjects. At 6 months, virtually all infants continued to turn in the familiar direction, while at 16 months all looked to the appropriate window.

It has been speculated that the onset of mobility via crawling, an activity with negative implications for the validity of a strictly egocentric location code, may be responsible for alerting the infant to the advantages of using readily identifiable and/or stable features of the spatial framework to code object location (Bremner, 1980, 1989). In the standard manual search task, for instance, the A$\bar{B}$ error is significantly decreased if landmarks in the form of distinctively coloured covers are used, *provided* the infant is moved before being allowed to search on B trials (Bremner, 1978). And infants are less likely to make the error if they crawl to the hiding place(s) rather than being moved passively (Benson and Uzgiris, 1985). Coming to 'foresee' the consequences of behaviour involves the infant in solving something like the *frame problem* in AI, simply defined as: 'keeping track of what is going on in the world while attempting to change some aspect of it' (Burge, 1974, p. 17). The development of voluntary locomotion may provide a naturalistic context within which the infant's initial solution to this problem is forced to change.

Bremner (1989) suggests that landmarks work for the infant by facilitating the updating of what remains a self-referential location code, rather than supporting an objective spatial code such as 'under the cloth'. Landmarks aid fixation and those infants who keep fixating them while they are moving or being moved prove least likely to make the A$\bar{B}$ error. However landmarks are being used, this approach needs to clarify what change(s) in mechanism enable the infant to begin exploiting them. Despite this, the approach is particularly interesting from the viewpoint of understanding the infant's pragmatic knowledge, since it concurs with the view of self-related activity, in this case description of object location, as a *positive* achievement, not a drawback for the infant:

> There is nothing egocentric in using a self-referent code appropriately; adults do so whenever they correctly locate an object relative to the self. Use of self-referent coding is only egocentric when the user fails to update the coding appropriately to compensate for a movement of the self or the object that is being tracked. (Bremner, 1989, p. 128)

Changing coordination and preconditions in action programs

The types of theory discussed above, alone or in some combination, are liable to provide only local alternatives to Piaget's theory, in the sense noted earlier. Analysis of a single performance phenomenon through dependent–independent variable methods offers limited help towards conceptualizing a system of *interacting* processes in which the 'cause' of performance requires a systematic explanation. Even an apparently complex looking theory runs the risk of being 'trivial' in a technical sense suggested by Priest and Young's (1988) attempt to devise quantitative criteria for evaluating certain classes of programmed model. They point out that what looks like an exact fit between 'behaviour' generated and the data on which the theory is based may be neither impressive nor surprising, since just those data points that actually exist in the data may be used as a basis for constructing the theory. Triviality comes about when a theorist handles new pieces of data that need to be explained by adding (often special-purpose) novel components or assumptions to their theory. An example of this kind of maneuver might include direct attribution of the lengthening delay that is required to obtain the A$\bar{B}$ error as infants get older to an age-related increase in the size or strength of some discrete cognitive component such as short-term memory rather than to the interaction of more primitive components (cf. Harris's (1989) eclectic model of multiple components that might be involved in the search task).

What must be kept in mind is the need to get behind the data to characterize deeper, 'non-trivial' principles of a system that could generate them. A non-local alternative to Piaget's claims should try to place the A$\bar{B}$ error in a broader context of the developmental sequence, including how search comes about at all; parallel developments in performance, such as the 'behaviour of the support' (grasping an out-of-reach object by first pulling the in-reach support on which it rests); and the issues of means–end behaviour, goals and intention featured in Piaget's original account.

My approach to this starts from the idea that basic action programs exhibit *differentiation*. Mounoud (1982, 1986) has drawn attention to the fact that initially tightly knit coordinations undergo a common pattern of dissociation during development. This 'decomposition' or 'segmentation' as he calls it is considered to be essential to new organizations. In the case of prehension, a significant organizational change appears to occur between 6 and 9 months of age. Hand and arm movements involved in reaching for and grasping an object become individualized, with the consequence that they are more adaptable to the range of circumstances in which they will be used. In part, such behavioural changes may be located at the level of movement execution, but what is of interest here is their significance in relation to changing program control of behaviour and the way that behavioural procedures are executed. While data and theory on the A$\bar{B}$ error go against (over)simple views of action as motor response, they prove compatible with a computational, procedural view of action.

This diverges from Piaget's (1953, p. 147) treatment of how actions become more complex, from which perspective *coordination* of actions is the central process and provides the criterion of *true intention*. The greater the number of intermediary behaviours involved in attaining an object, and the more interchangeable these 'means' to the 'end' are, the greater can be our certainty that the infant knows and experiences the object as independent of his or her actions towards it. However, Piaget himself is forced to admit that there is no 'sharply discontinuous' relationship between single and coordinated actions (Piaget, 1970, p. 19), and I shall argue that *number* of actions is an inappropriate criterion for demarcating developmentally significant distinctions in psychological mechanisms.

The kinds of infant action that were considered in previous chapters were very rudimentary, but the computational approach stressed how complex the processes involved in even such 'single' actions must be. From the Piagetian perspective, this might be considered irrelevant. Piaget suggested that analysis of the formal Structure of coordination could be taken back to levels of description 'below' the psychological level of organization, such as the movements involved in a single action, or even the neural level of organization. But this is not what the computational analysis of action is doing. It is concerned with the organization of processes that include but are not restricted to movements. Its emphasis on the organization of action programs and the causal embedding of the infant in the environment presents a more radical challenge to Piaget's model of what an action is.

Two broad aspects of the restructuring of action during the first year are especially relevant to clarifying what might underlie the differentiation that will contribute to, among other developments, the emergence of manual search. The first of these concerns changes in how perception and behaviour are *coordinated*, which may be particularly relevant to understanding the nature of the infant's 'motor' and 'egocentric' memories for object location in the $A\bar{B}$ error situation. In prehension, a transition from *visually elicited* to *visually guided* reaching is generally acknowledged, but it is rarely placed in a broader developmental context. To talk of early behaviour as being 'elicited' has unfortunate and probably inappropriate connotations of stimulus–response causation, but it captures the purely prospective way in which early action programs appear to use perceptual information. Elicited reaching is often termed 'ballistic' because, once it has been initiated, it is apparently executed without modification or correction. For example, vision and behaviour are coordinated in such a way that the location of a stationary object in the visual field provides all the information needed to program movement to that location. Once behavioural procedures have been invoked, visual information concerning either target object or moving limb will not alter their course. Change in the infant's or the target's location results in failure to reach the previously determined position.

In contrast with this, visually guided reaching uses vision in an ongoing manner throughout its execution. The development of this type of coordination is revealed in the infant's increasing success in situations where visually specified object location provides an insufficient basis for determining movement. Recordings of the structure of reaching in 4- to 10-month-old infants with and without displacing prism spectacles show that prism trajectories deviate from normal ones most in the youngest subjects; older subjects show an increasing ability to switch from a 'miss' to a 'hit' path after 'mis-guided' reaching commences in the prism condition (McDonnell, 1975).

These two forms of visual-behavioural coordination are related to notions of 'open loop' and 'closed loop' control from the motor skills literature (e.g. Stelmach, 1982), and reveal different ways of solving the problem of getting perception and behaviour into concordance (von Hofsten and Lee, 1985).

Open loop control involves movement that is specified centrally on the basis of information about environmental conditions. Examples include Schmidt's rapid, single behaviours controlled by 'anticipatory' motor programs (Schmidt, 1982a) and the 'prewired' behaviour patterns of many species that are executed without afferent feedback (Gould, 1982). Closed loop control involves feedback from peripheral receptors to detect and correct errors. An example is Bruner's (1968, 1973) 'feed-forward, feedback, knowledge of results' model, although his own data suggest the significance of both forms of control.

The non-trivial nature, from the infant's perspective, of the initial emergence of visually guided activity is vividly illustrated by one of Bruner's observations (1968) on the development of manual control. Films of infants at around 7 months reveal that noticeable effort is mobilized for the apparently simple task of reaching out to capture an object. One way in which the protracted problem of getting the hand to the object is resolved is through the paradoxical but expedient solution of firmly closing the eyes and completing the reach without visual feedback. This appears to show us the infant reverting to the earlier 'ballistic' form of coordination. Normally seen much earlier, during the prereaching period, this kind of reaching is characterized by its speed, force and pounce-like quality; its successor (from around 16 weeks), in contrast, is much slower and may involve the hands 'hovering' close to the object rather than immediately falling away if no contact is established (Bruner and Koslowski, 1972; Rutkowska, 1992b). It is interesting to note that designing early manipulators, as 'robot' arms are more appropriately called, encountered a whole new scale of difficulty when vision was included and the manipulator required simultaneously to coordinate seeing and moving its arm. The only partly successful systems operated by keeping the arm well out of the visual field while the target object was viewed, then switching off vision while the arm moved to the object's location.

The first type of coordination could underlie a unitary mechanism for the *motor memory* and *egocentric coding* of spatial location which may be available to pre-landmark using infants, in terms of behavioural procedures' role in coding object location. Its basis in concordance between a seen location and the end result of movement would clarify why and how maintaining a fixed orientation to a hiding place is effective. Additionally, the prospective way in which the elicited behaviour is programmed means that the behavioural procedures involved could provide the basis for an autonomous *position memory*.

On each of the A hiding trials, computation of values for the variables associated with behavioural procedures will yield an identical program specification. Maintaining these values following repetitions would make the program *anticipatory* at future A hidings, or *perseverative* if hiding were now at B. Pertinent here is Shapiro and Schmidt's (1982) review of the effects of variability of practice in the development of what they call *recall memory* for motor tasks such as throwing an object. A range of studies with 3- to 9-year-old children shows that *uniform practice conditions* result in greater error on transfer tasks such as throwing to a novel location.

The fact that searching infants will perseverate, even if they have not been allowed personally to uncover and retrieve the object at the A location, has been seen as problematic for action-based explanations. But, by contrast with Piaget's view of action, it is not essential to the present view of action that the infant personally intervene in the object's reappearance. Nor does it require that programmed movements be executed to completion and 'reinforced' by finding the object. On this view, the A$\bar{B}$ error is unrelated to action understood as motor response or motor skill, accommodating the often noted fact that much the same movements are needed to reach either of the hiding places. However, the A$\bar{B}$ error *is* related to behaviour in its procedural sense, and different *behaviours*, namely 'reach-to-A' versus 'reach-to-B', are required with regard to the two locations.

If this suggested basis for perseverative performance is correct, we should expect to find a common pattern across the early organization of action programs, and this does appear to be the case. Behavioural repetition in the looking behaviour of 6- and 9-month-old infants was mentioned in the earlier discussion of sensitivity to landmark information (pp. 154–5). At the much earlier age of 7 weeks, infants tracking a moving object have sometimes been seen abruptly to return to its starting point (Bullinger, 1977). In early prehension, 4-month-old infants who have repeatedly manipulated an object placed within easy reach will manipulate and fixate the now empty location when the object is moved to an equally accessible position (Willatts, 1979). Working with neonates, Butterworth and Castillo (1976) observed an oculomotor response that was spatially coordinated with the location of a repeated sound. Infants showed evidence of continuing to turn in the same direction when the location of the sound source was changed.

The perseverative component of performance in the A$\bar{B}$ error situation seems to be a general phenomenon of restructuring in action systems. It is neither unusual nor unique to stage IV, and it is an example of 'anticipation' as much as of 'perseveration'. The other side of this perseveration notion raises the issue of how *variability* of practice leads to restructuring that supports transfer. This issue will be taken up again in chapter 8's discussion of mechanisms of change, but two examples are particularly relevant in this context. With regard to search, 6-month-old infants are able to orient themselves after movement towards an event at a location that is not specifically distinguished by a landmark, if their own position relative to the location is varied during training trials (McKenzie, Day and Ihsen, 1984). Along similar lines, an early means–end behaviour of pulling a string to make a mobile move showed generalization to a novel context, if infants had previous experience of a variety of string–mobile combinations (Fagen, Morrongiello, Rovee-Collier and Gekoski, 1984).

The second aspect of the restructuring and differentiation of action taken up here involves the incorporation of *preconditions* into action programs, leading to an extended range of action and an increasingly complex *transaction* between infant and environment. A precondition is essentially a description or pattern (which may include variables) defining conditions that must be met *before* some procedure can be successfully executed. The significance of preconditions can be illustrated by Sussman's (1975) analysis of the general pattern of problems or 'bugs' that arise in his classical machine learning system, HACKER. HACKER acts in a microworld of spatially arranged blocks, under certain constraints such as being able to pick up only a single block at a time (cf. the infant's being able to grasp only objects within reach or of a certain size). Reciprocally, this gives rise to 'environmental' restrictions on behaviour – preconditions or prerequisites – such as a block having to have a 'CLEARTOP' before it can be picked up and moved. This means that the execution of even a simple task will involve an intricate interplay between the 'goal', information about the state of the world, and the selection and execution of procedures, which would be behavioural procedures in the corresponding human case. For instance, to make three bricks into a tower where 'A IS-ON B' and 'B IS-ON C', it is essential that the second subgoal is executed first. Otherwise, if A is placed on B, then B cannot be placed on C since it will no longer satisfy the CLEARTOP precondition necessary for pick-up.

HACKER is able to abstract the type of bug involved when it is forced to undo its first move of placing A on B. Subsequently it recognizes problems of a similar structure such that the goal of constructing an 'ON' chain of any length will be executed without error by starting construction from the bottom up. One might consider that a vast number of problems could be encountered by a system of this type. In the case of a human infant, say, s/he may drop a block, or lack the requisite degree of motor control for accurate stacking. Significantly, however, Sussman demonstrates that, at the level of HACKER's *plans* for getting from a start-state to the task-goal, many types of bug can be mapped onto a very limited number of patterns of program organization. Most involve *unsatisfied preconditions* or *protection violations* (in which the execution of one procedure makes execution of others impossible or eliminates the prerequisites they require).

With the early action programs that have been considered so far, if the perceptual component makes explicit appropriate information, the program can activate appropriate behavioural procedures. With the differentiation characterized by inclusion of preconditions, this type of program control of behaviour is superseded by one in which control may be passed temporarily to subprograms or subroutines.

For instance, the control of action in the behaviour of the support can be seen in terms of a program for grasping any object, X, that includes the precondition that it must be 'WITHIN REACH'. This means that before the goal procedure of grasping is activated, a check must be made to see if the system's description of X contains the pattern 'X IS WITHIN REACH'. Should this fail to be the case, control of behaviour will be delegated to a subprogram whose procedures attempt to satisfy the precondition; that is, to modify the environment so that the superordinate goal procedure of grasping can be run. Here, the appropriate procedure consists of pulling an object Y which is 'WITHIN REACH'. If it is to be effective, the activation of this procedure needs to be conditional upon satisfaction of its own precondition that 'X IS-ON Y', where 'IS-ON' is a fixed pattern and 'X' and 'Y' are variables.

Performance in manual search situations appears to involve a parallel aspect of differentiation in which a key precondition for success is that the object has a clear surface. If an object is made to disappear by plunging the room into darkness, then even 5-month-olds will reach for it (Hood and Willatts, 1986), but disappearance via a cover is quite a different matter. Empirical observations support the view that cover removal comes to be a distinct *subroutine* or procedure for satisfying this precondition. Gratch (1972) reports that virtually all 6-month-old infants will remove a transparent cover placed over an object they are already holding, but less than half succeed if the cover is opaque. Interestingly, the way in which the transparent cover is removed predicts performance with the opaque cover: if they simply shake their hand until the cover falls off, then they fail to remove the opaque cover; if they use their free hand to lift it off, then they do the same with the opaque cover.

The acquisition of 'support' and 'cover' subroutines based on these distinct precondition requirements appear to be independent and cumulative developments. For example, 9-month-olds can succeed in using a cloth support, even after the supported object changes location, in a two-support analog of the manual search task; but they will make the $A\bar{B}$ error on the search task itself (Willatts, 1989). If a composite task is constructed, with the object located on one of two support cloths, the ends of both of which are then covered with transparent or opaque cloths, the infants exhibit typical $A\bar{B}$ error behaviour of going to both locations and divide visual fixation between them.

Detailed observations of 6- to 8-month-old infants fit the idea that the acquisition of cover removal behaviour is a step in the development of the infant's pragmatic knowledge of objects' properties, rather than an improved ability to put to use ready-available knowledge of their continued existence. Behaviour when a hidden object is uncovered leads Willatts (1984) to endorse Piaget's claims of a transitional period in which the object may be uncovered or retrieved but is not *intentionally* sought by the infant. Differences in behaviour towards the uncovered object are observed depending upon whether the covering cup is grasped and raised or removed by swiping or moving the hand behind it. In the latter case, the infant may concentrate interest on the cup (for example, taking it to the mouth) and/or show little interest in the uncovered object. Many subjects fall into this category, leading Willatts to question the interpretation of findings that infants as young as 5 months 'search for' hidden objects in certain contexts. A similar transitional stage was found in support behaviour, when the supporting cloth would be grabbed and become the focus of attention, rather than the supported toy which had been brought within reach (see Willatts, 1989).

The developmental dimension of using 'landmarks' or 'background features' can be considered as an instance of the general way in which precondition-based programs exploit aspects of the environment to scaffold behaviour.

A distinctive feature of the location at which the object is hidden – 'X IS-AT Y' – may have much the same function as the 'X IS-ON Y' relation in the behaviour of the support. Would this reduce the chances of perseveration when the hiding location changes? The structure of precondition-based programs entails that they rely considerably on an ongoing use of vision, so a relation might be suggested between the development of such programs and visually guided perceptual-behavioural coordination. In the type of program that seems necessary for successfully attaining out-of-reach objects, for example, the structure of its distinct behavioural components suggests that the movements to be executed can no longer be computed *in advance* as a single autonomous unit. When the reach and grasp is eventually executed, the object will no longer be in the position it occupies when the program begins to operate.

The inadequacy of an 'elicited' or 'open loop' form of control for such behaviour suggests an explanation of why apparently simple detour reaching is non-trivial for the infant. For example, presented with a toy in a perspex box, infants of 6 to 7 months characteristically reach directly along the line of sight, although the opening of the box is to one side. By 12 months they spontaneously reach along the non-direct path to retrieve the toy. And, in an intervening phase, successful object retrieval sometimes occurs and is achieved by the expedient method of changing position until the infant can see through the opening and reach directly along the line of sight (Diamond, 1981, 1988). This latter observation looks much like systematic evidence for Bruner's (1968) observation of an infant 'falling back on' (developmentally earlier) open loop control of reaching in the face of difficulties.

What may look like a single act of reaching actually necessitates *two* distinct behavioural procedures: reaching to the opening; then reaching to the object. The unfortunate fact about the design of the manual search task is that the converse may be true: a single reaching movement is sufficient to get to both cover and object. Attempting to discover what relations between vision and behaviour are used on A trials is uniquely problematic because cover, goal object and landmark, if available, are all in the same location. However, better opportunities are offered by visual search for moving objects, a context that will be considered in chapter 8.

My approach is distinguished from the theories of the preceding section by its emphasis on *action*. But it differs from Piaget's action-based account in a number of ways. Especially important, in relation to the general view of the infant being developed here, is the attribution of the development of increasingly complex exchanges between infant and environment to *differentiation* of action programs as opposed to coordination of independent action schemes. This means that the functioning of ready-available perceptual-behavioural coordinations in the environment can provide a structured context from within which the further development characterized by acquisition of preconditions can occur.

Related to this is the idea that the infant is not 'liberated' from the restrictions of the environment, but becomes able to engage in increasingly complex *transactions* with it. In the manual search task, the object's cover is not so much a 'residual stimulus' as the key component of a *newly symbolized precondition* for the extension of action to a new range of situations. Through individual learning, the cover is assigned the function of indicating the invisible object's presence or 'findability' by the action programs that come to make use of it.

This is an instance of infant action coming to exploit the *scaffolding* offered by a structured environment. The task of searching for a hidden object, simple as it may seem, actually contains numerous degrees of freedom – totally unconstrained search might involve checking every possible location within the visual field. Preconditions provide a powerful way of exploiting environmental structure to constrain performance.

During the infant's passage through stage IV, search at an old location may deserve the status of 'mistaken anticipation' rather than 'residual error'. If we think back to Dretske's (1988) analysis of natural systems of representation, we might think of the 'errorful' infant as having a 'protobelief' that the object is at location A. On a traditional analysis, this belief would not count as knowledge about the world; leaving aside the issue of justification, it is simply not 'true', even in a broad sense of corresponding with the way things are in the environment. If, however, we agree to see infants' performance in terms of pragmatic knowledge, this conclusion no longer seems warranted. Nor need it follow.

Barwise and Perry (1984) have suggested a relevant distinction between belief as 'failed knowledge' and as 'successful strategy'. Their hypothetical example involves a blind squirrel in a nut-rich environment, repeatedly reaching out and trying to gobble whatever comes into its paws (sometimes a nut, sometimes nothing). We might describe it as conjecturing that there are nuts before it, but its unsuccessful conjectures (failed reaches) do not amount to failed knowledge; even if they are often wrong, they offer a good strategy for the animal in the right kind of environment. In developing search, then 'committing' the A$\bar{B}$ error, the practical infant is not on a one-way path of systematically declining egocentrism moving towards ideal 'objective' knowledge. Rather, s/he is developing strategies that underwrite increasing success in more of the environment.

Action differentiation and the structure of infant intention

Explaining the infant's understanding of his or her behaviour and its consequences is an extension of the problem of meaning in action that was discussed in chapter 4. The implications of the computational approach to action for analysis of the existence and nature of *intention* in infancy are important because they bear on a range of closely related issues, such as the problem of attributing *expectancy* to infants, and the status of the precocious anticipation that they reveal in, for example, visual search tasks.

In earlier chapters, infants' perceptual-behavioural coordinations were deliberately and cautiously referred to as 'functional', so as not to prejudge arguments as to whether or not intention is involved. As far as Piaget is concerned, he is willing to consider thumb sucking or early acquired habits as examples of *praxis or action*, defined as 'a system of coordinated movements functioning for a result or an intention' (Piaget, 1974, p. 66), but the 'or' in this definition is critical. To describe the activity of preconceptual subjects as controlled by intention would be using the term in a trivial sense.

The traditional philosophical background to Piaget's distinction and empirical criterion of intention is transparent, and it is relevant to understanding my view that intention is not usefully conceptualized as a property that only one kind of system 'really' has. For the kind of traditional philosophical perspective by which his position is informed, 'intentions', 'purposes' and 'goals' are all facets of intentionality. One property of intentional accounts, namely, their referential opacity, was discussed in chapter 4 (p. 102). Another significant property is the fact that one can 'think *x*', 'believe *x*', 'want *x*' or 'seek x' without there being anything in the environment that corresponds to *x* (sometimes technically referred to as 'failure of existential generalization'), whereas one cannot 'drink *x*' or 'find *x*' unless *x* exists. 'Intending' in relation to the control and execution of action is an intentional notion – as are 'expecting' and 'anticipating' – because an 'intention to *x*' always implies an end, result or outcome *x* that typically does not yet exist and might never exist. One can, for example, 'think about' or 'search for' a unicorn, but one cannot 'see' or 'find' such a creature. (Cf. the example of infants' progress from 'finding' objects to 'searching' for them, p. 161.)

These ideas have obvious links with Piaget's treatment of the relationship between intention and conceptualization, and with his attempt to operationalize them in terms of behavioural search criteria. Increasing search ability and decreasing perceptual reliance are interpreted as evidence of action being guided by concepts, viewed as the 'objects of thought' so central to traditional philosophical accounts of intentionality, rather than in terms of mere principled responding to things in the environment. And acting on the basis of intention becomes equated with action being directed by an explicit representation of the goal object.

The earliest actions certainly involve objects and events that are, so to speak, right in front of the infant, as when s/he reaches out and grasps a suspended toy. However, the analysis of infant action in chapters 3 and 4 suggested that it already exhibits some properties of intentionality, despite the absence of concepts in the traditional (and Piagetian) sense. Seeing the infant's causal embedding in the environment as a sufficient condition for meaning in action can be extended to imply that it makes sense to talk about an intentional 'object of action' under such circumstances, despite the absence of a traditional 'object of thought'. (Cf. ecological psychology's general position on intentionality: Turvey and Kugler, 1984; Turvey, Shaw, Reed and Mace, 1981.)

As with meaning or knowledge, we may be better off regarding intention as a property of a range of systems. The question then becomes: can developmentally significant distinctions in the nature and structure of intention be mapped on to the changing organization of action programs and the behavioural processes they control? The kinds of functional distinctions that need to be made are suggested by Boden's (1972) philosophical and computational analysis of the explanatory force of 'purpose'. The emergence of an increasing number of the characteristics of intention can be linked to changes in the procedural basis of action due to the differentiation of action programs.

In chapter 1, I noted that Bower, Broughton and Moore (1970b) and Trevarthen (1974, 1975; cf. Legerstee, Pomerleau, Malcuit and Feider, 1987) originally attributed intention to the manipulatory and communicative activities respectively of even the youngest infants on the basis of the same type of behavioural criterion. If an ongoing sequence of activity (for example, reaching towards an object, or communicative exchanges with an adult) is interrupted (because a virtual, untouchable object is employed, or the adult appears to stay oriented towards the infant but begins a communicative exchange with someone else) then the infant displays a negative emotional reaction characteristic of frustration. In my view, such phenomena are amenable to the same explanation. They illustrate that the sequence of movements observed is not simply causally chained. It is controlled by an action program whose operation has a *definite end-point* in the execution of the behavioural procedure(s) that it controls. Thus, there is an implicit kind of *prospective reference*, but it is structurally distinct from that which will be found following restructuring. At the earliest level of organization, the operation of the infant's undifferentiated action programs can be said to reveal a 'completion tendency'. This draws on Mandler's (1975) suggestion that the interruption of integrated or organized behavioural sequences leads to emotional reaction if completion of the sequence or substitution of alternative behaviours is prevented. It may be precisely because of the tightly knit nature of such early organizations that frustration can occur but substitution of behaviour cannot.

Contra the rigid nature of such early coordinations, some degree of *flexibility* or *variation* in the execution of behaviour is generally deemed essential to intention. Where this does not exist, Boden (1972) suggests that one might consider the behaviour to 'serve a purpose', but one cannot credit that purpose to the individual concerned (cf. Winograd and Flores's (1986) view of the 'reflex' infant). In the development of action programs, the transition from a prospective (elicited) to an ongoing (guided) coordination of vision with behaviour may play a significant part in this respect. The 'ballistic' nature of behaviour resulting from elicited coordinations means that it ends when the final behavioural procedure is executed, but this may well fail to result in the end-state or outcome that an adult would consider to be the objective of the behaviour sequence. In contrast with this, the guided form of coordination confers some flexibility on behaviour through its sensitivity to circumstances during the behaviour's execution. These two kinds of coordination might be characterized as *goal-terminated* and *goal-directed*, respectively.[1]

Goal-terminated action is designed, through evolution or development, so that its behavioural procedures will probably achieve a certain end in a normal environment (cf. Brooks's task-achieving behaviours, 1991b), but the acting system does not itself represent the environmental state that it will so often achieve. Even apparently sophisticated behaviour sequences in some species can be accounted for in terms of a set of 'sign stimuli' each activating a particular behaviour; if behaviour fails to alter the 'sign stimulus', or that stimulus is repeatedly reinstated by an experimenter, habituation offers the creature's only escape from continuous production of the same ineffectual response (Gould, 1982). For example, an insect that builds a funnel at the entrance to a burrow appears to 'recognize' a hole made in the completed funnel as a burrow entrance and proceeds to build a new funnel there. Boden's (1983) analysis of ring-doves' complex pattern of chick-rearing activities in terms of a sequence of independent behaviours would fit this notion of goal termination. In keeping with the mainstream AI view, Sloman (1983a) suggests that a sequence of concrete, short-term goals may need to be *explicitly* represented and sought to explain the control of such sequences of behaviour (for example, 'catch worm', 'sit on eggs', etc.). But evidence that at least some behaviour patterns can be appropriately employed in an 'open loop' manner without visual or afferent feedback suggests this need not necessarily be the case (Keele, 1982).

In goal-directed action, context-specific variation – as seen in reaching with prisms – increases the probability of reaching an end. The suggestion that some form of goal direction is involved stems from the fact that notions such as 'feedback' and 'success' begin to seem relevant when this ongoing adaptation to current environmental conditions is observed. There is a sense in which the infant appears to strive towards an end since, although the object to which the infant reaches is visible, the ultimate outcome of holding the object is not. This need not mean, however, as we shall see later in this section, that the infant's procedural representation includes anything like an explicit description of that end considered as a *state*. I am not claiming that visually guided behaviour is sufficient for attributing explicitly represented goals to the subject. There may be no necessity for a description or pattern 'OBJECT-IN-HAND' against which the infant's action program repeatedly compares descriptions of the environment until a match is achieved.

The use of *intermediaries* to reach some end can be considered a more explicit instance of the behavioural flexibility that is required for the ascription of intention to the subject. At the level of action programs, this can be related to their differentiation to include preconditions. This makes it possible to achieve an end from a broader range of initial states and through distinctively different means procedures: the end 'manipulating object X' can be achieved through procedures involving covers or supports, depending upon whether the object is out of sight or out of reach. As program structure comes to include preconditions, the teleological nature of infant action becomes more explicit and we become increasingly willing to talk about *the infant's* intentions or purposes in action, as opposed to the purpose served by the infant's action. But it is important to note the line of continuity between variation in a single means behaviour and the use of intermediaries – both demand an interactive coordination of vision and behaviour.

An increased number of individually identifiable behaviours will characterize programs including preconditions. Contrary to Piaget's position, however, their number, as such, is irrelevant to the issue of intention. A numerical increase could result from chaining and fail to satisfy *any* of the pertinent criteria discussed above. What is important is the structure of the program within which the procedures controlling those behaviours are embedded. The notion of a program that can pass control temporarily to subprograms or subroutines provides a way of understanding increasingly complex and explicit intentional phenomena in organizational terms. Of particular importance is the fact that this type of program can operationalize a basic notion of *planning*. It makes possible the selection and execution of some procedure(s) *in order to* execute others. Piaget's claim that the infant comes to look *in order to act* emphasizes this very dimension of organization. The differentiation of action programs to include preconditions provides the structural foundation for the infant's *nascent plans* and increasingly evident intentional activity.

Need early intentions towards persons be different?

The examples that introduced this analysis of the component properties of intention included the infant's behaviour in physical and social domains. It might be objected that my approach has considered only infants' pragmatic knowledge of the physical world, and has nothing to say to the issue of what *communicative intention* is or how it might develop. Views that are informed by traditional philosophical treatments of intentionality suggest that mechanisms qualitatively distinct from those that suffice for functional activity with inanimate physical objects are required for the infant to be treating some*thing* as a person.

The parallel that I drew between object- and person-related findings could be taken to suggest that both demonstrate a form of first-order intentionality on the infant's part; that is, that action programs underlie the infant's psychological understanding of something *about* objects and *about* persons, as contrasted with a possible zero-order intentionality interpretation in terms of purely neurophysiologically grounded reflexes. But understanding persons or 'intersubjectivity' can be taken to require at least *second-order* intentionality: an understanding that others have minds, and are themselves centers of intentionality, exhibited in an ability to represent the fact that the other is also representing things (for example, *intending* to affect the other's *intention* or what s/he *knows* about some state of affairs).

In this context, I can barely graze the surface of the many complex issues relevant to this debate, but I shall briefly sketch a viewpoint consistent with my other arguments about infant knowledge. This starts from the idea that using and seeking evidence for the kind of distinction proposed above may involve inappropriate category boundaries as far as the infant (and maybe in some respects the adult) is concerned, and that these generate spurious theoretical problems for the explanation of early abilities. What happens if issues about first- versus second-order (or indeed zero-order) intentionality are replaced with the general question: what mechanisms do infants need to behave *as if* they understand others' mental states? As Reddy (1992) has suggested, with regard to competing interpretations of why infants point, viewing the infant either as having the goal of sharing experience with another or as 'merely' trying to influence the other's observable behaviour may simply set up a false opposition. A viable hypothesis is that early social responsiveness and 'protoconversations' *can* be treated as a mode of action whose mechanisms obey similar general principles of organization to those employed in object-directed action. One of the outcomes of this approach is that it offers a way out of boundary-demarcation disputes that arise from attempts to interpret contemporary developmental data with the traditional categories and assumptions of intentionality in mind.

Sylvester-Bradley (1985) argues that infants up to 10 weeks of age (and possibly much older) do not distinguish people from things. Those who propose a person–object distinction on the infant's part are committing the psychologist's fallacy of drawing on their own experience to attribute to the infant awareness of the difference between people and objects, and knowledge of what that difference signifies. But the infant does not categorize people and objects differently and experience them as adults do. The infant's exchanges with adults do not involve 'communications of intention' or an ability to 'read others' intentions', but result from the interplay between 'social facilitation' (adults' oral and facial activity evoking broadly similar behaviours in infants); adults' attribution of communicatively meaningful content to infants' behaviour; and 'primary narcissism' (the infant's enjoyment of appearing to produce effects that reflect his/her actions).

This critique may read rather more than is strictly accurate into the claims actually being made by those who argue for early person–object discrimination, and owe as much to Sylvester-Bradley's conventional ideas of the necessary conditions for person understanding. The social facilitation mechanism that he proposes (although supposedly involving more general behavioural correspondence than imitation) does not seem far removed from the kind of ability that Trevarthen (e.g. 1992) sees at the heart of infant sociability: a preadaptation to display emotional states and to detect and respond to those of others in an unmediated way; that is, without recourse to rational thought or language. Might not such abilities suggest an embryonic understanding of persons? A positive answer seems to me to be in order for two interrelated reasons.

First, even those who advocate explanations involving some version of traditional second-order intentionality on the infant's part probably do not need such constructs to explain what they are observing. For example, as we have already seen, Premack (1990) suggests young infants may be capable of perceptually based discrimination of *intention*; hence, the prediction of others' behaviour which is the function of a Theory of Mind (Premack and Woodruff, 1978), based on specific movement parameters (internally generated changes in movement of a self-propelled object; cf. Mandler's (1992) ideas on the early appreciation of animacy). (For philosophers such as P. F. Strawson, it is worth noting, a person can be defined as an object to which we attribute intentions.) Premack stresses that he is not claiming that the infant can perceive second-order intentionality on this basis; that is, that the infant can perceive that the other they are observing can themself perceive that others have goals. This would amount to attributing third-order intentionality to the infant – representing another's representation of another's representation.

However, Premack does propose that 'in perceiving one object as having the intention of affecting another, *the infant attributes to the object a representation of its intentions*' (Premack, 1990, p. 14, italics added). This does not seem to me to follow at all. In terms of my preceding analysis, the intentionality in infants' earliest social understanding is grounded in action – the definite end-point and prospective reference implicit in their behaviour towards things that they describe as social, which could include things that are perceived as exhibiting intention. Such mechanisms enable the infant to represent people at a procedural level, without needing the kind of Theory of Mind to which Premack alludes.

For the young infant, people may be defined in terms of the 'social' properties towards which the infant is geared to act, not as individuated objects that happen to possess these properties rather than others. (Cf. chapter 3's questioning of the need to attribute an integrated object description to the infant.) Displays of negative affect when the adult partner in 'protoconversation' begins to exhibit 'irrelevant', unsynchronized behaviour, or no behaviour at all, may arise from sudden removal of the conditions for completion of the infant's own, already initiated, social behaviour. Legerstee's (1992) conclusion that infants start out with different mechanisms specific to social and non-social domains would also fit these action-based assumptions. These assumptions support the idea that there is different knowledge of people and objects, but do not require that this knowledge be 'processed in different ways' in the two domains.

The second reason for believing that infants can sensibly be credited with innate, pragmatic person knowledge, without the cognitive trappings implied by focussing on second-order intentionality, is the extent to which person properties that centralist accounts identify with mental states have an external, behavioural, and hence perceptible, component that can be exploited by infant action (Hobson, 1991; Neisser, 1988; cf. Premack, 1990 on intention). Sylvester-Bradley's study, it is worth noting, did not contrast behaviour towards the infant's chatting mother with that towards a stationary object. The structure of the two conditions may have been confounded by jerkily moving the ball that was used in the object condition, and by a means that would make it appear self-propelled from the infant's viewpoint. Despite this, some differences in infant behaviour between the conditions were found; for example, infants were more likely to open their mouths and less likely to turn away in the presence of their chatting mother.

Whatever discrimination is involved here is not as clear cut as Sylvester-Bradley feels it ought to be. In particular, some supposedly social behaviours can be elicited by objects and vice versa. For example, infants may exhibit mood switches that do not correspond with a change in whether an object or a person is present, as when they break off apparently 'social' behaviour with the mother to reach for her nose or hair; or they may exhibit features of a social mood, such as smiling, with an object. But what significance should we ascribe to this? It is not as if such cross-over effects cannot be found in the ways adults behave towards people and objects. A not insignificant number of sporting activities involve trying to push people around, or even picking them up and throwing them – behaviours usually reserved for inanimate objects – while social life is generally characterized by getting others to do what we want through negotiation. And adult rational thought, which some believe evolved to enable us to predict and handle the behaviour of other agents (Humphrey, 1976), is sometimes (mis)employed in attempts to control the inanimate world.

Two quite distinct mechanisms may underlie such cross-over effects, neither of which in infant performance would correspond to 'failure to know about the person–object distinction'. On the one hand, a behaviour can have different *meanings* depending upon the *action context* of which it is a part. Throwing someone out of the way in a wrestling ring can be considered as (a somewhat specialized form of) social behaviour, and has a meaning quite different from executing the same behaviour if you happen to meet someone coming the other way on a narrow path. On the other hand, *misrepresentation* in the sense outlined in chapter 4 (pp. 117–18) can be involved. This is likely to occur when an object exhibits some feature(s) of the movement pattern(s) characteristic of people's behaviour. (Machines such as cars are more likely to be the subject of our inappropriate adult 'social' remonstrations than unmoving inanimate objects.) As far as infants are concerned, some behaviours that an adult might deem to be 'confusions' may also stem from overlap between person and object properties; the mother's nose *is* an object.

Distinguishing which of these applies to the infant in any particular case is a fraught task. The relations between action and behaviour that are involved make it difficult to rely on isolated behaviours, even large numbers of them, to indicate when the infant is behaving in a person- or an object-appropriate way. Both the possibility of the 'same' behaviour having different meanings and that of misrepresenting persons/objects are predicated on an ability to represent them in the first place – the appropriateness or otherwise of the infant's behaviour depends on the action program controlling it, not on observers' objective assessments of its success or failure.

Whether the infant is able to *compare* people and objects, or to be *aware* of the distinction between them and of what it *signifies* is beside the point, especially since we would be hard pressed to map these notions on to computational mechanism proposals. As far as the infant's procedurally based, pragmatic knowledge is concerned, we can say that it extends to objects and to people: early action programs assign to visual descriptions of properties that are more often associated with objects or with persons the function of indicating 'things' for which social exchange or manipulation are the appropriate forms of activity. More sophisticated knowledge, involving, say, 'expectancies' about what people are going to do in response to one's behaviour, may develop by the same route as search for hidden objects or the behaviour of the support: through the discovery of the effects that one has become able to achieve.

Procedures and states in infant 'goals'

With the emergence of programs including preconditions and the rudimentary planning that they support, the infant's performance begins to take on the form that Newell and Simon (1972) see as typical of 'goal-directed' behaviour: there are correlations of behaviour over longer time intervals, in contrast with the 'stimulus bound' behaviour characteristic of individual productions. On Newell and Simon's view, goals are symbol structures with two aspects: they include a test to determine when some state has been attained; and they are able to evoke patterns of behaviour that have a rational relation to the goal. But to what extent need the infant's plans include a description of their outcome in order to be effective? Specifically, if we were to argue that the formation of a plan including intermediaries involves 'having a goal in mind', how would that goal need to be represented: as the procedure(s) that will eventually be executed when conditions have been made appropriate, or as the outcome state which should result from the execution of those procedures?

Emphasis on *outcome description* is implicit in studies of infant search that treat the extent to which infants can bear an object 'in mind' during its absence as a *functional component* of the mechanism controlling search behaviour, identifying this with the object concept. Empirical studies in this vein pose questions such as the effect of changing object features during a task (e.g. Evans and Gratch, 1972; Le Compte and Gratch, 1972). The idea of anticipation of specific content also appears central to Bruner's suggestion that 'patterned behaviour that initially occurs "reflexly" or "instinctually" is . . . converted into intentional action when the infant has had an opportunity to observe the results of his own acts' (1973, p. 2).

Two observations are relevant to the present discussion of action programs and intention. The first is that the procedural aspect of the goal is essential from a functional viewpoint. Adult intentions can vary enormously in the degree of detail with which the outcome aspect of the goal is specified (Boden, 1981, ch. 6). If this were not so, it would be impossible to formulate as everyday an intention as to see a new film. One can plan getting to the cinema and checking the title, but these are a preliminary to a viewing process of which the outcome content is more or less unknown. The earlier discussion of vision made clear that a single object may be described in many ways, and it does not seem necessary to assume that an infant activity such as searching for an object must be controlled by an exhaustive description of the object's features. Indeed, it may be naive to do so. Yet the organization of action may still fulfil important criteria of intention without such an individualized description. The act of reaching to a location in a search task, say, is a *behavioural prediction*, inferential by virtue of its role in a causally embedded perceptual-behavioural action system, not by virtue of guidance from an image or re-presentation of the end-state it achieves.

Second, action programs obviously require *particular contexts* and *contents* in order to function, but it is much more difficult than some studies suppose to explain how infant intention might come to include specific outcome predictions. For instance, what would be necessary for the goal of infant search – from the infant's as opposed to the observer's viewpoint – to be finding the particular object the infant sees hidden? Say the infant repeatedly retrieves a hidden object and is able to construct a description of its features. How can that affect *future* search activity? From a simplified memory viewpoint, we might suggest that such repetition will lead to recognition when the object is found on future occasions, surprise if search reveals an object with different features, as has been reported by Le Compte and Gratch (1972). There certainly seems to be a description of a specific outcome, but could it play a *functional role* in the selection and execution of program procedures? A negative answer is suggested by Evans and Gratch's (1972) finding that infants who uncovered a toy at location A continued to search there when a different, new object was used for B hiding trials. Presenting the infant with an object that had novel features evidently did not turn 'B hidings' into a completely new search task for the infant. This suggests that either the infant's representation of the hidden object does not individuate it by its specific features, or such a description is unable successfully to control the location of search.

The infant has a dual task: to learn that search procedures result in contact with an object; and to learn that the object revealed by these means is the same as the one they saw hidden earlier on. In chapter 3, I stressed that perceptual salience is independent of the 'familiarization' that results from repeated exposure to objects or events. For a goal to be represented as a specific outcome, there must be much more than a specific description of that outcome. Some kind of relation must be established between the sequence of hiding and finding events, and it is necessary that a *general relation* be established between hidden and found objects. These features were illustrated in Prazdny's (1980) hypothetical program for anticipatory visual search, part of which was sketched in chapter 5 (pp. 133–5). The complex of program procedures involved provides the subject with an integrated representation of the 'task', and the program's treatment of the 'disappeared' and 'goal' objects was based on general patterns against which a range of particular descriptions could be matched, not on featural copies. The idea of preconditions is similar in so far as it involves a piece of program that can check for a general relation between the specific objects involved in any particular run of the program.

In the Prazdny example, a temporary IF-ADDED procedure checked for partial-occlusion information in a new location. (If the pattern associated with this procedure was matched, visual following would be resumed.) The precise nature of the pattern associated with this procedure can underlie distinct differences in the system's 'expectancies'. For instance, it could be specified that it must include the previously bound values of the variables involved in the disappearance – [PAROCCLUDES ^OBJECT-Y ^OBJECT-X]. Then the procedure will be invoked only if the vanished object reappears at the new location; that is, if [PAROCCLUDES GREENBOX REDCAR] is added to the data-base. Alternatively, the pattern could include a new unbound variable whose value can be bound by any suitable object – [PAROCCLUDES ^OBJECT-Y *OBJECT-Z]. Then the procedure will be invoked whatever object appears, for example if [PAROCCLUDES GREENBOX BLUEDOLL] is added.

The first of these possibilities, together with the other procedures concerned, corresponds closely to what might be an adult's schematization of the search situation, a schematization which would be able to support a general search strategy. But the question that must be posed from our perspective with respect to the infant's perspective is: how could such a schematization develop? (From a more centralist or mentalistic perspective, we might ask: how does the infant acquire the *notion* that an object generally continues to exist at the last place it was seen?)

Trying to get at the mechanisms that underlie these kinds of changes is the aim of the remaining chapters. But enough has already been said about the ways that basic actions become more complex to support a provisional evaluation of the developmental potential of two of the proposals for computational treatments of sensory-motor coordination that were mentioned in chapter 5. Locating increasing complexity with the restructuring of action suggests an explanatory path somewhere between Churchland's (1986a, 1986b, 1987, 1989a, 1989b) claim that the explanation of sensory-motor coordinations can and should be reduced to a neurocomputational level and Kirsh's (1991) argument that a conceptual level of functioning is necessary to explain how a system like a Brooksian mobot would need to be scaled up to produce more complex activity.

Churchland's geometric approach to ability is unsuited to tackle either of the two underlying ways in which basic actions exhibit differentiation, suggesting that retaining action programs as a psychological level of explanation is preferable. His examples of coordinate transformations between (sensory and motor) state-spaces being used to account for sensory-motor coordination rely heavily on knowledge of the subcortical superior colliculus to establish their neuroanatomical validity. This region of the brain exhibits a topographical mapping between neurons involved in the sensory field and the direction of movement of parts of the body, and developmentalists have considered it to be the likely site for physical implementation of the innate, ballistic (open loop) form of sensory-motor coordination that characterizes early infant activities such as reaching to a location in the visual field (see, for example, Bronson, 1974; K. R. Gibson, 1981).

While Churchland's approach may offer a reasonable fit at this level of control, it seems unable to handle the routine developmental transition to a visually guided (closed loop) form of perceptual-behavioural coordination that is typical of human infants and of other higher primates, and generally associated with cortical functioning. Perhaps some story could be told in terms of new, multi-dimensional state-spaces, but that would offer only implementation details for the new kind of coordination that we are attempting to understand. As the dimensionality of the state-spaces increased, such an explanation would become increasingly opaque.

The development of programs that include preconditions and the beginnings of planning also provide problems for the reductionist neurocomputational framework, as they do for the connectionist models by which it is influenced. It has often been observed that they appear to be suited to explaining more 'automatic' sensory and motor processes but fall down when temporally extended, 'higher' sequential activities such as problem solving or planning are concerned (Norman, 1986). There is obviously a trivial sense in which the developing infant's ability to invoke one procedure *in order to* be able to execute another involves mapping between sensory and motor state-spaces, but the intervening portion of this 'state-space sandwich' is liable to be more complex and less directly amenable to neural reduction than Churchland allows. The kind of spatially localized neural–computational correspondences found in the superior colliculus are unlikely to exist for higher levels of organization of sequential processes. This again suggests retaining explanatory constructs at the level of action programs, viewed as components and organizational properties of a (modifiable) virtual machine.

Evidence for nascent plans on the infant's part is also an issue for architectures that forefront emergent functionality, such as the layered control of Brooks's (1991b) mobots. Kirsh (1991) concludes that the kind of situationally determined activity exhibited by such systems can be 'concept free' and work perfectly well so long as they have access to enough 'egocentrically perceptible cues' that act as local constraints on behaviour. This, however, will not be the case whenever task-achieving behaviours cannot be executed without taking account of some *future* desired state; that is, when 'anticipation' or 'prediction' is involved. In such activities, situationally determined control mechanisms are insufficient to account for how the *global* structure of activity is organized. For example, a good deal of the activity of a (to-date hypothetical) tea making mobot might be effectively implemented in independent task-achieving behaviours such as laying out cups, filling a kettle, boiling water, and so forth. But longer-range task properties contribute non-local constraints on efficient functioning and will need to be taken account of too; for example, how many people the tea is to be for will affect things like how many cups to lay and how much water to boil.

During the first year or so, the infant is already beginning to exhibit such globally organized activity. This includes cases where behaviour involves looking ahead to some future activity (for example, manipulation of an out-of-reach or out-of-sight object); and generalization or adaptation of a strategy to a new context (for example, increasing generality of successful manual search), which Kirsh considers a form of 'objective understanding'. Such a system, Kirsh suggests, needs recourse to mechanisms of 'conceptual' functioning (of the type outlined in chapter 4); for example, concepts can 'license inferences' about invisible properties.

From the perspective of developmental mechanisms, it is difficult to concur with this view, which illustrates the over-large gaps that sometimes characterize the paradigmatic cases of ability into which computational and AI work carves up the space of possible minds. At one extreme, we have systems that achieve something like moving around without bumping into things (concept free and, for an increasing number, representation free too); at the other extreme something approaching the all-reasoning chess player (conceptualizing the world and manipulating mental representations of it). Unsurprisingly, *rapprochement* between such extremes is problematic, and finding a developmental path that leads from one to the other is doubly so. In view of the preceding analysis of how infants proceed to more complex actions, a good option is first to consider further what can be achieved by restructuring of action to novel kinds of organization. This, as we shall see in chapter 7, ends up with a new way of looking at the process of 'conceptualization'.

Notes

1. I am continuing with my earlier use of the labels 'goal-terminated' and 'goal-directed' (Rutkowska, 1988), despite Dretske's (1988) distinction between 'goal-directed' and 'goal-intended' behaviour which is similar in some but not all respects. In particular, the description 'goal-terminated' seems potentially less misleading than 'goal-directed' for referring to behaviour that is considered to involve neither an explicit goal nor intention on the subject's part.

Chapter 7

Looking for developmental change and its constraints

Previous chapters rehearsed some features of two key examples of changing ability in infancy: the infant's appreciation of size–weight covariation and the acquisition of manual search for objects. On the face of it, the kinds of processes involved in grasping and lifting objects might easily be construed as ('mere') improvement in motor skill, whereas search is a quintessential index of cognitive development. However, a computational perspective on infants' action-based knowledge made it possible to begin bringing them within the scope of similar mechanisms; and so to appreciate something of the cognitive aspect of object manipulation and the behavioural-motor aspect of object 'conception'.

In this chapter, I look explicitly at what these and other examples of infant development might have in common. Is it possible to characterize a *general form* to infant development? If so, what is it? And in what sense(s) can we usefully talk about *constraints* on that development?

A major concern must be to clarify whether the type of *radical reconstruction* of ability suggested by Piaget is either feasible or necessary. Keil (1990) locates the primary developmental constraints suggested by Piagetian and other stage theories, such as those of Bruner and Vygotsky, in the *domain-general acquired* cell of his heuristic classification of constraints on development. Notably, this cell is distinguished from the others by being occupied solely by a '?', due to the controversial status of claims for the requisite across-the-board changes in knowledge organization. Keil suggests also that a place among *domain-general innate* constraints would be apposite for any specific competences that arise due to maturation; indeed, Piaget could be considered an extreme empiricist to the extent that domain-general acquired constraints emerge from domain-general innate ones (cf. classical learning theory). As far as infancy is concerned, locating the subjective developmental constraints of Piagetian theory primarily within the domain-general innate cell would indeed seem to be the right move. (Since the infant has, supposedly, yet to achieve the first level of global reconstruction, this purportedly domain-general acquisition is as yet irrelevant.) But seeing this as grounds for the conclusion that an empiricist direction for development would be implied reveals a broad problem with the way the task of development is typically being defined.

From the viewpoint of Piagetian theory, classifying stage structures as domain-general acquired constraints could be seen as giving the outcome of development precedence over its all-important process. At the level of mechanisms of change, it is the endogenous processes which underlie *equilibration* that are central, and they are most appositely classified as domain-general innate restrictions (grounds for relocation but certainly not deletion of Keil's '?'). Innate schemes are, of course, also considered a crucial innate component of the infant's processing repertoire. But their initial form has a domain-general flavour, since Piaget (1953) characterizes their primary mode of interaction with the environment as 'functional assimilation': simply trying to operate on anything and everything until, with experience, discrimination of circumstances is established.

This may look like an empiricist programme for general-purpose learning about the external environment, but that is not, of course, what Piaget had in mind. Varela (1988) locates Piaget at the enaction pole of his cartography of cognitive science precisely because he aimed to explain knowledge acquisition in terms of an *epigenetic* process that arises from a type of subject–environment interaction that is *neither* extremely nativist or empiricist *nor* validly seen as a shade in between (representing a direction that Varela feels has gone missing in mainstream PSSH-informed styles of cognitivism). The Structures of knowledge were certainly not, according to Piaget, 'in' the environment, nor did they originate 'in' the subject; they achieved their necessity only after they had been developed through action (Piaget, 1972). Admittedly, Piaget did a far from compelling job in establishing this, not least because his innate allocation to the infant seems inadequate for bootstrapping to any level of organized action. But the important point to highlight here is the nature of his proposal for a significant *constructivist* dimension in development. This goes beyond the straightforward idea that organization must be acquired through exchanges with the environment – the input to the developmental process emerges from a mutual interplay between subjects' behaviour and environmental structure that cannot meaningfully be teased apart into independent contributions.

Nowadays, it would be hard to find any developmental theorist who would deny that interaction between infant and environment is essential for development, but this does not rule out most views remaining closely allied to an empiricist or a nativist stance in the way they locate the primary constraints on developmental outcomes. A theoretical reason for the difficulty we seem to have in escaping from characterizing the form of development on an empiricist–nativist scale may stem from seeing the problem that development tackles as one of mapping between environments and mental structures. This view has built into it a subject–environment, internal–external dichotomy that supports a nativist versus empiricist style of thinking. Epigenetic notions require a systematic style of explanation. Despite doubts about radical reconstruction, it remains worthwhile to explore the possibilities raised by an approach based on the systematic account of action that has been outlined in earlier chapters.

This generates a second major question: how far can we take action in explaining the performance phenomena that emerge during infancy? As suggested by the analysis, in chapter 6, of changing object understanding in terms of the restructuring of action programs, the answer may be considerably farther than is generally thought.

Action need not be seen as a 'precursor' of more sophisticated abilities in the sense of being a mere stepping-stone to representation, a stopgap until it is transformed into mechanisms that liberate the infant from the restrictions of perception and of the environment. Support for this viewpoint comes from comparing computational structure, function and adaptive change in infant action with relevant work dealing with other ages in a number of domains of ability. Such age–domain comparisons are made possible by conceptualizing the infant's action-based abilities in computational terms, and they offer a particularly powerful route to highlighting what is unique and what is common in different cases of functioning and adaptive change.

The reasoning behind reconstruction – and some alternatives

As we have seen, Piaget's view of the *reconstruction* of infant knowledge through development is characterized by suggestions for a number of now familiar transitions, such as object perception to object conception, action to re-presentation, subjective practical to objective logical-conceptual knowledge, and elementary awareness to cognizance (Piaget, 1953, 1955, 1976, 1978a, 1978b). As far as the developmental reconstruction of knowledge is supposed to go, distancing from the environment underlies both of its hallmark acquisitions: re-presentational ability in the sense of imagery, deferred imitation, symbolic play and language; and content-free, domain-general cognitive processes, ultimately applicable to even totally novel or hypothetical material. It is instructive to note, however, that even Piaget's own work carries hints that reconstruction to a representational level of functioning may not be the only or primary mechanism of change during infancy. In particular, the distinction between his notions of 'conceptualization' and 'representation' may involve more than two different ways of talking about the same acquired mechanisms.

Progressive conceptualization and representational development are inferred from the same performance phenomena, but Piaget's publications after the infancy texts no longer used these notions interchangeably, and evocational representation came to play a lesser role.[1] Sometimes, infants' object permanence is viewed as a six-stage development based on *true representation* in the sense of mechanisms for evocation. Often, however, Piaget maintains that object concept is formed in the fourth stage, placing emphasis on the *conceptualization-cognizance* dimension. It is not at all evident that, as much of the infancy literature has tended to assume, having a 'notion' of the object = conceptual ability = a capacity for re-presentation.

Piaget's 'reconstructionist' view of the aims and mechanisms of development hangs on his distinction between learning and development. A process of *empirical abstraction* is said to underlie learning about isolated object properties (such as shape) through action. Development of new structures for knowing, on the other hand, is said to involve *formal* or *reflective abstraction* from general properties (such as ordering) of the coordination of actions. This bears on Piaget's view of interiorization as a process that frees the general form of knowledge from its specific content to reconstruct it at a new level.

In chapter 5, I argued that an account of diachronic process needs to start from a new approach to the relationship between content and process in general, and between perception and behaviour in particular. Piaget's versions of these distinctions were neither clear nor useful, and this has implications for his learning–development dichotomy. Increasingly flexible application of procedures to a wider range of contents and contexts may not be equivalent to constructing general-purpose, content-free, thought-like processes. Indeed, Piaget (1972) himself eventually suggested that even the ultimate 'stage' of formal operational thought might not be as completely dissociated from subject matter as he had initially insisted. Rather, its development might be specific to the domains of knowledge with which the subject had greatest involvement. As far as developmental mechanisms are concerned, concepts such as 'interiorization' may be no more than metaphors for the Structural isomorphisms and experiential distinctions that Piaget believes to be central to adaptive change. And discussion of these phenomena in the following sections will suggest that they are inappropriate metaphors.

Two alternative frameworks challenge notions of discontinuous development and have implications for the computational view of infant action. The first is due to ecological psychology. The view from direct perception theory is that development involves the *differentiation* of *perception*, through which an increased number of environmental properties become salient, and increased sensitivity to the affordances of the environment; not *enrichment* and the construction of *representations* (E. J. Gibson and Spelke, 1983; Gibson and E. J. Gibson, 1955).

The view presented in chapter 6, of the infant coming to make explicit new action-relevant information is sympathetic to ecological psychology's claims. But agreeing *that* perceptual learning occurs does not answer the question of precisely what it entails, or how it occurs. We can assume that innate capacities to perceive and act in accordance with environmental affordances should constitute an important (domain-specific, innate) source of constraint on functioning – hence on possibilities for learning; and that maturation may be called upon to play a role (for example, the emergence of visual cliff 'avoidance' with crawling). But there is no strong contender for an account of the process(es) through which the infant might, in the language of ecological psychology, become 'attuned' to new information or 'discover' further affordances of the environment. Nor is it clear how far-reaching the perceptual differentiation model is supposed to be. Sometimes the idea that perception and conception are distinct abilities is alluded to, but the developmental relation between them is ambiguous and consensus on the form that it takes is currently lacking.

One line of argument sees the ecological approach as *replacing* any need for a separate account of the development of purportedly cognitive phenomena. Wilcox and Katz's (1981) 'radical environmentalism' takes an uncompromisingly empiricist stance and rejects any form of developmental discontinuity, by arguing that questions of internal state, and of how and why differentiation takes place, can be dispensed with if it is assumed: performance can be predicted solely from input; and the input *itself* unfolds over time, as more of the environment is discovered. They draw an analogy with a series of photographs of the same scene with different exposure times (and film speeds appropriate to those times). Very short exposures will capture only moving objects; longer exposures will capture more permanent features, while transient events will now be blurred. What we call development – here they include prototypical cognitive, conceptual achievements such as the quantitative conservations of the middle years of childhood – is the subject's *Umwelt* coming to contain more and more of the permanent features of the environment. Conservation of volume, for example, amounts to using an input 'exposure time' that allows the child to perceive an invariant property in transformations of shape.

This position seems unsatisfactory in so far as it is significantly more behaviourist than mainstream ecological psychology. Wilcox and Katz's central claim is that the deficiencies of Piagetian-style cognitivist concepts establish that any notion of psychological mechanisms is unnecessary for *prediction* in developmental psychology:

> if input is lawfully related to state and state to output, then input is lawfully related to output. Since state merely serves as a shorthand for input anyway, and is not anything that is itself observable, there is no need for it . . . it would only be useful if access to it were somehow easier than access to past inputs. But is there an *it* over and above past inputs? (Wilcox and Katz, 1981, p. 254)

Prediction may be a goal of explanation, but *understanding* is equally if not more important, and this perspective offers little in that direction. In particular, it does not amplify ecological psychology's key notions such as the 'education of attention' and 'effectivities'.

Butterworth (1981a, 1989) also expresses commitment to direct realism, but his concern with the complexities of infancy data lead him to propose a more eclectic approach that considers direct perception as a *precursor* of conceptual-representational functioning. On this view, 'to perceive' is not 'to know'; and infants' changing abilities increasingly suggest a change involving information innately detected by direct perception being overruled by objective knowledge. For example, experience with a posture such as standing reduces or eliminates effects in the 'moving room' that is used to assess infant's use of visual proprioception. Initially, infants fall over, but then become capable of quick compensation for their 'perceived' loss of balance (Butterworth and Cicchetti, 1978). Also, studies of 6- to 18-month-old infants following the line of visual regard of an adult are thought to reveal an initial 'ecological' mechanism, which cognitive development supplements with a 'geometric' mechanism and eventually a 'representational' mechanism, which enables the infant to represent space as a 'container' of self, others and objects (Butterworth and Jarrett, 1991).

Evaluating this proposal is difficult, since the mechanisms proposed are specified primarily in terms of properties of the tasks on which infants are increasingly successful. Thus, the 'ecological' mechanism of 6-month-olds is one that supports looking in the infant's visual field, in the same general direction as the adult; which of several available objects will then be fixated depends on their attention-attracting features. By 18 months, infants are credited with accessing a 'represented space', because they will follow the adult's gaze to an area behind them – hence, outside their visual field – provided there are no objects in the visual field to compete for their attention.

The plausibility of this suggestion depends on identifying perceptually based abilities as restricted to handling whatever is going on in the immediate visual field, outside of which representational ones are called into play. If, however, we locate perception within action mechanisms, this distinction is eroded. Coordinated with behaviour, perception can be seen as an aspect of action-based knowledge that is not subject to the same spatio-temporal restrictions as vision alone. (Cf. the possible relations between spatial landmarks and behavioural inferences introduced in chapter 6.)

The view of development that I shall propose suggests that it must involve both differentiation – of action programs and hence of perception – and enrichment in the sense that it consists of constructing novel action-based representations. On this view, we are looking at developing pragmatic knowledge, and need not propose a discontinuity between perception and objective knowledge. This is compatible with ecological psychology to the extent that its arguments for the validity of pragmatic knowledge, and for information specifying the real world, question the idea of perceptually based knowledge being 'overruled' in favour of something more reliable. Seeming to take account of more than the immediate visual field is not equivalent to knowing how things 'really are' as opposed to how they 'appear to be'.

From a cognitivist direction, a distinctive and significant alternative to ideas of developmental reconstruction comes from the perspective of Keil's (1981a, 1981b, 1985) *formal constraints* framework, which has many similarities to predominantly nativist directions in theorizing about the development of infants' object knowledge. On this view, mental representations are attributed to the infant from the outset, and how they change is the developmental problem. The solution is informed by Chomsky's (e.g. 1965, 1980) nativist and rationalist position that it is only where the mind has no special structure to deal with the properties of a task situation that new acquisitions will have to rely on processes such as trial-and-error, conditioning, abstraction and induction. Chomsky's view of 'natural development' actually involves little by way of development, or learning either for that matter; rather, he speaks of: 'an innate endowment that permits the growth of rich and highly articulated structures along an intrinsically determined course under the triggering and partially shaping effect of experience, which fixes parameters in an intricate system of predetermined form' (Chomsky, 1980, p. 1). While Keil's perspective proves less extreme in certain respects, he argues that development should be seen in terms of the differentiation of *domain-specific* knowledge structures guided by constraints, and that we should not expect qualitative discontinuities.

What is the force of the 'formal' in the claim that there are formal constraints or formal restrictions on the development of natural knowledge structures? Despite its apparent precision, the term's range of application was initially hard to pin down, but it has been clarified during the course of this framework's evolution. The question needs to be answered both synchronically, in terms of the subject's initial mechanisms, and diachronically, in terms of how those mechanisms are assumed to change, in order to clarify in what sense these constraints can be said to be 'guiding' the acquisition of knowledge.

Initially, 'formal' constraints brought to mind Marr's (1977, 1982) idea of an abstract computational theory, which aims to spell out the constraints or principles of operation in the known domain (Rutkowska, 1988). They were defined as: 'formal restrictions that limit the class of logically possible knowledge structures that can normally be used in a given cognitive domain. Constraints reduce such a class toward a limited class of naturally learnable concepts' (Keil, 1981b, p. 198). And making clear the restrictions on the set of knowledge structures that can operate in a domain is what Marr's computational theory purports to do.

But a developmentally useful account of psychological mechanisms is going to need to be at the process-based algorithmic level, and a clearer distinction is now made between formal constraints as properties of an external *task domain* and properties of the subjects' initial knowledge structures that contribute *psychological* constraints or restrictions to development in natural conceptual domains (Keil, 1990, 1991). These psychological constraints are envisaged as *domain-specific*, *abstract*, *structural properties* of the subject's representations, and as playing a role akin to cognitive constraints such as memory restrictions. The central developmental claim is that initial and outcome representations share the same abstract properties, which ensure convergent development despite experience of varying environments.

Sometimes, an abstract form for possible strongly constrained mental representations is spelt out very precisely. In Keil's (1979) work on children's ontological knowledge, for example, categories are assumed to have a hierarchical structure that is independent of their particular content. This structure is said to restrict what spanning relations can hold between the predicates and terms of natural language, and thus to limit childrens' hypotheses about new word meanings, in a form of abductive inference to the best hypothesis. (For example, if predicates and the terms they span are depicted in a tree diagram, they cannot form an 'M' structure, where two predicates span intersecting sets of terms, or a 'W' structure, where two terms are spanned by intersecting sets of predicates.) In other cases of cognitive constraints, however, admissible constraining structures are increasingly thought of as networks of concepts or interrelated principles, whose 'general form' is not precisely specified in an abstract sense.

At both the computational and algorithmic levels, Marr's explanatory framework is predicated on notions of 'input', 'output' and 'intervening process' at a purely synchronic level. Chapter 2's reinterpretation of infant vision in terms of Marr's framework showed how analysis at the computational theory level could offer valuable insights; for example, the computational theory of the 2½-D sketch provided a novel perspective on Spelke's (1988, 1990) principles of infant object perception. So it is worth considering what a developmental version of perception of the physical world might look like from this viewpoint, and what place the innate knowledge of domain constraints that is embodied in the algorithmic level's early/low-level visual processing might have in it.

The most direct developmental analog to Marr's style of analysis would link innate, domain-specific initial-state representations ('input') with outcome representations ('output') through a *special-purpose* process that could itself be viewed as embodying *domain knowledge* ('algorithm(s)'). This kind of 'knowledge acquisition device' is, in fact, close to Chomsky's (e.g. 1965) heavily nativist view of a predetermined developmental pathway (cf. Fodor, 1975). With Keil's (1985, 1990) idea of structural differentiation, the analogy is more subtle, although the implication of a predetermined outcome is maintained. The transformation between input and output representations is thought of as mediated by relatively few, probably inductive, 'feeble' domain-general processes, whose constraining role is 'little or none for natural domains in which domain-specific innate constraints are heavily influential' (Keil, 1990, p. 163). The significant constraint on development is thus the subject's initial state, viewed as a domain-specific mental representation.

For infancy in particular, Keil's position concurs with Spelke's (1988, 1990, 1991) recent views on the development of object perception and physical reasoning in infants. These consider the principles that are believed to underlie infants' object perception – cohesion, boundedness, rigidity and no action at a distance – as constraints in both of the senses that are of interest here: as domain-specific knowledge inherent in initial organization, and as the primary restriction on subsequent development. 'Core principles' are assumed to be unchanging, and development to involve only their *enrichment*, a process with properties similar to Keil's 'differentiation of structure' (and hence *partly* similar to the view of 'enrichment' that direct perception theory rejects, since extending ready-available mental representations is involved rather than constructing ones that are qualitatively distinct or that generate novel domains of knowledge). Therefore, no knowledge revolutions or radical conceptual changes are expected. Infants perceive the world as adults do; and the principles they exploit also underlie the origins of reasoning about physical objects, such as the understanding of their causal interactions.

As we have already seen, this view of development is grounded in mappings between theory and much impressive performance data. Caution is needed, however, to ensure that the question of development's overall direction is not prematurely foreclosed. In particular, it is important to be sure that the formal analysis of mental representations, whether of abstract structures or networks of principles, is getting behind the question of 'how' processing mechanisms operate in an appropriate way.

In terms of synchronic analysis, preceding chapters have already shown how complex is the distinction between accounts of task structure and those of subjects' underlying mechanisms; and how the difficulty of using and maintaining such a distinction can be exacerbated when abstract/formal descriptions are invoked. We saw in chapter 2 how some of Spelke's abstractly stated principles would indeed count as knowledge of constraints on the physical world in a 'strict' Marrian sense – the processes underlying them, and the role they play in computing structural descriptions, could be spelt out, and they fitted the computational theory. But other principles seemed more akin to intuitive descriptions of the computational task/problem that infant mechanisms solve, and required additional clarification of the computational processes underlying them. In chapter 5, we saw the problematic relation between Piaget's Structures and scheme mechanisms. And chapter 3 noted problems that arose with the very possibility of a formal analysis of the everyday task structures with which infants learn to cope, once perception was grounded in action and the transactional nature of action-based abilities was acknowledged. Such issues carry over into developmental concerns.

If the output of the developmental process is domain knowledge – a view whose appropriateness at some level is hard to dispute – it is harder than it first appears to answer even the most fundamental questions such as 'What and where *is* a domain?' and 'What has the infant acquired once s/he has come to grips with one?'

Defining a domain *seems* straightforward when object perception is concerned; it is generally taken to encompass things 'in' the external environment, their properties and interactions. But we need not go far afield in natural conceptual development to find far less clear-cut examples. 'Number', for example, can be thought of as a similarly strongly constrained domain of knowledge, but it is most difficult to locate independently of the subject's activities. For example, Keil (1981b) questioned whether principles such as one-to-one counting that are obeyed by even very young children are either abstract or general enough to be constraints. They might, he suggested, be closely related to 'more powerful, abstract constraints that are yet to be discovered' (1981b, p. 212) that guide the child's acquisition of knowledge about counting. In my view, the reason that this issue arises is because both the boundaries of the number domain and subjects' knowledge of it are dependent upon subjects' activities. Several areas of research now support the conclusion that subjects do not develop a unitary representation of what a non-computational perspective might consider to be a single concept or piece of knowledge.

Sloman (1978) suggests explaining the child's developing knowledge of number as the acquisition of numerous procedures. He develops a conceptual analysis of multiple procedures that might be needed to acquire and perform even as apparently basic a numerical ability as counting: counting forwards; counting backwards; answering questions such as 'What is after N?'; obeying instructions such as 'Count from N'; etc. This type of analysis is essential, Sloman argues, to understanding exactly what it *is* that a child acquires in learning about numbers. Striking support for this viewpoint comes from Lawler's (1981) case study of mathematical understanding. His 6-year-old daughter employed quite different procedures to add the 'same' (from our viewpoint) two numbers if the sum was presented verbally, on paper, or in financial terms (as when the number of cents concerned corresponded to the price of two packs of her favourite chewing-gum). Influenced by computational concepts concerning knowledge and understanding, Lawler (1985) develops in detail the notion that mental development is founded on disparate, active structures that he calls 'microviews'. (In language, compare children's use of different procedures for comprehension and production of the 'same' word (Clark, 1983).)

Even with an artificial puzzle such as the 'Tower of Hanoi', which has a circumscribed state-space that can be clearly formalized, it is difficult to view its recursive structure as external to the subject, and awaiting their discovery. It is intricately bound up with a means–end analysis based on action–environment transactions featuring behaviours such as 'moving' and 'placing' discs that span the dichotomies subject–environment and internal–external. As with the synchronic process of perception, domain boundaries, and the information used for their developmental construction, may be a mutual product of subjects' perceptual and behavioural activity and of the environment that supports it.

None of this is intended to mean, contrary to Keil's (1991) assessment of this position, that a functionalist, action-based alternative to the treatment of domain knowledge in terms of structural properties of mental representations entails abandoning all attempts to contrast or distinguish different kinds of constraint, in favour of collapsing them into a holistic model of subject–environment interactions; or that questions of innate initial conditions are considered irrelevant. It does mean two things. First, carving up significant kinds of developmental constraints along the lines of internal/external or subject/domain dichotomies is, at best, contentious (despite its intuitive plausibility). And second, confounding the aspects of a subject plus environment system's functioning to which proposals for different kinds of constraints are assumed to refer (for example, by inadvertently collapsing them) can produce a misleading picture of what the system is doing.

Keil (1991) suggests that Marr's distinction between computational and algorithmic levels of explanation might well be useful for developmental models, even if a formal analysis is difficult to arrive at. He notes that the computational theory encompasses a functional, teleological analysis of the goals and strategies associated with a computation, and suggests that 'one might well wish to further divide the computational level into functional analyses and analyses of formal structure in such areas as geometry and logic' (1991, p. 241). This seems to me to be the maneuver that accounts of developmental constraints will, in fact, need to adopt, with their focus on the functional analysis. We should bear in mind, however, that this 'unformalized' approach does not provide the crucial element that Marr was looking for in a formal theory: a logical proof that the proposed computation between initial and outcome structures is possible. So we will need to fall back on empirically based arguments concerning how effectively any input–output mapping proposal seems to account for available developmental data.

At this level of analysis, considering infant knowledge in functionalist pragmatic, terms suggests a more open-ended view of the 'task' of development than does the structural mental-representation perspective. For the latter approach, the task tends to be 'to acquire structure X', which entails predetermination of knowledge to the extent that evidence of the same or similar structures is believed to be uncovered in the infant's initial mechanisms. By way of contrast, the action-based perspective comes to look closely at infants' preadaptations for functional activity and at the *ways* functional outcomes may be achieved during the developmental sequence.

A general direction for development is associated with the two synchronic senses in which constraints have been encountered in the preceding chapters. First, there is early/low-level visual processing's implicit knowledge of physical constraints, into which the studies of Spelke, Baillargeon and others appear to be tapping. More significantly, as far as the 'task' of development is concerned, there is the idea of the infant being attuned to constraints or regularities in the world, in the sense of information being explicitly represented (via action programs) for use in action (cf. Israel's (1988) 'attunement' and Dretske's (1988) 'implicit belief'). In this sense, constraint knowledge is embedded in the processes that underlie viable activity in the environment. This establishes the subjective context within which vision's domain-specific innate knowledge of the physical world can operate; and it may be a key locus of developmental change.

It can be suggested that infants are equipped with preadapted processes that satisfy (that is, implicitly embody) some constraints. *What* may develop are new processes for exploiting structure in the world; these processes could be characterized as satisfying further constraints. *How* this development occurs becomes a problem of how one psychological organization is mapped into another. This is a question of how the infant's initial mechanisms, their operation in the environment, and reorganizing principles and processes result in a novel psychological organization. Each of these factors can be viewed as playing a role in constraining – in an 'unformalized' sense – the developmental process and its outcome.

This view keeps open the issue of whether an account of the significant source(s) of constraint on development must choose to attribute it *either* to the domain-specific content and structure of knowledge *or* to general-purpose processes of transition; *either* to elaborate representations *or* to many complex computational routines (Keil, 1985).[2] There may be processes involved in reorganizing knowledge that are domain-general yet specific to natural human learning and quite unlike those proposed by traditional learning theories. This viewpoint, as we shall see later (pp. 196–207), need in no way imply the simple mapping from environment to knowledge structures that Keil so rightly rejects. For there need be no conflict between proposing powerful domain-specific innate constraints *and* significant domain-general innate processes.

What does continuity of Structure mean?

Piaget's constructivist theory and the predominantly nativist formal constraints perspective invoke formal analysis of psychological organization to justify orthogonal views of the relation between initial and outcome psychological organizations in development. For Piaget, Structural analysis supports not only *continuity* within stages ('horizontal decalage'), but also *radical reconstruction* between them ('vertical decalage'). The abstract logico-mathematical level of description is epistemologically useful to him because it enables him to demonstrate Structural isomorphisms between what he considers to be autonomous yet ontogenetically related levels of knowledge. His account of the sensorimotor period claims that the infant first comes to act in accordance with, then to perceive, and finally to represent the 'group of displacements'. For this *ontogenetic* link to be feasible, there must be an explicit relationship between Structure and the mechanisms underlying performance. But, as we saw in chapter 5, it is unlikely that useful links can be established by exploiting the very general formalisms that Piaget adopts.

What, then, do Structures describe, if not mechanisms? Piaget's Structures are descriptions of *tasks*, from an observer's viewpoint, in the *most general* terms possible. The group of displacements, for example, is primarily a formal description of the set of environmental transformations with which the infant is deemed able to cope. Object movements and spatial relations can be described in terms of the group; hence, the infant is said to perform increasingly in accordance with this Structure as performance on search and manipulation tasks improves. This might be considered as akin to Marr's computational theory. But it lacks Marr's (psychologically useful) decomposition of complex information processing problems into a number of functional components, each of which can then be treated as a 'problem' that merits its own formal analysis. Unlike Marr's concern with specific task analysis, Piaget's formal analysis is directed at very general properties that are shared by a wide range of tasks.

Piaget has tended to suggest that Structures describe mechanisms in addition to tasks, but significantly different psychological mechanisms can underlie tasks that are identical in terms of his abstract logico-mathematical descriptions. An example is provided by demonstrations of 'early competence' in logical reasoning, following modifications of the content and/or presentation of tasks dealing with 'transitivity' (Bryant and Trabasso, 1971), 'conservation' (McGarrigle and Donaldson, 1974) or 'classification' (McGarrigle, Grieve and Hughes, 1978). Piagetian and other critiques of such counter-experiments argue that the tasks have been altered so that mechanisms other than those of concern to Piaget are brought into play, and this is extremely likely to be the case (Neilson and Dockrell, 1982; Smith, 1982; Youniss and Furth, 1973). But it is important to stress that, at the Structural level of description, the tasks *are* identical to those used by Piaget.

What this demonstrates is that Piaget's logico-mathematical level of description is so abstract that it can establish correspondence between tasks that may have little or nothing in common in terms of the psychological processes they involve. Piaget (1971) himself noted that Structural similarity did not provide intuitively obvious support for reconstruction; although a flea is a partial isomorph of the moon, nobody believes that one evolved from the other. Furnishing evidence for his ideas about the origins of logical Structures is considered to be the key reason for studying ontogenesis, but the kinds of developmental studies noted above have conspicuous difficulty in offering the requisite support. In fact, the intimate relationship between the content and process even of adult reasoning challenges the view that abstract logical rules are ever a component of psychological mechanisms; they may be better viewed as an emergent property of the operation of those mechanisms (e.g. Johnson-Laird, 1983, 1988).

It is interesting to note that such early competency studies are one source of evidence for what Keil (1981b) calls 'increased access' to a common structure and suggests as evidence (*contra* Piaget) for *continuity* in the development of mental representations. But the same conclusion follows; that is, similar abstract task descriptions need not mean overlapping psychological mechanisms.[3]

Nor need evidence of success on a formally described task implicate psychological restrictions on performance in general; for example, that Keil's (1979) structural M constraint on children's ontological knowledge may reside in specific properties of the task materials employed. Carey (1985) suggests that subjects' pattern of performance is due to the use of hierarchically related materials, not necessarily to any structural constraint on the way children can conceptualize language. Thus, she disputes claims that ontological categories form a strict hierarchy; predicates do not span different ontological types; and challenges the status of the M constraint. Other materials do not provide support for this constraint: it fails to hold for comparative adjectives (for example, of spatial extent); and, rather than being structured to avoid ambiguity, natural languages regularly use single terms for different ontological types (for example, the same word for a country and a people/government).

Attempts to explain infant abilities in formal terms attract the same danger of drawing inappropriate conclusions about psychological mechanisms on the basis of task structure. Valsiner (1987) has argued that developmental psychology is particularly vulnerable to being ill-served by mathematical formalisms, which 'may lead to an increasing confusion in those areas of science where their application helps to change some essential aspects of the phenomena which are being studied' (Valsiner, 1987, p. 234). In the case of perceptual-behavioural action, for example, there is a route to devising a Marrian computational theory by seeing its goal as specifying transformations in initial and outcome information about spatio-temporal relationships between subject and environment that drive action and are changed by it (for example, see Young (1988) on a computational theory of the control of running). Such descriptions, however, would collapse together the different forms of performance that infants reveal in basic actions like lifting objects, and thus appear to offer no route into their developmental significance.

As far as Piaget's Structures are concerned, even if they were accepted as (very abstract) descriptions of mechanisms, they are even further removed from the process models that a developmental account requires. An important point is that they are applicable only at a few restricted points in a mechanism's development. Thus the infant's performance is repeatedly ascribed to *absence* of the group of displacements. This does not tell us anything about what the infant *is* doing at formal *or* process-mechanism levels of description. Nor does it place restrictions on the developmental path in a *prospective* manner. Piaget's preoccupation with *closure* in Structures and systems can be considered a major example of his drawing limited and inappropriate demarcations in the 'space of possible minds'.

As an alternative, characterizing *types* of computational structures and processes may provide a rigorous yet psychologically more profitable route to understanding organization and change, by spanning many of the distinctions that Piaget attempted to make between logico-mathematical Structures and psychological systems. In particular, it supports a unified treatment of his distinction between 'mechanisms common to all subjects at a certain level' and their 'effective realization and operation' (Piaget, 1968b, p. 69). The outline, in chapter 6, of programs that include preconditions illustrated how we can make explicit a program's organizational properties independently of particular applications. (Cf. Simon's (1962) point that the broad structural properties of programs such as GPS might produce phenomena like those Piaget tries to explain in terms of operational structures.) Thus, we can discuss general organizational principles at *any* point in development, and effectively compare different task domains and age ranges – as both Piaget's constructivism and the formal constraints framework need to do.

Action and concept: a discontinuity in experience?

While continuity of Structure has been one key theme in attempts to grasp the direction of development, the intriguing notion that infant development leads to a radical discontinuity in subjective experience is also central. This is captured in Piaget's (1955) claim that the acquisition of object permanence marks a Copernican revolution in the infant's world. His experiential account of cognizance and qualitative discontinuity in knowledge squares with our adult intuitions about how we know things, and still emerges in the infancy literature, whether or not Piaget's views are agreed with; for example, Carey (1991) opposes his idea of 'ephemeral sensory experience' against conceptual understanding. I have already questioned traditional interpretations of adult introspections and intuitions concerning how we 'know' objects and 'refer' to the world; and similar issues are raised by ideas concerning consciousness and its relation to conceptual knowledge.

Butterworth's (1981a, 1981b, 1981c) suggestion that development reveals an ability to overrule perceptual information by knowledge of the objective situation draws on something like the Piagetian view. The type of change he has in mind is exemplified by infants who are able to regain their standing posture after stumbling in the 'moving room' situation, and whose behaviour and emotional responses suggest some form of understanding of the anomaly involved. The increased persistence of object search in the second year is also noted, as if the infant now 'believes' that an object must continue to exist somewhere, no matter how it disappears. Such developments are assumed to involve the acquisition of mediating representations, and of an ability to *reflect* on the information they contain: 'In order to make the transition from a perceptual to a conceptual mode of intellectual functioning, the infant must acquire *self-conscious awareness* and *awareness of the permanence and the individual identity of physical and social objects*' (Butterworth, 1981b, p. 165; italics added).

Introspectively, this view appears plausible, and it points to interesting phenomena. But it is important that experience terms should not be substituted for, or confounded with, accounts of mechanisms. Consciousness, like knowledge and meaning, may be better seen as an emergent property of mechanisms. As with chapter 4's discussion of notions like 'belief', how clear and useful the analysis of such a mentalistic construct is depends on how successfully it is tied in with plausible mechanism proposals.

In developmental psychology, such a mechanism-related approach to experience is found in Mandler's (1984, p. 79) definition of recall in terms of 'accessing (bringing to awareness) a cognitive structure pertaining to a past experience not currently available to perception', coupled with analysis of what type of *retrieval* mechanism would be sufficient to support it. As we saw in chaper 5, it is proposed that such *conscious access* depends on an ability to retrieve explicitly represented concepts and facts (which Mandler equates with a declarative coding of information). Relevant to possible views of the relation between evocation memory and re-presentation, Mandler distinguishes between two kinds of recall. No evidence supports the idea that infants are capable of 'deliberate' attempts to recall, using 'queries' and 'search' to guide the retrieval process. By around 8 months of age, however, they may be capable of 'being reminded', a form of recall where external or internal cues automatically lead to information about something being retrieved and brought to awareness. In the infant's case: 'It may be that such cues are effective only if they are external or part of an action schema' (Mandler, 1984, p. 82).

A similar style of proposal that features a different view of mechanisms (discussed more fully in chapter 8), is found in Clark and Karmiloff-Smith's (in press) account of conscious access to knowledge. Their primary concern is to clarify what mechanisms might be associated with a valid attribution of *belief* and *conceptual knowledge* to a subject (of clear relevance to talk of the infant's 'belief' in object permanence). Performance phenomena suggestive of increasing *flexibility* and *explicitness* of knowledge are highlighted, and explained in terms of mechanism changes arising from redescription of underlying representations.

The significance of such proposals about experience comes from their embedding within a framework of explanatory psychological mechanisms. Looking directly from the mechanism side of the equation, it may be invalid to assume that there is a single major discontinuity in the infant's experience of objects. I noted earlier that Piaget's theory establishes an ambiguous relation between 'concept' and 'representation'. Indeed, representation, in the sense of evocation, may prove to be a side-issue in respect to the infant's increasing success in and understanding of action. The *extent* of cognizance is supposed to be determined by the *degree* of conceptualization that the subject has achieved in the domain concerned. Thus, for the infant, experience of, and belief in, a permanent object is made possible by new mechanisms – whatever the mechanisms underlying conceptualization might be.

The computational perspective is consistent with the view that the infant's beliefs about objects are a function of his or her procedures for interacting with them, not a reified component of mechanism upon which the infant can reflect (self-consciously or otherwise) in order to decide upon a course of behaviour. The pre-permanence infant, for example, might experience finding a displaced object as 'creating it', to the extent that this is the experiential concomitant of a system operating with a restricted range of procedures for interacting with objects. Nevertheless, s/he cannot be said to activate a procedure such as looking to a previous object location *because* s/he believes this will recreate the object. Similarly, the infant may eventually show extremely persistent search for an object no matter how it disappears, but not *because* s/he now possesses a belief in the object's independence which overrules perceptual information. In this respect, the present framework concurs with Harris's opinion that 'any belief in object permanence is no more than an epiphenomenon of the fact that the infant's search strategies will eventually be sufficiently accurate to ensure success on most occasions' (1983, p. 730).

It is unlikely that acquiring the notion of the object marks a distinct developmental discontinuity. The underlying development of understanding in any domain can be viewed as the acquisition and organization of a wide range of procedures, as suggested for the earlier example of number knowledge. It takes many years for children to master even a fraction of the procedures that, taken together, we call the number system. It is immensely difficult to conclude that the child has or has not got a notion of number at any point on this path. If we include professional mathematicians in our analysis, then probably we would conclude that no school-age child has a notion of number. Presumably, there are subjective experiences associated with each type of procedural organization, and the collective experiences that correspond to the child's notion of number will be different at different points on the developmental path. Our 'pass criterion' connotes not that the child now has *the* notion of number, but that the child has *a* notion of number corresponding to those procedures whose acquisition characterizes a criterion level of numerical ability – a level that is, from some perspectives, arbitrary. (The same might be argued for 'consciousness' in general.)

This is surely also true in the infant's development of the notion of an object. From this perspective, the experiential phenomena addressed by the metaphor of cognizance moving from the periphery to the center of action may not be due to a progressive reconstruction of action. They may be the experiential concomitants of restructuring in action programs, such as that which chapter 6 began to describe. Take, for example, infants who know 'in advance' of activity that the 'IS-ON' precondition must be met for the behaviour of the support to be successful for object retrieval. Their knowledge of this constraint on action suggests that their awareness of the situation must differ from infants who simply pull the support no matter what. And they have, to some extent, moved beyond 'practical success' to 'understanding'. But they are still functioning in terms of action.

Action as 'non-trivial' representation

The present chapter has included discussion of two ways of looking at mental representation: Piaget's re-presentation, with its allegiance to quality discontinuity in mechanisms; and the formal constraints view, with its focus on innately determined continuity. Chapter 4 discussed one sense in which mechanisms underlying action-based abilities could be thought of as a natural system of representation. This section explores further action's representational characteristics, and their implications for understanding development.

If we forefront mechanism considerations, we can think about representation generally as the means by which knowledge at different ages and in different domains of ability is 'coded'. This allows for the possibility of different codes, each of which can be viewed as a *naturalistic representational system*. From a functionalist viewpoint, subjects can then be seen as possessing or developing a range of ways for representing the world, each of which offers different possibilities for interacting with it.

As with 'domains of knowledge' (Keil, 1990), the boundaries of 'systems of representation' are hard to draw precisely, and their scope varies from theorist to theorist. For example, as I have already noted, Mounoud (1981, 1982) talks of a stage-like sequence of three major coding changes during development, with the construction of new domain knowledge within a *perceptual* code occurring during the period from 6 months to around 2 years of age. Along similar lines, Bruner (1966, 1970) originally talked of three codes, with infancy as the period where an *iconic* code (involving something like mental imagery) took over from an *enactive* one (where knowledge was coded, metaphorically speaking, 'in the muscles'). Karmiloff-Smith (1984, 1991), on the other hand, alludes to multiple modes of representation operating simultaneously, including *kinaesthetic*, *spatial* and *verbal* codes as examples, and talks of each input system (in Fodor's sense) having its own representational code. Yet another proposal of relevance to infancy involves Leslie's (1983, 1987, 1990) claim that the appearance of pretence during the second year is due to the all-or-none emergence of a *metarepresentational* system, not to advances in object understanding or to the gradual construction of symbols or representation. A special-purpose metarepresentational operator is assumed to be capable of decoupling 'literal' primary (perceptually driven) representations from the input–output relations in which they are normally embedded; they can be manipulated in 'non-literal' ways, without the pretence involved being confused with misrepresentation.[4]

A general working definition, in computational terms, could consider any naturalistic representational system to be a set of programs. Each has its own procedures, which accept certain types of input and are suited to a particular range of purposes.

In principle, there should be no inherent incompatibility between considering knowledge development from the vantage points of the 'domain knowledge' that the subject acquires, or of the 'representational system(s)' that underlie this knowledge; they are simply different aspects of a single knowing system. But it has been suggested here that, in practice, preoccupation with domain considerations directs attention to descriptions of 'what' the subject knows, or comes to know, in an observer-oriented way that runs the risk of reifying knowledge and of divorcing its properties from those of the knowing subject. This undesirable consequence is less likely if representational systems are the focus, since we are forced to keep sight of the process-basis of 'how' knowledge in whatever domain is organized and used.

Taking language as an example, viewing it primarily as a domain of knowledge tends to be associated with a focus on the syntactic Structure of linguistic productions. But if language is viewed from the functionalist perspective of how it is used, the broader context that is established influences what it means for language to be a 'domain' or 'what is known' from the subject's viewpoint. Language is equally a *representational system* and a *mode of action*. It can support selective correspondences between subject and environment that include those traditionally associated with representational ability and the use of conventional symbols; and it can achieve a range of purposive effects in the environment that are different from those within the scope of purely perceptual-behavioural action, whether those linguistic actions are directed at other persons or, as emphasized by Soviet psychology, directed at the self. From the subject's viewpoint, 'what' they know in the language domain is grounded in the computational structure of their procedures for using language (including any procedures whose structure is able to support what is normally considered to be metalinguistic ability).

From this perspective, infant perceptual-behavioural action can be considered the first naturalistic representational system. From the Piagetian perspective, the mediation and selective correspondence established by action programs that rely heavily on perceptual inputs can never be more than *trivially* representational: such context-bound mechanisms *are* knowing; they do not stand for or in place of what is known. Establishing which of these approaches is more useful for explaining and understanding development depends on comparison of how infant action and the abilities of older subjects work.

The Soviet framework provides a useful route into highlighting the line of commonality between functioning and change in infant action and more mature abilities. Throughout its developmental analysis, the focus remains on a functionalist concern with activity, while nevertheless proposing one of the most blatantly stage-like dichotomies between *lower* and *higher* mental processes (Vygotsky, 1978, 1981). In this it shares Piaget's fundamental assumption that perception and behaviour are tied to the environment, whereas higher processes (such as planning and voluntary memory) are considered able to transcend its restrictions.

Two of this framework's features are significant in this context. One is the idea that the development of all higher mental processes involves a process of *internalization* (in which speech plays a central role), whereby initially interpsychological processes become intrapsychological. The other is that higher mental processes and consciousness involve self-produced *mediators* – from a knot in a handkerchief to internal speech – that support voluntary control of activity, in place of the direct control by environmental stimulation that characterizes 'natural' (unconventional) abilities. (Cf. Dennett's (1982, 1985) view of *self-stimulation*, including talking to the self, as a method for establishing new links between functions with a limited number of 'hard-wired' connections, such as the 'computational reflexes' of the sensory and motor modules in Fodor's model of the mind.)

Soviet psychology, and infancy studies that are informed by it, consider adult intervention to be a key source of constraint on the development of higher mental processes (e.g. Kaye, 1982; Rogoff, Malkin and Gilbride, 1984; Valsiner, 1987). A key explanatory construct is learning in the *zone of proximal development*, which can be defined as 'the set of actions that the child can perform when helped by another person, but which are not yet available to the child in his individual acting' (Valsiner, 1987, p. 233). For example, Kaye's (1982) detailed studies of infant–adult interaction (from the first feed to language use) lead him to conclude that the possibility and organizational sequence of many early abilities stems as much from the adult partner as from the infant. He finds Goffman's notion of 'frames' useful to describe regular patterns of adult behaviour that provide the spatial and temporal structure within which the infant develops and displays increasingly sophisticated voluntary skills. These ideas intersect with the Soviet-inspired notion of *scaffolding*, which was introduced in chapter 3 (pp. 78–9).

It is significant for our understanding of action as representation that the extension of abilities through a phenomenon like scaffolding is not only displayed in interpsychological social interchanges; it is also found in the infant's natural transactions with the physical world, where it facilitates the spontaneous (that is, untutored) restructuring of action. What are, from the infant's viewpoint, serendipitous features of their activity's causal embedding in the environment (in the sense outlined in chapter 3) can result in the infant's experiencing an outcome that they did not and could not achieve through intentional action. This is the primary aim of social engineering of the infant's action–environment exchanges: 'What he could previously do only with adult help, he is now able to do unassisted. This fact becomes the basic law in a child's development' (Luria, 1961, p. 2). In the case of 'manual search' acquisition, for example, we saw how the intentional activity of reaching for a cover resulted in unintentional finding of a hidden object.

This highlights a general feature of representational development. A prominent aim of social scaffolding is *marking critical attributes* of a novel task (Wood, Bruner and Ross, 1976). Outcomes achieved by virtue of the causal fit between behaviour and its embedding circumstances can support the infant's acquisition of novel intrapsychological processes that we might associate with 'becoming aware' (at some level) of the significance for action of new environmental features, and consequently being able to exploit those 'critical attributes' in future voluntary behaviour on what is, for the infant, a novel task. Thus, the infant's behaviour towards a cover underlies its becoming a 'perceptual index' for an object's location, enabling the infant to *search for*, rather than merely to *find* out of sight objects. It is along the same lines that patterns of grasping can lead to size becoming a 'perceptual index' of the invisible object property of weight (Mounoud and Hauert, 1982a, 1982b).

Environmental scaffolding is not only part of the 'means' of such developments, but also an outcome. It can be seen as an active principle of organization that is increasingly intentionally exploited by infants (as well as children and adults) to support novel transactions with the physical and social environment. As has already been noted (in chapters 3 and 6) a key property of scaffolding is reduction of the number of degrees of freedom in a task. Socially, this is achieved by restricting the range of environmental structure with which the infant can interact; and it can be viewed as what infants achieve on their own behalf when they extend their range of activity through learning to exploit environmental structures such as covers and landmarks to restrict search, or generally to anticipate future events or actions.

Similar processes can be found in more mature abilities and conventional forms of representation, which, as we saw in chapter 3, involve transaction between subject and environment in much the same way as infant action does. Relying upon procedures that involve recognition rather than evocation memory is a good strategy during infancy, and it remains so. We can think of tying a knot in a handkerchief or reminding oneself to mail a letter when one sees the mailbox as temporary IF-ADDED procedures that rely on recognition. Conventional 'higher' forms of representation may actually *project* new external structures into the environment, such as drawings or words that aid the organization and execution of further procedures. In perceptual-behavioural action, the subject's mediating processes *assign significance* to descriptions of environmental structure, which thus becomes able to serve as an external memory or guide for the execution of behaviour.

Both physical and social environments can and do come to be exploited in this way. For example, Levin and Kareev's (1980) study of 10-year-olds' problem solving at a computer club emphasizes the way these children draw on each other as a 'rich external resource'. In one computer game, the problem is to direct a screen harpoon at a screen shark. A typical episode revealed a division of labour in which one child provided the 'bearing', a second provided the 'range', and a third acted as typist for these suggestions. Everyday situations are frequently social situations, and a comprehensive theory of problem solving will need to include procedures for 'help seeking' and 'help using'. Human infants soon come to exhibit such adaptations, which appear to be species-specific; unlike gorilla infants, they readily demonstrate an ability to use other individuals as agents of action by, for instance, handing a mechanical toy back to an adult for a repeat performance following a demonstration (Redshaw, 1978).

In chapter 4, I argued that the meaning of things for the infant is determined by the structure of infant action and its causal embedding in the world. Preadaptations ensure that some action-based meanings are soon operative, together with processes for extending the range of action. We can view social interventions, such as social scaffolding, as exploiting the organizational form of infant ability to determine the direction in which action is extended. Social and technological cultural tools serve to embed infant, child and adult in a network of culturally prescribed meanings.

Absence of formal tutoring has been noted as a possible criterion for distinguishing strongly constrained knowledge acquisition (Keil, 1981b). But a tutored/untutored distinction may prove unhelpful in clarifying the locus of constraints on development, or how strong the restrictions they impose are, to the extent that tutoring takes advantage of the same principles of organization, functioning and (internally motivated) change that infant action spontaneously exhibits. For this type of reason, Mounoud (1986) has argued that no compelling distinction can be drawn along these lines. Say we ask why children across the world are tutored in reading and writing in an alphabetic system from the age of around 6 years, we find that this is not merely a convention but capitalizes on an untutored reorganization in the child's knowledge.

The type of *adaptive change* involved in such reorganizations appears common across domains of ability and systems of representation considered at widely varying ages. It is revealed in 'natural' acquisitions ranging from object manipulation, through language to reading and writing. There is an interesting sense in which the organizational changes involved reveal important properties associated with 'interiorization' in Piagetian theory, and with socially guided 'internalization' in Soviet psychology; but have no counterpart among developmental proposals that eschew qualitative change, such as strongly nativist accounts that forefront the role of domain-specific innate structural constraints on knowledge representation.

As we have already seen, Mounoud and Hauert's (1982a, 1982b) studies of the infant's pragmatic knowledge of weight revealed three main levels of organization of action. Infants began by successfully grasping objects of any size–weight combination by means of a one-off organization of movements suited to current circumstances. They progressed to a second level, where they appeared to anticipate appropriate behaviour, and errors were made if, say, a heavy object was replaced with a lighter one of identical size and appearance. At a third level of performance, they appeared able to coordinate the two earlier levels of organization, first making errors in the trick-object context, then rapidly compensating for them. It is notable that the second and third levels of organization correspond to the two levels of functioning described for Butterworth and Cicchetti's (1978) 'moving room' study.

Children well past infancy show similar patterns of change in object manipulation. For example, Karmiloff-Smith and Inhelder (1974) presented 4- to 9-year-old subjects with blocks that were either of uniform density or weighted, visibly or invisibly, at one end. Even the youngest child was able to balance any of the blocks, apparently making use of proprioception in order to move them back and forth along the bar until their balance point was found. At a second level, however, they tried to balance every block at its geometric centre, including the weighted blocks. When these failed to balance, they were put to one side and designated as 'impossible' by the children. At a final level, the geometric centre was tried first, but failure with weighted blocks now led to a readoption of the first-level proprioceptive strategy.

The changing performance of the older subjects does not seem to implicate representation in a manner significantly different from that of their infant counterparts. And this is not merely an artifact of particular task content or of the knowledge domain involved. Similar data exist for numerous domains, including a functionalist analysis of the child's changing knowledge of *language* (e.g. Karmiloff-Smith, 1979, 1984), a domain that is associated with representation in the 'strictest' of Piagetian senses.

For example, there is an initial level at which young (French-speaking) children use the term *même* correctly in its two adult contexts (that is, where it demarcates 'same kind' or 'same one'). But the apparent appropriateness of this successful usage masks the fact that the child may simply be employing two unrelated, situation-specific procedures in these two contexts. From the perspective of the child's representation, s/he may not, as an adult might assume, know that *même* is an element of language, whose meaning is determined by its linguistic context and which has two different meanings when used in these two ways. The sentences in which the term occurs are not represented as general patterns or program procedures with variable slots that can be filled by multi-purpose words for particular runs of those procedures. Support for this view comes from evidence for a second level of organization, where a term *même de* is spontaneously introduced to mark cases of the 'same kind'; this term never occurs like this in adult usage, so it gives the appearance of being an error. At a third level, the child again uses the single term *même* in both contexts, but now, Karmiloff-Smith suggests, a single 'plurifunctional' representation underlies it.

Reading and writing provide uncontroversial examples of representation in the sense of understanding and producing externalized symbols, and fit this general pattern. Mounoud's (1986) analysis of the literature on reading highlights the fact that children before the age of 6 years are unable to read in an alphabetic system because they lack the ability *explicitly* to segment speech into phonemes. (Cf. Clark and Karmiloff-Smith's (in press) analysis of the idea that children do not know what a word is until they reach around 6 years of age, as indicated by, for example, their inability to discriminate the number of words in a sentence.) He suggests that there are two aspects to this development, which arise from a qualitative shift in the child's representation of whatever domain(s) s/he encounters. First, there is a general capacity for segmenting wholes into abstract units whose representation has a status different from that of the 'same' elements when they were initially 'embedded' in a larger whole. Second, there is a consequent general capacity to integrate these abstract units into novel wholes, in which their meaning now depends on the part they are playing in the new structure. (Cf. the second and third levels of the preceding language example).

Preceding examples of adaptive change have a number of things in common. Most basic is their support for *qualitatively distinct* levels of organization of what might superficially appear to be the 'same' knowledge. This contrasts with the resolutely *single-level* systems implied by ideas of structurally constrained representations and networks of concepts and principles, wherein knowledge is embedded in the same uniform representation and remains at that level as the structure or network differentiates. On the basis of such domain-system comparisons, Karmiloff-Smith (1984) has argued that it is legitimate to speak of *phases* of representational development. If one considers simply the outcome of task performance from an observer's viewpoint, we find two phases of 'success', separated by what looks like a phase of 'error' or regression. Performance in both 'success' phases would be indistinguishable at a formal-Structural level of description of the domain knowledge they appear to reveal. But, as we saw clearly in the object manipulation examples, subjects with lesser or greater experience in the domain concerned do not achieve their seemingly equivalent success in the same way; and are thus representing differently what looks like 'the same task' to the observer. In the first phase, we can consider the infant's or child's behaviour as achieving 'successful' outcomes in a series of contexts that are (from the perspective of their representation) unrelated as far as critical 'task' attributes are concerned: they grasp a series of objects, balance a series of blocks or use a word in a series of different contexts. Only following a phase of representational change, marked by their apparent regression in terms of successful performance, is there evidence that their representation is, in some way, now relating these individual problems in what is, from the subject's perspective, a novel task domain.

A further generalization, concerning the changing object understanding exhibited in the manipulation tasks, is that the way perception is coordinated with behaviour in the control of action may *change* with the subject's changing representation, but its role is by no means *eliminated* or even necessarily reduced. For example, if we recall Bruner's (1968) subject who overcame difficulty in visually guided reaching by closing her eyes and pouncing in more primitive 'ballistic' mode, it is intriguing to note that second phase subjects in Karmiloff-Smith and Inhelder's block balancing task had no difficulty with balancing otherwise 'impossible' weighted blocks if they were asked to perform the task with their eyes closed. The subtle interplay between context and behaviour in more 'mature' phases of successful performance is reified and rendered mechanistically opaque if we consider the subject to be 'overruling' perception in favour of 'objective' knowledge.

Probably the most significant generalization that can be made about these and similar examples is that changes in the subject's representation, including those that occur within infants' control and organization of action, give rise to abilities that begin to exhibit properties generally associated with *conceptual* functioning in the (technical) sense discussed in chapter 4. We are seeing evidence for increasing *interrelatedness* and *systematicity* of subjects' knowledge, as they appear to develop an *explicit representation* of the conditions that underlie their initially purely practical success across a range of individual contexts. The changes seen suggest an ability, which my earlier discussion of concepts suggested was missing from the innate knowledge implicit in isolated low-level visual processing, to represent a regularity across a range of experienced situations as a *property* that is not only common to those situations but also applicable to others. Changes between systems of representation are not implicated, but qualitatively different *levels of domain knowledge* within a single system are. Domain regularities that are initially implicit in the outcome of subject–environment exchanges are eventually brought to bear on the 'anticipatory', prospective control of action. New mechanisms are certainly not 'content free' in the sense that Piaget sought, but they are applicable in a *foresightful* way to *novel* contents in the specific domain concerned. And we also see increasing *flexibility*, in the infant's ability to shift between two ways of representing the situation (two forms of action) as and when the situation requires.

The second phase of this sequence of representational change also provides evidence of subjects coming to appreciate the *anomaly* inherent in 'trick' experimental trials; that is, that some examples depart from a (newly represented) 'norm' or 'general rule'. (This, as we saw in chapter 1, has been treated as evidence of violation of 'strongly constrained knowledge structures' (Keil, 1981b); here 'violation of newly acquired knowledge of a constraint' would be an apposite description.) Prelinguistic infants obviously cannot label 'counterexamples' to their newly established knowledge of domain characteristics, but Mounoud and Hauert's (1982a) 9- to 10-month-olds certainly distinguished 'trick' trials, repeatedly shaking and closely inspecting objects whose weight violated the size–weight relation their action system had recently come to represent.

Additional infancy examples along these lines could include the appearance of surprise on discovering a toy different from one seen hidden, increasingly accompanied by intense exploration of the trick box used (Le Compte and Gratch, 1972); and 40- to 50-week-old infants orienting significantly more intently to an object's disappearance if it is made to look as if it vanished instantaneously – hence, 'impossibly' – whereas 20- to 30-week-olds behave in the same way that they do with progressive 'normal' disappearances over a range of speeds (Bower, 1967). The type of qualitative restructuring of mechanisms that is being suggested here may be associated with the idea that search tasks seem to reveal the infant's 'sense of object permanence' – their awareness of 'something out there' – getting stronger towards the end of the first year, despite so many measures suggesting that they 'know' a good deal about objects long before that time. As noted earlier, infant's knowledge of their own (standing) posture may also follow this general pattern, and we can note that Butterworth (1981b) assumed a 'conceptual' mode of intellectual functioning must come into play to explain the infant's development from falling over in the 'moving room' situation to a new level where they stumble but their balance is rapidly regained.

This suggests a view of *conceptualization* as a restructuring of representations that is decoupled from the stage theory assumption of an across-the-board change in the subject's knowledge. Essentially, you can have a conceptual level of understanding without your transactions with the environment depending upon what are generally thought of as re-presentational modes of representation. From the viewpoint of subjects' mechanisms, what is involved is a change in representation and in the control of behavioural exchanges with the environment. This qualitative change is domain/task specific, and it occurs irrespective of what system of representation is mediating the subject's exchanges, whether that is perceptual-behavioural action or, say, language. This is not to discount any idea of developmental change in what systems of representation are available to subjects, only to propose that this need not be implicated in the key performance phenomena with which studies of the first year or so of life are concerned (and that understanding the latter is an essential prerequisite to clarifying the former). Although Mounoud (1981, 1982, 1986) generally describes change during this period to a transition from a 'sensorial' to a 'perceptual' system of representation, my view is that both of these levels are encompassed by the functioning and restructuring of action-based representations.

Significant changes in infant object knowledge need not, therefore, be due to a reconstruction of action to a novel kind of system of representation. They may be the outcomes of restructuring in what remains a perceptual-behavioural action system. Talking of infants (or older subjects) acquiring concepts, and of the relations that hold between those concepts, may for some purposes be a convenient way of describing the current content of their knowledge – at an abstract level and from an observer's perspective. It is, however, a description of limited use for clarifying what the subject has acquired from the mechanism viewpoint, or how they managed to acquire it.

'What' is being acquired, as far as the infant is concerned, can be thought of as a representation of *novel constraints* on *successful action*. To acquire, or to become attuned to, novel constraints means constructing new processes whose synchronic functioning can be thought of as embodying principles that support an extended range of viable action exchanges with the environment. For the infant, the increasing range and flexibility of performance that is underwritten by this construction is served by restructuring action programs, whether it is described as a consequence of 'interiorization' (Piaget), 'internalization' (Vygotsky), or the acquisition of a 'perceptual code' (Mounoud).

Infants' changing behaviour gives the impression that their knowledge has become more objective, but this is not because they have transferred their allegiance from perceptual to conceptual knowledge. It is due to their action-based representational system taking on board an increasing number of constraints on viable action. Hence, they give the appearance of getting things 'right', and of understanding what is going on, in more contexts more of the time. The direction of development need not be seen as a move from being deceived and misled to appreciating how things 'really are'; it can be from more- to less-restricted pragmatic knowledge.

'How' these changes occur implicates a domain-system neutral process that can be thought of as a powerful *domain-general innate* constraint on development. This operates at a diachronic level, taking as input an initial state of functioning of a representational system to output a new form of organization that makes explicit synchronic constraints on its pattern of exchanges with the environment. Such an innate constraint may be the most significant answer to the vexing question of what human infants start out with that makes their naive/intuitive physics appear different from the knowledge of the physical world that seems to be implicit in the web-spinning skills of spiders and the technically proficient performance of similar creatures (Keil,

1990). The way in which the earliest abilities work may turn out to be similar for infant and insect, but a human-specific process of change may equip the infant with both an initial endowment and a range of possible outcomes that are qualitatively distinct. Piaget may turn out to have been right in his delphic pronouncement that human infants' development takes longer than that of other species (see, for example, Redshaw, 1978) because they are going farther.

Initial state revisited

The view of the infant's initial state that has been advanced so far suggests two central constraints on development: the preadapted action programs that underlie domain-specific, innate pragmatic knowledge of synchronic constraints on viable action in the physical and social environment; and an innate, domain-general process directed at representational change. Coupled with the idea that one form of restructuring of action programs can be characterized as differentiation (cf. chapter 6), the proposed developmental pathway may look, on the face of it, relatively inflexible and close in its implications to the structural constraints framework. The organization of action programs might be considered analogous to the structure of initial mental representations; and the process of representational change as analogous to abstraction heuristics for mapping between environments and mental representations.

But a closer look at the changing organization of early action reveals that it is more difficult than it seems to specify the action-based equivalent of the infant's 'innate endowment' or 'domain-specific knowledge', and hence innate constraints (structural or otherwise) through which the organization of initial-state mechanisms might be thought to predetermine the development of more mature forms. Two related phenomena are of special interest: what I shall call *substitutability*, and the infant equivalent of 'exotic' behaviours.

The (very predictable) development of forms of locomotion can serve as an example. Detailed analysis of biomechanical contributions to the development of walking have been used as evidence for a phylogenetically determined maturational sequence, with no signiificant place for cognitive factors (Thelen, 1984; contrast Zelazo, 1976, 1984). For crawling, there is also a straightforward picture of predetermination through 'prewired' or maturationally acquired coordinations between specific perceptual and behavioural components. For instance, the age of onset of crawling has been suggested as the crucial variable in predicting performance on the visual cliff. Rader, Bausano and Richards (1980) demonstrated that infants who began to crawl after 6.5 months did not cross, even with very limited experience of crawling. They suggest that failure to cross the 'deep' side of the cliff involves neither avoidance nor experience; it is failure to crawl in the absence of specific visual information. Maturationally determined crawling, they suggest, is based on 'a visual-motor program that guides crawling in response to visual information specifying a surface' (Rader *et al.*, 1980, p. 66).

Such conclusions are complicated by substitutability: neither the perceptual nor the behavioural components of everyday action structures appear to be 'fixed', despite their normally predictable 'maturation'. First, different perceptually provided information can support infant locomotion: infants who crawled before 6.5 months tended to cross the 'deep' side of the cliff, because they appear to use tactile information in place of visual information to guide their behaviour. And second, different motor components can be employed: while refusing to cross the cliff by crawling, many infants nevertheless seem willing to do so while in a baby-walker. The important point here is how easily infants spontaneously learn to execute the behaviours that enable them to exploit a cultural artifact, the baby-walker, for precocious locomotion.

A striking example of substitutability is found in Dennis's studies of locomotion. His work on the apparently negligible effect of cradle-boarding practices on the onset of walking in Hopi Indian infants is a staple of numerous textbooks (Dennis and Dennis, 1940), but little or no attention has been paid to the corresponding studies from impoverished institutional environments. In one Teheran orphanage, only 15 per cent of 3- to 3.9-year-old children were able to walk unaided (Dennis, 1960). Of particular concern here is these children's spontaneous acquisition of a unique kind of locomotion: 'scooting'. This involved pulling themselves into a sitting position, then propelling themselves forward by pushing with their arms. Dennis argues that maturation cannot explain these data: 'experience affects not only the ages at which motor items appear but also their very form' (1960, p. 57). Scooting clearly differs from walking in significant ways, and there is no simple sense in which the action structure that underlies it can be thought of as innately specified. (Cf. rats kept in hot and overcrowded conditions, who learn to sleep hanging by their teeth from the wire grid at the top of the cage. Stewart (1989) notes that the idea of such structure being 'transmitted' by genetic means or otherwise, is most unlikely.)

Such 'exotic' behaviours, and culturally specific ones such as using a baby-walker, seem to me to offer a window on to the kinds of mechanisms with which the infant is equipped for 'normal' acquisition of action structures. They reveal that it is difficult to establish that initial action coordinations share (abstract) structural properties of their mature successors; and that the amazing regularity of development may be far more fragile than it seems. Dennis's conclusion that the 'very form' of locomotion is variable might be questioned if a very abstract analysis is employed; scooting and walking are both ways of getting from one place to another, as are crawling and moving in a baby-walker. But this level of abstraction obscures most of what is psychologically interesting about what is being acquired here, and about the acquisition process. It would be tempting to define the infant's innate endowment for getting to grips with the domain of locomotion in terms of various sensory and motor capacities plus a guiding 'goal' of locomotion (cf. Bruner's (1969, 1973) view of the development of skilled action in general). Then, of course, there are many ways to reach a goal. But that is the very problem – what goal? What kind of mechanism could possibly substitute new (perceptual and behavioural) elements in the action-path to a goal that has never previously been achieved? We must ask how innate endowment and environment interact so that it *looks* as if the infant has the 'goal' of locomotion, which then controls a search for appropriate 'means'.

A first step in making sense of both 'normal' and 'exotic' developments is to see them as evidence for the kind of flexibility suggested by Rumelhart and McClelland's (1986) connectionist proposal for something like a 'test-' or 'confirmation-period' for interconnections between input (sensor) and output (effector) units that were 'prewired' into the human infant when they proved useful during the course of evolution. There is no need to believe that the infant will be 'free to modify or completely reverse any of these behavioural dispositions' (Rumelhart and McClelland, 1986, p. 141) to suggest that the infant's mechanisms can support the acquisition of alternative forms of activity, if evolutionarily specified input–output connections do not prove viable (or additional forms of functional activity even if they do).

This would see the initial level of the development of preadapted, perceptual-behavioural action programs as a form of 'tuning' to the individual's particular environment. A well-documented example of this in another species is Knudsen's (1988) work on young barn-owls' ability to turn to look in the direction of a sound source, which has been suggested as a good example of the role of innate constraints on the development of perception in action (Keil, 1991). Very young owls display a preadapted connection between sound and vision by turning towards the approximate location of a sound source. The system gains in accuracy through *simultaneous experience* of visual and auditory information about objects, and distortions of input result in a predictable pattern of mislocalization (cf. a similar outcome in the pecking behaviour of young chickens – which might be considered an analog of human neonatal reaching – when they wear distorting prisms (Hess, 1956)). Physical implementation of such changes is mediated by the optic tectum, the equivalent of the mammalian subcortical superior colliculus (cf. the neural basis of the type of sensory-motor examples used by Churchland). For human infants, a similar idea concurs with Morton and Johnson's (1991) work on face perception, with its proposal for a preadapted subcortical structure that 'locks' neonatal vision on to approximately facelike inputs, if and when they are available. Such a phenomenon may well account for difficulties in reliably demonstrating certain (purportedly) neonatal abilities. For example, in the case of avoidance behaviour directed at approaching objects, the period in which the ability is hard to demonstrate unequivocally may parallel the period in which the type of directional selectivity processes that could underlie it (see chapter 3) are being 'tuned-in'. (As noted in chapter 2, clear discrimination of directional selectivity is unlikely prior to 1 month).

In an average environment, this kind of tuning can be viewed as setting the stage for subsequent developments that have a more flexible and constructivist direction, such as the kinds of representational change that were sketched in this chapter's preceding section. Thus, in the case of face perception, subcortical and (subsequently functioning) cortical mechanisms should perhaps not be viewed as independent 'modules', so much as two developmentally related levels of organization of domain knowledge.

A second step in making sense of 'normal' and 'exotic' developments is to suggest that preadapted action mechanisms may in fact be organized in terms of the *emergent functionality* that characterizes Brooks's (1990, 1991) idea of layered control (see chapters 3 and 5). That is, some basic programs underlying *independent* perceptual-behavioural coordinations may be in place, rather than the kind of hierarchically organized, goal-directed programs that Bruner (e.g. 1973) has suggested. Steels (1991) has clarified a number of features shared by systems that are organized to exhibit emergent functionality, such as the greater ease with which *new environmental dependencies* can be added, when contrasted with ('brittle') hierarchical control structures. Such features map closely on to requirements that are satisfied by, and/or would be developmentally useful to, the earliest preadapted infant actions:

> systems with emergent functionality are useful when there is a lot of dependency on the environment and it is difficult to foresee all possible circumstances in advance, when the functionality is not so complex that a hierarchical approach is needed to decompose it hierarchically in many different subfunctions, and when robustness is very important. (Steels, 1991, p. 459)

The key gain that comes from putting together these assumptions is the possibility of a system that offers an account of development involving *preadaptation without predetermination*. It could support the 'confirmation', 'disconfirmation' or 'recombination' of a range of preadapted task-achieving behaviours.[5] This meshes with Valsiner's (1987) argument that valid accounts of psychological development need to separate subject and context in an 'inclusive' manner; that is, to view infant action as an *open system*, whose structural organization depends on, and is enhanced and maintained by, its continual exchanges with the environment. Biologists' doubts about 'programming' notions stem from a concern that the idea of structure being 'transmitted', by genetic or other means, is a denial of the development that it purports to explain (e.g. Oyama, 1985, 1988; Stewart, 1989). However, the way classical computational ideas are being elaborated here does not deny the central argument that development is something more than an instantiation of preexisting form.

Scaling up from such basic actions to more complex ones, and the construction of new forms of activity, has been encountered in several guises, all of which may implicate the same underlying mechanisms of representational change. First, there was *differentiation* of basic action programs, exemplified by prehension coming to incorporate preconditions that heralded the extension of adaptive action to new contexts and the beginnings of plan-like control. Next, there was *reorganization* of what look to an observer to be already successful activities, such as the distinct levels of performance found in grasping and lifting objects. Both of these are examples of the restructuring of ready-available action programs. And there are also instances of *concatenation* of perceptual and behavioural capacities into novel forms of activity, of which the 'exotic' behaviour described above is one kind of example. A more 'normal' instance comes from Valsiner's (1987) detailed studies of feeding; in particular, how infants acquire the use of utensils and the eating rules of their culture. This is characterized in terms of initially adult imposed constraints on the interplay between the infant's sensory-motor possibilities (especially motor control) and properties of the environment. The infant's *zone of free movement* (determined by the range of currently available behavioural possibilities and the constantly varying environmental context) is narrowed by various interventions, to push the infant towards a *zone of promoted action* (the desired action or range of actions). For example, a high-chair simultaneously acts as a physical restraint and brings the infant to the level of the adults' table. Special nursery utensils (for example, blunt-pronged forks or cup-shaped juice containers that substitute a narrow spout for the usual larger openning) restrict the dangerous or messy possibilities of their adult equivalents in infant hands, while simultaneously supporting approximations to adult activity. Infants have the behavioural potential to do things like grasping objects and bringing them to the mouth before this kind of tutoring begins, but acquiring the ability to use a socially meaningful tool such as a spoon involves a significant developmental achievement (as everyone who has played the role of 'tutor' will be painfully aware).

All of these kinds of change involve the construction of novel forms of activity, and of new 'goals' for the infant system, following conditions where the infant has been led (by the structure of the physical environment and/or the structuring activity of adults) to experience a new form of 'successful' activity. (Cf. evolutionary examples of changing function, such as the respiratory processes of breathing and panting coming to be exploited for temperature regulation (Satinoff, 1978).) Infants' cover-directed activity comes to be used in order to locate objects, although *search* may not initially be its goal. Behaviour directed at lifting objects is present months before it is directed, from the viewpoint of infant action programs, at lifting objects of variable *weight*. And the 'goals' of exotic forms of locomotion and the social rituals of human mealtime activity are more appropriately viewed as an outcome of relevant perceptual-behavioural coordinations having been acquired than as the motivating force behind their acquisition.

Whether we are addressing the confirmation and extension of preadapted action possibilities, or more novel concatenations of perceptual and behavioural functioning, the *constructive* aspect of this representational development is due to the dependence of its input on the *causal meshing* between infant action and environmental structure; its basis is not simply an *inductive mapping* from the (objective) environment to the changing representational system. Thus, contrary to empiricist thinking, environmental structure is not the significant constraint on the direction of development; but mutual structures established by action exchanges with the environment are. An exploration of *how* patterns of successful activity can alter the organization of infant action can have reciprocal benefits for AI work on learning.[6]

Any psychologically valid model of ability must be capable of adaptive change, and the computational analysis of infancy suggests what kind(s) of change it will need to be able to implement.

Notes

1. My discussion has maintained the use of both 'concept' and 'representation' because it is central to the original theory, and because Piaget continued to stress a qualitative discontinuity between action and later knowledge through ideas such as 'reconstruction' and 'overtaking'.
2. Of course, the formal constraints approach necessitates processes too, though not as a primary source of restriction on the acquisition of new mental representations. Gauker's (in press) philosophical arguments claim that the network theory has so far failed to provide any convincing account of the *abstraction heuristics* that it requires to explain how abductive inference from experience to appropriate mental structures occurs. If it cannot offer an account of the process of development, it is unlikely to be a useful characterization of the outcome of development.
3. Keil's (1981b) endorsement of Fodor's (1972) and Rozin's (1976) suggestions that, during development, preadapted computational processes or subroutines become accessible for new purposes and in new contexts will prove more apposite.
4. As typically 'deviant' metarepresentations, Leslie (1983) includes object substitutions (for example, the shell is a cup); pretend attribution of properties (for example, it is full of tea); and pretend presence of absent objects. This view sees evocation as an outcome of this general metarepresentational ability, not a cause; and the metarepresentational system as the first step in a 'Theory of Mind'.
5. Systems displaying emergent functionality may look like Production Systems. However, P-S often feature far more interplay between their situation-action rules or 'productions', including a central control process that may make use of a database to keep track of where the system is in a particular task (Thornton (1987).
6. Valsiner (1987) suggests 'social canalization' as the key developmental mechanism, but this labels a key source of constraint on certain kinds of acquisition, not details of internal mechanisms involved (cf. Kaye's (1982) similarly socioculturally motivated account of 'frames' that adults establish for infants).

Chapter 8

How does developmental change work?

If perceptual-behavioural action is to be viewed appropriately as a naturalistic representational system in its own right, without assuming that it is reconstructed to a qualitatively distinct type of psychological mechanism during infant developmment, it is necessary to address the question: *how* could *action* develop to give rise to novel performance phenomena, such as goal-directed search, that emerge during infancy? A computational approach to mechanisms of representational change is needed, to establish that actions mechanisms are, in fact, sufficient to explain these acquisitions. But there are methodological problems involved in bringing together the methods and concepts of developmental psychology and AI in order to achieve it.

As Johnson-Laird (1983) stresses, we need theories that both cohere and correspond to the facts. But which 'facts' should those be? I shall suggest that 'microanalytic' studies of representational change, which take behavioural processes seriously, make a unique contribution towards the goal of constructing powerful theories of development. In this chapter, I consider how this approach helps to support the idea of *representational redescription* (Karmiloff-Smith, 1987, 1991) as a mechanism of development, its potential relevance to infancy, and to what extent ideas from the AI literature in machine learning can clarify the mechanisms underlying it.

To look at conceptual links between AI and developmental psychology, I introduce the notion of a *learning program*. What is intended by this is a program that operates on other programs. The idea is not new – Simon proposed it as early as 1962 – but it has not previously been exploited in this context. In our current state of understanding, I think of this notion as something of a 'black hole' concept; that is, it will be employed (metaphorically) as a useful way of mapping out the problem of development: what kinds of adaptive change are involved, and what kinds of computational processes would be required to produce them? Bringing these concerns together through the notion of a learning program makes it possible to draw on and extend the computational analysis of the preceding chapters. Program procedures express the infant system's potential for process, in particular for viable action. Its learning program expresses its potential for adaptive change.

Programmed models of object concept development: data and assumptions

One approach to bringing together developmental psychology and AI might suggest that developmental psychologists have provided data on what infants do at different points in development and AI researchers should now offer theories of mechanisms which could explain observed changes. But both the data that are generally provided by developmental psychology and mainstream AI ideas about learning make this suggestion simple minded.

The issues involved can be illustrated by two attempts to build programmed models of data concerning infant performance on the visual tracking task. This task is highly relevant to developmental concerns because it has traditionally been used to index the infant's re-presentational status and degree of object concept development. And the models are of special interest because it is extremely rare to find competing attempts to model identical data. Hence, comparative criteria can be used to evaluate them. The models concerned help to clarify two closely related points. One is the type of *behavioural data* that we need in order to construct powerful (that is, precise, detailed, useful and psychologically valid) theories. The other concerns very basic assumptions about the strategy we need to adopt in order to get a computational explanation of *developmental change*.

Prazdny (1980) and Luger, Bower and Wishart (1983) offer programmed models of developmental data from the tracking task, which uses head and eye movements ('visual search') to assess infants' object understanding. Its logic is deceptively simple: if infants 'believe' that an object continues to exist when it moves out of sight, they should 'anticipate' its reappearance by turning to look at the opposite edge of a screen or tunnel, just as or before the object becomes visible. Typical data on which the programmed models are based involve infants of about 8 to 12 weeks revealing early competence by anticipating the re-emergence of a moving object at the opposite edge of an occluder behind which it disappears from view. However, they turn to look at the occluder's opposite edge even if the object stops in full view before reaching it; and they visually follow an emerging object whose shape and colour are quite different from the one that initially disappeared (Bower, Broughton and Moore, 1971). In contrast with this, 20- to 24-week-old infants stop following when the object stops, and they track back to the occluder if a featurally distinct object emerges.

Working from an AI perspective, Prazdny (1980) devises a computational theory (in the general, process sense) which is implemented in a working program. His model includes parallel channels for processing movement and feature information, and modular computation of motion, trajectory and features. In contrast with this, Luger, Bower and Wishart (1983) use a computer program to test the (cognitive-developmental) 'identity theory' that Bower (e.g. 1974/82) constructed to explain his group's empirical findings. This proposes that infant performance is governed by a series of conceptual rules, which interpret perceptual information and generate behaviour. Each rule embodies different criteria for attributing identity to successive appearances of an object. The younger infants' rule takes account of an object's place or movement, but not of its features. Older infants use a rule that employs features too.

Elsewhere (Rutkowska, 1987), I provide a detailed comparison of these attempts, and their relative strengths and weaknesses need not be repeated here. Two points are relevant to present concerns: neither model is truly *sufficient*; and neither shows *development*.

As far as sufficiency analysis is concerned, the closeness of the programs' behaviour to that of infants is determined by how concerned we are with the details of the behavioural process. If we adopt – as a standard psychology experiment well might – a 'pass/fail' criterion of whether the programs exhibit a successful or errorful 'search' outcome, then *both* models match up for *all* studies. If we move beyond this to take account of the *microstructure* of behaviour, we find differences between the programs, and discrepancies between programs and infants.

For example, when a moving object stops in full view, the younger infants pause and fixate it for up to half a second, then continue tracking along the now empty trajectory (Bower and Patterson, 1973). But Luger *et al.*'s program does not pause, even though the original empirical study placed great stress on this as evidence that the infant's following is not simply overshooting or an inability to stop head movement. Additionally, both programs 'look back and forth' more than infants do between the now empty movement path and location of the now stationary object. Do such fine behavioural details and this degree of discrepancy matter? I believe that they must, if we are to get beyond superficial theories.

Pylyshyn (1984) distinguishes between weak and strong equivalence of processes – a psychologically valid theory (or computational model) needs to exhibit strong equivalence with human performance. There is *weak equivalence* if two systems show the same input–output relationship. Establishing *strong equivalence*, however, entails showing that this is achieved through the same computational method or process.[1]

To demonstrate strong equivalence between a model and psychological processes, it would be necessary to reveal – at an appropriate, non-trivial level of aggregation or generalization – all of the basic cognitive or representational states that underlie regularities in behaviour. Pylyshyn (1978, 1979, 1984) proposes several ways of using empirical data to tease out these basic components and processes of a system; each imposes an empirical constraint on psychological validity. These include the *relative difficulty* the model and human subject have with a range of problems in the area of ability concerned; the likelihood of making an *error*; and the relative *time taken* for the tasks.

Error data are not straightforward in the infancy context because a great deal of the data with which we are concerned consist of wall-to-wall 'errors' when viewed from the perspective of mature performance. Thus the *likelihood* of an error will prove less pertinent to analysis than will the *nature* or *form* of that error. This meshes with a further important dimension of comparison: *intermediate state information*. A goal or problem solution is rarely achieved in a single step. Subjects' task or problem performance involves various stages of computation. Techniques that go beyond capturing performance on a single dimension – whether that is pass/fail, proportion of trials correct, or time taken – to reveal *intervening states* that subjects pass through may provide a powerful basis for the construction of strongly equivalent, psychologically valid theories. Without such detail, a theory will remain *superficial*: its explicit component processes will be too molar because further necessary and important processing must be contained within them.

Data suited to such purposes are unlikely to be the outcome of psychological studies that are restricted to dependent–independent variable experimental designs and use predetermined criterion measures for assessing performance. For example, Bower *et al.*'s descriptions of infant behaviour on tracking tasks are far more detailed than those of many experimental studies, and seem to move beyond simplistic pass/fail criteria. Nevertheless, the research conforms to the conventional logic:

> If subjects do/do not possess process(es) *p*
> then they will/will not produce behaviour *b* in situation *s*.

Even when the criterion behaviour, *b*, appears to be a fairly complex pattern of behaviour, it generally derives from the experimenter's prior analysis of the demand characteristics of the task. As such, it may implicitly embody a premature theory of what subjects are doing and offer a very selective account of their actual performance.

Because behaviour is not of interest in its own right, criterion measures devised in this framework will rarely offer the precision required for analysis of error patterns or intermediate states. In the tracking studies, for instance, the way in which 'anticipation' is operationalized involves a very molar behavioural measure, which allows many different behaviour patterns to qualify as achievement of the criterion (Bower, Broughton and Moore, 1971). When it comes to comparing such behaviour with that predicted by a verbally stated theory, or generated by a computational model, a genuine problem emerges. There is no way of telling whether an imperfect fit results from deficiencies of the theory or from incomplete, selective reporting of relevant aspects of subject's behaviour.

Of the two models of infant visual search, it is interesting to note that Prazdny (1980), who uses computational concepts, acknowledges the insufficiency of his theory. Also, he bemoans the 'macroscopic' level at which infant visual tracking behaviour has been described. Luger, Bower and Wishart (1983), on the other hand, assume that their program's behaviour is close enough to that of infants. And they assume that quite different behaviours can provide evidence of the same conceptual rule: either visual or manual search offers access to the subject's 'object concept'. While sympathizing strongly with their intention to emphasize the complexity and significance of infant mental processes, the plausible intuition that what subjects 'know' can be expressed through quite different behaviours may be misleading except at a superficial level of description.

Computational ideas have been an important motivating force behind developmental psychologists' *microanalytic studies* of the interplay between action sequences and changing cognitive representations. Microanalytic studies of representational change have much in common with the data-collection techniques exploited by cognitive scientists whose aim is cognitive modelling and who believe that behavioural observations (as opposed to their own intuitions) should provide their data. These rely on subjects' 'thinking aloud' protocols while tackling problems (Ericsson and Simon, 1984; Newell and Simon, 1972), or on their non-verbal behaviours, such as eye and hand movements (e.g. Young, 1976). (Karmiloff-Smith's (1979, 1984) studies of children's problem-solving, which were introduced in chapter 7, provide a good example, as does Mounoud and Hauert's (1982a, 1982b) analysis of 'kinematic' properties of behaviour.)

Microanalytic methods take behaviour seriously and they mesh with two basic assumptions of the computational model of action that was outlined in chapter 3: that of behaviour as an *externalization* of process; and the notion that even a single trial may involve a complex *transaction* between subject and environment, giving rise to intermediate processing states in task execution. The broad aim is to move away from criterion measures that embody the experimenter's view of a task to a subject-oriented methodology. The focus is on distinguishing between outcome and process, and an intimate relationship is assumed between *what* subjects can be said to have achieved and how they achieved it. A primary goal is to describe the range of *behaviour patterns* shown by subjects – that is, *how* they attain an end, outcome or goal. Commitment is to the computational and functionalist view that ('external') behavioural processes offer us access to the subject's ('internal') program for acting in the world.

The type of hypothesis testing characteristic of the experimental method may be prematurely and inappropriately constrained unless a microanalytic analysis has preceded it. The collection and analysis of data through case studies, on the other hand, may be underconstrained, resulting in anecdotally fascinating transcripts of very limited theoretical significance. Microanalytic studies based on individual and group observations offer a principled balance between these two approaches.

Besides their lack of sufficiency, it is especially disappointing that neither Prazdny's nor Luger *et al.*'s model exhibits *development*. Pylyshyn (1978, 1979, 1984) highlights several ways in which developmental considerations can impose empirical constraints on the psychological validity of computational theories. One is that a system's behaviour should be malleable; among other types of change, it *should* be able to cope with new environmental demands and 'learn by doing'.

Prazdny suggests that his model might 'develop' if the calculations of the trajectory component were given access to feature information. But the structure of the model seems to militate against this. These independent computations of trajectory and features are assumed to be executed by hard-wired modular processes, so any new connections between them would have to be mediated by some additional unknown process. Bower (1974/82) proposes that conflict between partially successful rules concerning 'place' and 'movement' is resolved by amalgamating them and constructing a new conceptual rule, which incorporates 'features' as a criterion of object identity. But what mechanisms are involved to 'bring in' features is not addressed. In any case, this new 'conceptual rule' seems to be no more than a redescription of the new infant performance – a further instance of what Pylyshyn (1984) terms a 'generalization stated over properties of the represented domain'. Bower's (1982) expressed hope that AI research may soon provide a precise specification of the rules determining such changes seems over optimistic.

Despite their differences, Prazdny and Luger *et al.* share two fundamental, common assumptions about the nature of change and how to set about explaining it. In my view, both are questionable. The first is that the infant must be trying to 'solve the problem' and that change involves improved performance towards the experimenter-posed goal in the task situation. The second is that we should first model each level of ability encountered. Then, it ought to be possible to work out what processes are needed to change the structures of one level into the those of the next. An alternative strategy, and the one I shall emphasize in the rest of this chapter, is that elements and levels of ability cannot be understood independently of each other. Looking at *behaviour patterns* can aid interpretation of their component elements. And looking at *how* a system is transformed offers unique insights into *what* is being transformed and why.

Requirements for a learning program

To begin addressing computational ideas about how change occurs, I shall look first at classical, PSSH ideas, before considering how they might fit particular phenomena of developmental change, such as those sketched in the preceding chapters and the specific example of changing performance on the visual tracking task. Sussman's (1975) HACKER makes a useful base (ideal) model for considering requirements for an infant learning program. HACKER is not supported by a computational theory in Marr's sense. But it offers a precise process model of improvement in problem-solving ability, and shares with traditional mainstream AI work on *machine learning* a number of assumptions about the nature of change and the mechanisms underlying it.[2]

One of Sussman's important contributions to how a learning program might operate is the notion of 'debugging'. This refers to engineering improvements in a program by locating, diagnosing and rectifying 'bugs' or points at which procedures fail to run smoothly, and it is relevant here for two reasons. One is what Sussman calls the 'virtuous nature of bugs' – to locate a bug, the system must already have *some* knowledge of the problem domain concerned. Of equal significance is the fact that 'bug' is a *semantic* concept, which must be defined in terms of what the system 'knows'. For example, when HACKER fails to achieve some goal, it checks to discover what type of bug exists – such as the unsatisfied preconditions and ordering problems mentioned in chapter 6 (pp. 160–1) – and can express the difficulty in terms of a *mismatch* between the consequences of a plan's operation and the goal it was supposed to achieve.

This (subjective) notion of 'error' is potentially much richer than that of failure to attain what the experimenter assumes is the goal of a task situation (although in HACKER they amount to the same thing). It serves the same theoretical function as the concept of *discrepancy*, which precomputational model-schema theories adopted to account for the developmental role of mismatches between the subject's current cognitive structure and the environment (see chapter 3). Degree of discrepancy, however, is a non-semantic, quantitative concept whose application is fraught with difficulties. Being able to articulate the *type* of mismatch is critical. In this respect, it is worth noting that Piaget's (e.g. 1978a) notion of 'disturbances', which he believes provide the motivating force for developmental change, cannot be equated with experimenter-determined failure. He stresses that there are different types of 'disturbance', not all of which foster development.

One role a learning program must be able to play, then, is detection of *mismatches*. It also requires some way of determining what to do in the light of such detection. (HACKER has program 'critics' which suggest changes of plan when they detect known bugs.) But development, including that of infants, is clearly more complex. There are minimally three important additional requirements for a model of adaptive change in *naturally intelligent* systems.

One requirement relates to HACKER's ability to learn to 'diagnose' the structure of a problem, such as forming an ON-chain, as needing a particular ordering of procedures (see chapter 6). HACKER can also learn to 'recognize' what *circumstances* demand a particular type of strategy for success; for instance, it learns that any stack of bricks that is structured in a certain way must be taken apart before certain goals can be reached. These examples of improvement in skill are possible because HACKER already has an *explicit representation* of conditions in the world that must be met if its activity is to be successful. It has built into it the knowledge that CLEARTOP-'X' is a precondition for picking up 'X', and its learned improvements permit HACKER's plans to satisfy this constraint between its behaviour and the environment. Thus, HACKER illustrates how knowledge of the constraints on successful action can increase a system's power to modify its capacity for 'voluntary' or 'purposeful' activity. But HACKER does not *discover* the significance of CLEARTOP for itself. An infant learning program needs to account for how an action system could gain knowledge of general prerequisites or preconditions that it does not already have, such as the MUST-BE-ON relationship in the behaviour of the support.

A second consideration is that Sussman's system is predicated on a definition of learning as improvement in getting to a goal. HACKER is given a series of problems and gets better and better at reaching the goals specified in them. This is a common enough notion in the AI field of machine learning, and it was shared by both Prazdny's (1980) and Luger *et al.*'s (1983) models of infant visual search. Indeed, Mitchell (1983) argues, from an AI standpoint, that learning is made *unnaturally* difficult unless the subject knows what the goal is. But infants patently do acquire new goals, besides becoming more efficient at achieving old ones. We saw this in the range of examples of normal and exotic acquisitions that were brought together in chapter 7. The operation of a learning program should help to explain how this could occur.

Finally, the notion of a learning program will need to avoid the limitations of considering only 'failure' and 'disturbances' as motivating forces for change. These notions are central to the task-/goal-oriented view of learning found in HACKER. But, as Winograd and Flores (1986) stress, an AI system cannot solve a problem by searching rationally through a problem-space until that problem-space has been created.[3] In the present context, emphasis on 'failure' is questioned by the studies of children's and infants' changing performance on tasks such as block balancing and lifting objects. We have seen that subjects reveal what look like two levels of task success. They begin by being able to balance or to grasp an object; from an observer's viewpoint, they succeed on the task and achieve the goal. Why should they change?

The design of HACKER, in keeping with most machine learning systems, is based on its creator's intuitions about problem solving and learning. Sussman notes that his anthropomorphism in referring to HACKER as 'he' reflects that the system's development was guided by introspection on his own problem solving behaviour, not that he claims the same mechanisms operate in people. Winograd and Flores (1986) make the straightforward yet non-trivial point that programmers can program only what they are conscious of. This suggests an important role for the detailed comparison of psychological data on behavioural change at different ages and in different domains of ability, such as that undertaken in the preceding chapter: through such comparisons we can become aware of regularities that escape our adult introspection.

Representational redescription

Within developmental psychology, a computationally informed view of possible mechanisms of knowledge acquisition is offered by Karmiloff-Smith's *representational redescription* framework. This framework focusses on the middle years of childhood, but, as chapter 7 showed, the kinds of change that it aims to explain have many features in common with patterns of development in infant abilities. My main concern is to highlight its ideas about *how* these changes might occur, especially as they might apply to the restructuring of action programs that I believe gives rise to novel infant abilities.

Representational redescription has been defined as follows:

> It involves an endogenous process whereby the mind exploits the knowledge that it has already stored (both innate and acquired), by re-representing recursively its own internal representations. This process may, of course, be triggered by external constraints but it can also be self-generating. (Karmiloff-Smith, 1991, p. 179)

The essential idea that underlies it is that knowledge can be represented at a number of different levels, each of which has different functional properties and is constructed through a conservative process operating on the preceding level (that is, constructing a new level of representation of knowledge of some domain does not destroy the lower-level representation that it is redescribing). This process is viewed as one of *explicitation* of knowledge; with each redescription, knowledge that was previously implicit at the lower level of representation becomes explicitly available to a higher level of the system and can thus be used in novel ways and/or by parts of the system that previously lacked access to it.

Karmiloff-Smith has devised two main versions of this framework (which differ in the range of developmental phenomena that they encompass). The initial model referred to 'procedural', 'metaprocedural' and 'conceptual' levels of organization (e.g. Karmiloff-Smith, 1984), which correspond closely to the recurring three levels of infant performance that were earlier highlighted for a range of tasks. The more recent model distinguishes an initial 'Implicit' level (cf. the previous 'procedural' level) from three different levels of explicitation of knowledge that are a consequence of its re-representation (e.g. Karmiloff-Smith, 1986, 1987, 1991). In so far as my concern is with representational change in infant action, some of the directions and preoccupations of my own ideas about representational change differ from those of Karmiloff-Smith, so I shall note these in the context of briefly describing the multiple levels of representation and redescription that explanation of her own work with older children is currently thought to require.

The first, *Implicit-level* representations are seen as underlying initial successful functioning in context – the first level of 'success' that was encountered in the earlier account of three phases in action organization. Karmiloff-Smith refers to this as 'I-level' or 'procedural' since the child's domain-knowledge at this point is thought to be embedded in procedural functioning and (computationally) unavailable to other parts of the system. Karmiloff-Smith has noted (e.g. 1987) that her framework does not address how newly effective, functional procedures get organized in the first place, an issue of special relevance to understanding the development of early infant action. This will, therefore, be one focus of this chapter's following discussion; that is, what mechanisms might be needed to get infants' activity even as far as the Implicit level of knowledge organization?

The essential idea behind the first level of explicitation (the E-1 level) is that knowledge that is part of the subject's procedures is said to be redescribed so that it achieves the status of a data structure and can thus be accessed computationally (without actually running the procedure) and related to find its 'common thread'. (Cf. the second phase of action organization, and the three-phase model's 'metaprocedural' level, wherein common elements in successful activity become explicitly represented.) It is at this level that we see 'error' and 'regression', a phenomenon that can be linked with a move to a new representation that supports prior analysis of activity and hence, exhibits the 'brittleness' that is avoided by systems organized in terms of situationally bound emergent functionality. Karmiloff-Smith (e.g. 1987) suggests that the procedure concerned is redescribed in a 'higher-level language'. However, my following interpretation does not assume that this is necessary, provided some 'record' of functioning is available to the system that can provide a basis for comparison and analysis that generates novel organizations of the procedures that constitute the infants' action programs. That record might be in the same kind of 'language' – for example, a copy of the symbol structures associated with procedural functioning in a range of contexts. As discussed in chapter 5, the symbol structures that are thought of as procedures are not only able to generate processes, but can also be treated as part of its data-base to be operated on or reorganized by other procedures.

Karmiloff-Smith's second (E-2) level of explicitation is assumed to redescribe E-1 representations so that they become not only computationally available to processes dedicated to their reorganization but also *consciously accessible* within the system of representation concerned. The final (E-3) level of 'tertiary' explicitation is thought to underwrite translation between different codes or systems of representation, in particular between non-linguistic codes and language.

Neither of these ideas figure significantly in my analysis of infancy. The E-2 level is difficult to map to infant mechanisms, and may not be required to explain the phenomena concerned. In keeping with chapter 7's discussion of discontinuities in experience, I consider that some form of consciousness may be associated with construction of E-1 level representations without additional redescription to conscious access being essential. And the E-3 level's focus on intersystem processing, bringing in language, is beyond the scope of infancy. Overall, the three-phase model and current I and E-1 levels appear to provide an appropriate space within which to locate the intrasystem representational changes needed to account for infancy. It is notable that the computational implementation of these ideas via a connectionist system mainly focusses on the distinction between I-level representations and E-level representations in general, which are referred to as E+-level (Clark and Karmiloff-Smith, in press).

One of the important implications of the representational redescription model for our appreciation of the development of infant action concerns *why* this adaptive change occurs. Karmiloff-Smith stresses that it is not error-driven, and proposes that such reorganizations provide the child with greater *control* over both the environment and his or her own processing. These conclusions converge with ideas about machine learning.

Scott and Vogt (1983) argue that machine learning has paid too exclusive attention to *task-oriented learning* (TOL). A complementary approach should deal with *knowledge-oriented learning* (KOL). KOL can be defined as 'the construction of an organized representation of experience' (Scott and Vogt, 1983, p. 432), and it is not motivated by the attempt to improve performance on any particular task. Scott and Vogt's own model is based on ideas such as increasing its *certainty* about the *possibility* and *consequences* of applying particular behaviours in its (microworld) environment. As with other systems in this vein, it is based primarily on its designers' intuitions and logical analysis. They do argue that children's play is a naturally occurring form of KOL, worthy of investigation by the cognitive scientist, but this is unduly restrictive. Microanalytic studies of representational change offer wider insights into the interplay between TOL and KOL in human interactions with the environment.

Developmental studies suggest that two phases of apparently successful performance, bridged by one of apparent regression, may be a general property of KOL. In chapter 7, we saw how formal analysis of performance would collapse these two successful levels of ability. In cognitive science, two mechanisms are considered formally different, even if they use different computational processes, only if they lead to different sets of possible results (e.g. Dresher and Hornstein, 1976; Marr, 1982). But, as Winograd (1977) has pointed out, this assumption is related to a theorem in computer science stating that any machine with a certain minimal set of mechanisms can perform the same set of computations as any other machine, provided there are no limitations on memory or processing time. These *resource limitations* are important when they result in differences in the properties of the systems being studied and those differences can be related systematically to differences in the representations the systems use. This may well point towards the kinds of (mechanism-based) reasons behind why the successful subject recurrently comes to solve the task in a different way.

Suggestions that fit this line of reasoning come from Draper's (1987) analysis of machine learning. This stresses the relevant question: what is learning for? Two 'motivations' found in AI systems (independently of the particular tasks they learn about) mesh with the ideas and data discussed above, and they are associated with two relevant learning processes.

Inductive summarization involves constructing a general rule (concept or procedure) that captures several already known instances. It results in simplification and data compression. Its advantage is that memory is saved, at the cost of learning time. This process is comparable with one dimension of representational redescription – the first steps in establishing what initially independent procedures have in common.

Inductive generalization involves finding out what class of (similar) problems a method will also work for. It results in the extrapolation of known examples, through analogy and generalization, to similar but as yet unexplored cases. The performance of the second phase of representational redescription is comparable with this 'theory-driven' activity. Inductive generalization confers the advantage of saving memory space by maximizing the number of instances that are potentially covered by a single 'rule'. The cost (apart from learning time) is that generalizations will not always be accurate.

Viewed in this way, the apparent 'regression' of phase two loses the paradoxical quality it holds from a task-outcome perspective. If induction is 'motivated by' simplification and generalization, the accuracy and reliability of its conclusions will not be the only criteria of its effectiveness (cf. Karmiloff-Smith and Inhelder's (1974) 'creative simplification'). Draper notes that complete accuracy is not essential *provided* the system can fall back on some other strategy. For example, Mitchell's (1983) LEX learns to solve integration problems, discovering what sequence of operators to apply for particular types of problem. If a generalization to a new problem proves to be 'false', LEX recovers by returning to its initial method of exhaustive search and recalculation, trying all sequences of methods until it succeeds. This process of falling back on strategies tailored 'on the spot' to a specific context is comparable with the flexible phase three performance of infants and children.

Representational change in classical vein

The classical computational terminology that has been developed and used so far offers a route to clarifying the kinds of processes that are needed to support proposals for representational change, such as the notion of representational redescription, and how they can be applied to the infant in particular. What, for example, would redescription need to achieve? And what would it need to redescribe to achieve it?

In chapter 7, I suggested that the idea of the learner's representation *making explicit* previously implicit aspects of their activity, and of the situations in which it occurs, can be viewed as a change in procedural organization and control that enables them to satisfy novel constraints on successful operation in the domain concerned (pp. 200–1). Subjects came to *anticipate* that blocks should balance at their center, or to take object weight into account *before* grasping. This is highly relevant to the questions that have been posed about how infants could gain knowledge of *preconditions* for action, and construct novel *schematizations* of tasks. As far as infant action is concerned, it can be suggested that a process like representational redescription underlies reorganization at the level of *action programs*, and that it affects the structure of both the perceptually provided information those programs use and the behavioural procedures they control. The goal of this procedural reorganization is greater control over the environment and action in it.

Consider first the case of the relationship between action programs and the perceptual component of the action system. I have previously noted how the infant's embedding in the environment gives rise to *scaffolding* of activity, and can give rise to serendipitous 'success'. If, for example, an infant tried to grasp blocks without regard to their size, s/he would sometimes succeed and sometimes fail. However, in order to be used *voluntarily* in the control of activity, the pertinent information must be made explicit at the action program level.

The nature of the general problem that must be solved can be illustrated through the behaviour of the support. Say, for example, an infant pulls a green blanket and discovers that this brings an attractive red toy within reach. The behaviour succeeds for any object that 'IS-ON' any other reachable object, but this invariant relationship is embedded in a vast amount of situation-specific information about both the environment and the infant's own activity. How is the infant to know that this action brought about an 'interesting' result not because of the blanket colour, or because of the toy's shape or colour, but because the toy was on the blanket? Initially, as we have seen, they do not – when they first demonstrate this ability, they will pull the 'support' even if the 'goal' object is beside or several inches above it.

Three features of the computational mechanisms that underlie action are relevant here. First is the distinction, made in chapter 3, between the perceptual component's construction of descriptions, which make explicit potentially usable information, and that information's availability for use at the level of action. Second is the fact that descriptions produced by the parallel, modular, preattentive processes of the perceptual component will make explicit far more information than action programs seek for, require or use in their operation. Finally, chapter 5 showed how, given a suitably structured action program, the activation and form of behavioural procedures could be thought of as determined by (symbolic) patterns, whose fixed and variable elements were matched against perceptually provided descriptions.

One important way in which a learning program can extend the infant's action potential is by incorporating, into the control mechanisms of action, elements of descriptions that are initially 'implicit' or 'ignored'. To do this, it will require processes for *comparison* of the system's operation on various occasions; and for *variablization*: the construction, at the action program level, of novel, general patterns that simultaneously define fixed and variable characteristics of the *range* of contexts in which particular behavioural procedures operate. ('Variablization' underlies HACKER's ability to learn to generalize available solutions to new problems.) Comparison requires some form of 'record' – what is 'stored'? There is no difficulty in conceptualizing how a program like HACKER, operating with limited knowledge in a restricted microworld, can store a trace of 'everything' to do with environmental conditions, plans, operations and outcomes.[4] In the infant case, we cannot articulate with the same precision the computational structure of the infant's 'knowledge', hence what there is that might be 'stored' for comparison. Certainly, in order to make explicit new conditions for action, infants would seem to require some record of perceptually provided descriptions, even if these were not being 'used' by the action program level at that time.

In the comparison process, elements of descriptions that are matched repeatedly against themselves can be viewed potentially as fixed elements of a new general pattern at the action program level. This does not mean that fixed pattern elements must refer to static situations in the world. Descriptions referring to a type of motion, such as 'looming', or a type of disappearance, such as the 'kinetic occlusion' of the tracking task, could be fixed elements. Fixed elements could, however, refer to static environmental features, such as the spatial relationship of one object to another.

An element of a description that is not matched against itself will be a candidate for variablization; the pattern slot it occupies becomes a variable, which can be bound by a range of suitable objects or properties. Put very crudely a series of instances might include:

[REDTOY IS-ON GREENBLANKET]
[BLUEBOX IS-ON BLACKCLOTH]

The elements REDTOY and BLUEBOX will be candidates for variablization, whereas the repetition of IS-ON makes it a candidate for a fixed element in a new general pattern at the action program level:

[^OBJECT-A IS-ON ^OBJECT-B]

This is close to the level at which HACKER would describe examples and a general pattern. As chapter 3's discussion of human vision should make clear, it is most unlikely that the primitive elements of descriptions from which infants construct new patterns will be expressed in this kind of very molar vocabulary (or, it is worth stressing, that they will be straightforward to express in language-like form). Discrete, verbal labels are simply a convenient way of describing the system's ability to represent and respond to regularities in input.

Contrary to Fodor's (1975) 'mentalese' argument, the idea of inferring general 'patterns' from descriptions that are produced by hard-wired and unmodifiable perceptual processes does not commit the present perspective to the view that all possible information is already 'in' the perceptual component. The organization and explicit representation of information at the action program level involves an 'addition' to the system, and it is functionally very different from its implicit form.

This type of inductive inference would be necessary to explain Karmiloff-Smith's notion that a first level of 'explicitation' enables subjects to discover and mark semantic distinctions in the use of similar yet initially unrelated procedures. The initial comparison process is directed at the positive outcomes of repeated functioning, rather than at successful versus unsuccessful instances. This clarifies why *success* is so important: it enables the inductive process to make explicit how certain outcomes or ends can be achieved. These ideas begin to spell out mechanisms that could explain the 'social scaffolding' perspective's assumption that *experiencing* reaching a goal is a vital prerequisite for learning *how* to attain it. Additionally, they show why *variability* of experience is needed; as noted in chapter 6, variable practice seems an important prerequisite for restructuring and transfer to novel contexts, (for example infants' use of landmarks to recall event-locations, and their early means–end activity).This can now be clarified as avoiding inappropriate 'summarization'; hence, inappropriate 'generalization'.[5]

The (cognitive) processes suggested here begin to offer a potential mechanism for ecological psychology's notion of perceptual differentiation. Making explicit new constraints can be equated with 'attunement' to new information for action, in the computational sense outlined in chapters 4 and 7 – hence, with the 'education of attention'. But the 'record' that supports this reorganization is not the evocational 'memory' or 'image' for interpreting new inputs that Gibson (1979) rejects. It is consistent with his view that: 'An image of the past, if experienced at all, would be only an incidental symptom of the altered state' (1979, p. 254).

The preceding discussion focusses on the relationship between action programs and perceptually provided information, and considers the example of making explicit *conditions* for action. But similar processes could make explicit the general properties of *outcomes* of activity that are attained initially with the support of environmental scaffolding. Considering outcome knowledge in this way could contribute to an explanation of how individual infants can develop different solutions to a 'problem' like locomotion, and exotic yet viable behaviours, without needing to assume that they represent, from the outset, the 'goal' of getting from one place to another.

What of a learning program operating on the *behavioural* component of the action system? Previous chapters have argued that an action system's pragmatic knowledge, grounded in its causal embedding in, and structural coupling with, the environment, owes as much to the procedures of its behavioural component as to those of its perceptual equivalent. Perceptual processes may embody physical constraints to construct a description of the environment; behavioural procedures control processes of movement that must satisfy its physical conditions. There are two ways in which new features might be considered to become explicit at the action program level: change within the structure of behavioural procedures themselves, and change in the circumstances and order in which they are activated.

At present, the first of these is harder to conceptualize, computationally speaking, than the second. (HACKER is fairly typical of machine learning systems in so far as its behavioural procedures or operators do not contain variables; for example, it can execute 'PUTON' but not with differing forces, speeds or whatever.) Contrast this with the conceptualization of the behavioural-motor component of infant action within Mounoud's framework:

> The subject will become aware of the object's weight through the characteristics of his action. When he is holding or lifting objects, his movements will vary in amplitude, rapidity and regularity. Variations in these different parameters result simultaneously from properties of the object and from the motor commands

> initiated by the child. The child has to discover and control the *variations in the characteristics of his own actions* (force, amplitude, velocity, etc.) on the one hand, and the *variations in properties of objects* (weight, height, etc.) on the other hand. It is only by taking hold of objects, lifting them and so forth that the child can explore weight perceptually and process proprioceptive and tactile information as they relate to visual information. (Mounoud, 1981, p. 107)

A range of phenomena point to the question involved: how does the achievement of success on a series of related 'problems' lead to the development of *anticipatory* programming of behavioural processes based on new general patterns for procedural motor control? (In the terminology of the motor skills literature, the three-phase model illustrates development from 'feedback' to 'feedforward'.)

The case of behaviour seems comparable with that of perception. The level-one infant is able to grasp objects of different weights in a data-driven, feedback-governed way. The action program level need not explicitly represent the particular forces that will be needed. For such behaviour to be programmed in an anticipatory way, this information needs to become explicit. Could such a development be expressed in terms of comparison and variablization processes similar to those suggested for perception? These would need to operate over records of the behavioural component's operation, to make explicit fixed and variable elements of behavioural processes. Rather than specifying patterns used to 'look for' or 'respond to' perceptually provided descriptions, they would specify patterns to be used for generating behaviour. But what would be 'stored' as a basis for a 'behavioural' comparison process?

In a sense, what we need here is a computational model of the behavioural equivalent of preattentive vision. I argued, in chapter 4, that action programs offer a consistent level of analysis for behaviour, and that the behavioural processes which action programs control can be conceptualized in an implementation-independent way. But, as I have also noted, this does not discount the possibility of the behavioural component including 'lower' levels of computation, analogous to vision's 'autonomous computational reflexes'. It seems plausible that there is a corresponding level of analysis at which preadapted movement mechanisms, which provide the causal basis for behaviour, have their own computational principles of operation, distinct from the action program level. But we do not have a clear picture of what those principles are. Theorists disagree about the 'units' involved in motor activity, their functional organization and appropriate level(s) of explanation (Kelso, 1982; Whiting, 1984).

Hinton's (1984) perspective on movement supports the view that notions of central control and autonomous peripheral organization may prove to be compatible within a computational model. His detailed arguments draw on mathematical advances in robotics to show how central 'executive commands' could be related to distributed parallel computation concerning the joints of the periphery. While this is applied to a limited case (reaching and balancing), it does point to what may eventually be a promising way of thinking about levels of computational control of the body. In general, however, robotics has not yet provided an integrative framework for motor activity of the type that Marr proposes for vision (Brady, 1985; Brady and Daniel, 1986).

This 'gap' is illustrated in the asymmetry of Harnad's (1992) model for solving the symbol grounding problem. Since the way in which both sensory transducers and effectors mesh with the environment will determine what robotic activities work, both components need to have a counterpart in the kind of 'supervised learning' that he considers necessary. Harnad, however, focusses exclusively on a *sensory category* learning mechanism:

> it is clear that a learning mechanism is needed that can extract invariant features from sensory-projections on the basis of supervised learning – learning in which there is a right or wrong of the matter, and in which the consequences of sorting and labelling incorrectly are sensed somehow so as to guide the error correction process. (Harnad, 1992, p. 81)

A key goal for future analysis in AI as well as in infancy needs to be clarification of the computational structure of this effector side of activity.

Machine learning systems like HACKER are concerned with high-level planning and can make a clearer contribution to our understanding of how a learning program could affect the *circumstances* and *order* in which procedures operate. Say HACKER discovers that a sequence of steps planned for achieving a subgoal (for example, 'ON A B') work equally well for any variables X and Y, it constructs a subroutine ('MAKE-ON X Y'), which is essentially a sequence of procedures for achieving a general goal. The subroutine can be called on future occasions, eliminating replanning of how to achieve a subgoal of the same type.

HACKER's *concatenation* of procedures in this way is predicated on the task-dependent notion that the procedures are components of a plan to achieve a known goal. In the case of extending the potential of infant action, comparison processes might also be based on spatial and temporal criteria. This could offer a different way of changing the *coordination* of procedures with one another, in order to achieve novel accessing rights for 'old' procedures. For example, procedures that come repeatedly to co-occur in a fixed order – each being invoked by particular information in a situation – might become coordinated into a single program or subroutine that no longer requires each of their associated patterns or inputs for their sequential activation. This would produce a 'short-circuiting' of perceptual inputs, which could be interpreted as evocation or re-presentation at the performance level. In the following discussion of 'visual search' development, I shall suggest that this type of change may be particularly relevant to the development of goal-directed action.

To summarize this far, I have suggested that a learning program will need to include processes for detecting and handling (semantic) 'mismatches'; knowlege-oriented learning (in addition to task-oriented learning); comparison and variablization; and procedural recoordination. The similarities that were noted between change on different tasks and at different ages suggest that the learning program could be considered usefully as *domain-* and *system-general* – applicable across representational systems, including (perceptual-behavioural) action, and across domains of knowledge within those representational systems. So far as we know, such reorganizational processes may also be implicated in general learning.

These ideas offer some clarifications of a number of concerns that have previously been discussed in relation to knowledge development. The mechanisms outlined suggest how, as Kaye (1982) has argued, contingencies or regularities in infant behaviour acquire the status of 'rules' underlying the infant's awareness of how s/he is expected to act. And they highlight the kinds of processing mechanisms that Valsiner's (1987) notion of 'social canalization' requires. In the same vein, the notion of making explicit information (provided by preattentive vision processes) at the level of action programs is compatible with Dennett's (1991) view of the relationship between mediation techniques and voluntary, conscious activity. Acquiring such techniques, he suggests, is equivalent to implementing a von Neumannesque virtual machine (that is, a serial central processor) on a hard-wired, parallel computational system.

The notion of a domain- and system-general learning program offers an alternative to Piaget's 'interiorization' account of why supposedly general Structures are often far from general in their range of application *within* a developmental stage, yet similar forms exist *between* stages. Within the sensorimotor period, there may be no single 'stage-like' development of an 'object concept', but a cumulative restructuring of action in numerous areas of activity. The same may occur for other naturalistic representational systems. Structural isomorphism between 'stages' may be a consequence of similar organizational developments within the representational systems that characterize them, not evidence of reconstruction. These abstract Structural correspondences reflect the fact that the learning program will make explicit very general spatial, temporal and causal constraints that must be satisfied by any *adaptive*, *embedded computational* system.

Connectionist insight?

A problem for 'classical' AI machine learning systems is that their learning is restricted by the description language that they use to describe the examples they are offered (that is, the vocabulary of elements and possible relations in which inputs are described and examples stored) and its fit to the task domain. Draper (1987) has suggested that PDP learning networks could be characterized as computing a new description language, since they adjust their units in a way that enables them to discriminate hidden regularities about which they have not been 'told'. Thus, they can 'discover' more basic atomic descriptive dimensions behind a pattern of data.

This makes such systems look ideal for the role of implementing redescription of representations in a way that can make explicit new information and generate novel forms of organization. Indeed, their growing range of applications sometimes includes learning to categorize examples within a domain whose properties are too poorly understood for a classical rule-based system to enjoy any significant degree of success. Churchland (1989a) discusses this as the 'pay-off' of a PDP form of organization:

> There are various procedures for adjusting the weights so as to yield a network that computes almost any function – that is, any vector-to-vector transformation – that we might desire. In fact, we can even impose on it a function we are *unable to specify*, so long as we can supply a modestly large set of examples of the desired input/output pairs. This process is called 'training up the network'. (Churchland, 1989a, p. 202)

The limitation of this approach is that it not only looks too good to be true but probably is, as far as the requirements of the kind of developmental account that is being discussed here are concerned. Unless we agree to move in the reductionist direction of 'explaining' how such a system is working purely in terms of primitive connectionist concepts, we are left with little more than an example of material instantiation. In any case, three issues concerning connectionist systems suggest their 'learning' ability is still less powerful than it needs to be to tackle human development.

The first point at issue is that a good deal of the organization revealed by impressive examples of self-organization in connectionist systems may lie in the experimenter's preselection of the set of input examples from a domain – whether 'modestly large' or not. Say, as in many training studies, a network is presented with examples from a number of sets, each of whose members is characterized by a fixed number of features; each set will have distinctive features, some but not all features will overlap between sets. With training, the internal structure of the network comes to reveal a separation between the location of the network's mediating hidden units that are associated with the features from each set; the more distinctive features may be further apart than the overlapping ones; and the network becomes able to 'recognize' which set any new example falls into via the location to which it is 'allocated' by its particular feature combination. How impressive is this? And how similar to the human case? van Brakel (1991) suggests that it is neither of these. The number of features that are typically used is remarkably small, and the features are highly selected/structured by the network's trainers, in contrast with the apparently limitless choice of inputs that might be available in a 'real' environment. He suggests that such systems are acquiring a category system that is defined by their trainers, not one that is of their own or the environment's making.

This parallels one of Pinker and Prince's (1988) objections to the explanatory status of a typical connectionist system due to Rumelhart and McClelland that succeeded in acquiring the past tense on the basis of training with a set of linguistic examples. It exhibited levels of performance similar to those of developing children – for example, discrimination and overgeneralization of the regular *-ed* ending – without having built into it the kinds of rules through which grammatical knowledge is often described or any constraints on learning approximating to the kind of innate knowledge with which infants are increasingly being credited. However, Pinker and Prince suggest that inspection of the example set shows how significant changes in the structure of inputs during training may underlie what looks like the spontaneous discovery of a 'hidden' pattern in the data. The experimenter-selected input examples shift from (a small set) of mostly irregular verbs to (a much larger set) of regular verbs. Thus, any endogenous change in the system plays a minor role in its changing 'representation' of the verb domain; its overgeneralization stems almost entirely from reflecting the highly structured inputs with which it is presented – the kind of 'simple' environment to representation mapping that seems implausible for human development.

A second limitation with this style of learning mechanism relates to a general difficulty, raised in chapter 3, concerning 'pure' connectionist systems: that typical networks may be better viewed as components or fragments of a system whose global functional decomposition is more usefully understood in terms of classical computational structures and processes. Support for this view comes from Norris's (1991) attempt to train a network to perform a simple date-calculation task of reporting on which weekday a particular date falls. Idiot savants achieve remarkable success on this; and it seems they must do so by inducing underlying regularities in the calendar since their explicit knowledge of its regularities and their mental arithmetic ability tend to be poor. Norris used a back-propagation method (that is, giving the system explicit feedback concerning its outputs' matches to the successful solution) to 'supervise' his network's training on 20 per cent of dates sampled over a period of 50 years. The network exhibited no ability to 'discover' underlying regularities: 'The net was trained many times and it always failed to generalise. It is not that it did not generalise much. It didn't generalise at all' (Norris, 1991, p. 295).

The problem for the net, with its input, output and mediating hidden units, is that the simplest algorithm for solving the task involves three basic steps, each of which is a simple, table look-up, procedure on fixed relations between days, dates and years (for example, step 1 is to determine the day on which each date falls in an arbitrary month of an arbitrary year). Things are no better if a single layer of hidden units is supplemented with additional layers (in fact, its learning of specific instances gets slower). The only connectionist solution that Norris found was to train three separate nets, each devoted to one step of the algorithm. After the first net was trained on step 1, its learning was switched off while the second net was trained on step 2 (to prevent the first net changing its connection weights and 'unlearning'), and this procedure was repeated to train the third net. By this method, each step 'learned', and the three nets functioning together were about 90 per cent successful. But its mode of success indicates a significant pitfall of large unstructured networks – even with feedback, they do not learn: 'The network generalises because it was told about the structure of the task and the form of the algorithm' (Norris, 1991, p. 295). Unstructured connectionist networks thus seem to have little chance of spontaneously coming to display even the basic planning skills involving preconditions and subroutines that infants are able to acquire. Rudimentary as infants' plans may be, they still involve organization of component activities and their management across time.

Finally, much connectionist thinking suggests a preoccupation with TOL that diverts attention from significant sources of constraint on KOL revealed by developmental studies. The kinds of learning that have been implemented in connectionist systems have led to their performance being criticized as both under- and overconstrained in terms of the processes they employ; for example, contrast Partridge and Paap (1988) and Minsky and Papert (1988). Their algorithms range from designer-tutor restricted 'backward propagation' to the untutored, unidirected learning of the 'Hebbian learning rule' (see chapter 1). But, as we saw in chapter 7, the tutored/untutored distinction, focussed on environmental feedback, is of limited relevance to understanding the endogenous process that operates on subjects' representations of their own apparently successful activity.

Clark and Karmiloff-Smith (in press) offer some suggestions for implementing an appropriate endogenously motivated process of representational redescription in connectionist systems. Of interest here is the distinction between, and possibility of combining, *first-order* and *second-order connectionist systems*.

As a typical example of a first-order system, they consider Sejnowski and Rosenberg's NETtalk, which takes text as input and outputs speech (that is, 'learns to read aloud', according to its trainers). This is characterized by exhibiting example-driven learning, through which it becomes 'trained' to statistical regularities between inputs and outputs. The only 'rules' it comes to know are implicit, emerging from interactions between its units. We might describe some of NETtalk's successes in text-to-speech transformation as coming to 'know about vowels', for example, but this abstraction concerning what its hidden units are doing is nowhere available to the system itself. Such systems, Clark and Karmiloff-Smith suggest, provide a basis for the acquisition of example-driven, I-level knowledge. However, unless the problem domain stays stable, they are limited by being unable to change strategy without retraining, which would destroy the organization they have built up (cf. the preceding discussion of Norris's program).

Flexibility in the form of 'systematically manipulable components' that support example-independent generalizations might be achieved via a second-order connectionist system. Far fewer examples of candidate systems exist for this category, but a possible contender is Mozer and Smolensky's (1989) *skeletonization* technique. This is able to use a procedure that operates on a (first-order) network that has already been trained successfully. The procedure operates to identify the input and hidden units that are most important in determining performance, keeps them and deletes those that are least relevant.

The 'skeletonized' network proves to be better at generalizing, learns and its pared-down structure can be described 'in terms of a small num... rules instead of an enormous number of parameters' (Mozer and Smolensky, 1989, p. 15). This, of course, does not yield a conservative redescription of the original net, but Clark and Karmiloff-Smith suggest it might have the requisite properties for explaining children's 'explicitation' of knowledge if the skeletonization operated on a copy of the original rather than on the original itself.

Two observations about these ideas are pertinent here. One is how much what skeletonization is doing in a connectionist network looks like the novel pattern construction via classical comparison and variablization that I adopted above as a way of characterizing the process through which the consequences of activity might yield novel general patterns for perceptual-behavioural control by action programs. This suggests potential compatibility between PSSH and PDP views of computation, with the learning algorithms of the skeletonization network seen as an implementation of a classical learning program.

Second, work suggests caution about the extent to which first-order connectionist systems can be endorsed as a possible basis for the acquisition of procedural, I-level knowledge. Sensory-motor coordinations that compute a single input–output function, such as reaching to a location or turning to look at the source of a sound, would seem plausible candidates (cf. the examples of (subcortical) 'tuning' discussed in chapter 7). From a classical perspective, however, even the procedural or I-level representation underlying an ability like drawing a house (Karmiloff-Smith, 1990) involves the coordination of multiple procedures. Once temporally extended abilities involving sequences of coordinations are involved, it seems highly probable that more structure must be imposed on the untrained system and its exchanges with the environment even to reach novel I-level functioning (an issue pursued in classical vein in the following section).

Constructing the problem of object search through action

This section brings together the learning program ideas sketched above and the notion of action program restructuring to sketch a possible answer to the question of how a schematization or general understanding of a situation such as a search task could develop.

My approach to the tracking task is microanalytic in the sense outlined previously. I ask not 'When can infants do this task?' but 'What do infants do in this situation and what can this tell us about what coming to do the task entails for the infant?' (Rutkowska, 1983, 1988). The tracking task is of particular interest because it involves a *transaction* with a complex yet highly structured event. A single trial includes a sequence of visible object movement; object disappearance ('kinetic occlusion'); invisible movement;

object reappearance; and, again, visible movement. But, unlike the manual search task considered in chapter 6, the locations of the disappearing object and of the reappearing object (from an adult viewpoint, the 'goal' to which search should be directed) are not coextensive.

For the Piagetian approach, this is a visual analog of the manual search tasks that index the infant's ability to represent a 'goal' object in its absence. Here, I shall extend chapter 6's argument that the notion of acquiring a concept in the sense of a 'mental object' or 'image' does not offer an appropriate characterization of the developments involved.

From the direct perception perspective, Gibson's (1968) analysis of object understanding proposed that kinetic occlusion is a higher-order invariant that specifies the continued existence of an object that has disappeared. He noted that sensitivity to this invariant might be a developmental phenomenon. Following the arguments of earlier chapters, I shall suggest that development of action forms the basis for this perceptual learning. This touches on chapter 7's questions for ecological psychology, of how the 'differentiation' of perception works, and the complementary nature of affordance-effectivity construction.

For views of infancy that see contemporary data as evidence for an innate knowledge of objects, such tasks present a paradox. Thus, Baillargeon (1986) suggests that infants' inability to exhibit 'visual search' across the full range of situations so far tested means that this task is not a test of their 'belief' in 'object permanence', rather it involves building 'expectations' about regular spatio-temporal sequences. This is probably close to what is going in this situation. For the perspective developed here, however, such explanations are not alternatives – tracking tasks are tapping infant's developing pragmatic knowledge; hence, their implicit protobeliefs about objects.

When infant behaviour is described microanalytically – in terms of *behaviour patterns* describing the sequence of looking and fixation during each trial – at least 10 patterns are needed to capture infants' performance on a basic tracking task. These patterns subsume the criterion measures that have been used by various researchers, and they make clear what an immense amount of information relevant to interpreting the infant's understanding is lost if data collection is restricted to predetermined criterion measures.

Several behaviour patterns characteristic of 3-month-old infants involve failing to watch the object approach and disappear into the tunnel used, or failing to refixate it when it re-emerges from the tunnel. Paradoxically, however, these young infants also show patterns of *continued tracking*. As the object they are following disappears, their head and eye movements continue, as if the object were still visible. Needless to say, they have sometimes reached the object's exit point as it emerges, and they continue following it as it moves away from the tunnel. This pattern can be labelled *Type-I anticipation* because it fulfils the operational criteria of 'anticipation' that are conventionally used on this task. However, we do not see the increase in such behaviour that most theories would predict. Instead, it shows a linear *decrease* in frequency with age. This cannot be explained in Piaget's (1955) terms. There is no 'extension of acts of accommodation' in the sense of progressively increased extension of tracking movements. In fact, this functionally adequate Type-I anticipation appears to fade out. This also suggests that Prazdny's (1980) AI model of early anticipation assumes far more object knowledge than the data warrant.

Considering the behaviour of these young subjects in the context of later developments does not support the view that Type-I anticipation is motivated by *search* for the object or predictive *interpolation* of its trajectory. Nor is it evidence of inability to arrest head movement; it is a *behavioural* phenomenon, not a purely *motor* one. Infants bring to the situation a basic yet viable action program for exchanges with moving objects (it is not, however, the same action program that we shall see after development, even though some of its motor components are 'reused'.) Initially they are equipped simply with an action program whose behavioural procedures cope with visible, moving objects. This provides the infant with an essential 'interpretation' of the situation, without which further development could not occur. For the infant, however, object disappearance is not yet described as a *problem*. Observed behaviour is controlled not by what is present in the environment, but by the way in which the system's descriptions make information explicit and by the behavioural procedures that action programs can coordinate with it. To change behaviour when the situation changes (the object disappears), the change must be noticed, the situation redescribed and alternative behavioural procedures activated. If the information in a new situation is unavailable to the system at the action program level, two things might occur: it might do nothing (look away on the object's disappearance), or maintain the current interpretation of the situation (continue tracking to Type-I anticipation) – as 3-month-old infants do.

The behaviour pattern most characteristic of 6-month-old subjects – as contrasted with their younger and older counterparts – involves continuing tracking as the object disappears and moves through the tunnel, as if making a Type-I anticipation, but then turning the head and eyes back to the disappearance point. This *backtracking* pattern is of special interest. For criterion-based experimental studies and Bower's identity theory, it demonstrates that the infant has noticed a change in the identity of the emerging object – based on change in movement or features – and is looking back to the disappearance point where the original object was last seen.

But backtracking seems characteristic of this age group, even when a single object moving at a consistent speed is involved. It occurs primarily at the beginning of sessions, following which it is replaced by increasingly persistent fixation of the disappearance area. In the most persistent form of this pattern, infants intently fixate the disappearance point through the entire period of invisible movement, until the object re-emerges. Contrary to Piaget's argument that 'action-dominated' infants often return to the position at which an object was found, the infant increasingly concentrates on the place where the object is 'lost'.

This offers an example of how microanalytic investigation of *changing* behaviour patterns supports novel interpretations of what subjects' performance means. Performance levels that are isolated in describing changing behaviour may have different functions in the developmental process. The prevailing view, which was illustrated by Prazdny's (1980) and Luger *et al.*'s (1983) perspective on the tracking task, is that the behaviours characteristic of a given level reflect *qualitatively* distinct representations, strategies or hypotheses for tackling the *task* concerned. So, backtracking reveals object representation changing to include 'features' as an identity/search criterion. But some regularities may demonstrate 'errors' due to no interpretation of the task or situation being available to subjects (cf. Cromer, 1980). More significantly, certain levels may reveal characteristic *transitional*, exploratory patterns in the subject's construction of their understanding in the domain concerned. At this point, I suggest, the infant who exhibits backtracking or preoccupation with the disappearance point is not 'searching' at all.

Continued tracking is not motivated by 'search', and backtracking is not motivated by 'failure' to find the object. The infant's backtracking to the disappearance area can be viewed as evidence that the system detects and attempts to handle a *mismatch*: the absence of the input appropriate to visual following. The consequences of this are that a *redefinition* of the situation is needed, and this can be achieved by a strategy of checking what other information is available among the descriptions constructed by the perceptual component. The most recent information is the description of an event (the object's kinetic occlusion) in the disappearance area, to which the infant now attends.

Even in the arid laboratory situation, how is what is attended to limited? Why, for example, does the infant not look at an adult in the room, or at the video-camera? (3-month-olds, it should be noted, very often do.) As we saw in chapter 7, one effective way of limiting attention is by *social* means (cf. Kaye, 1982; Rogoff *et al.*, 1984; Valsiner, 1987; Wood *et al.*, 1976), distraction reduction (by removal of irrelevant or distracting cues) and marking or accentuation of a task's critical features being key features of the social scaffolding of activity. But another means is through inbuilt processes that impose spatio-temporal restrictions on what connections can be established. The well-known limits on infant conditioning can be interpreted as evidence of such restrictions. I do not suggest here that conditioning is the sole general-purpose learning mechanism through which development occurs. What I do mean to imply is that the reason why there are restrictions on what organisms can be conditioned to do is because processes for conditioning may be one essential component of a learning program directed at KOL and the restructuring of action.

We can view the infant's increased attention to the object's disappearance point as a *self-generated* process of marking the critical features of a novel and changing situation. It is important to note how the joint structure provided by the environment and the action program in operation at the time of mismatch detection will restrict the range of possible information to which the infant will attend. This suggests a systematic and action-related way in which new information can become salient; that is, relevant to action. As has been mentioned previously, repeated exposure to information is not sufficient to make it significant to the subject. The backtracking phenomenon most certainly is not a progressive extension of action in the Piagetian sense, but it may mark an important first stage in action program differentiation through restructuring.

Most distinctive to older, 9-month-old infants is a pattern which can be called *Type-II anticipation*. This is quite different in structure from Type-I anticipation, although they will be classified as identical if only outcome criteria are considered. In Type-II anticipation, the infant watches the object's disappearance, briefly breaking the previous tracking pattern to do so. After this, s/he turns to the re-emergence location and waits there until the object again becomes visible.

The development of Type-II anticipation can be viewed as a second type of application of a learning program: a *recoordination* of initially independent procedures which makes possible a new form of transaction between infant and environment. A striking aspect of infant behaviour is the way in which they get faster and faster at refixating the emerging object across trials, yet often without ever showing Type-II anticipation. This supports the view that they are not yet interpreting the situation as one of search for a particular object, in which the object that disappears is linked with the one that is found. Initially, the infant's 'finding' behaviour is *scaffolded* by the spatio-temporal sequence of transformations in the environment. Visual attention to the disappearance area and visual orientation to the reappeared object involve independent procedures whose activation and ordering is determined by the occurrence, sequence and location of environmental events. As with 'social scaffolding', the infant is led to experience *achieving the goal* without actively *seeking it*. The repetition of this particular sequence provides the basis for a recoordination in which detection/description of movement in the reappearance area ceases to be necessary for activation of turning.

At the level of perceptual-behavioural action, this establishes an ability to infer or predict an event (the reappearance) from the newly salient information to which the infant attends (the disappearance). This activation of turning, *prior* to an object's appearance in vision, can be seen as a form of *behavioural inference*. It can also be seen as an example – at the *mechanism* level – of the 'increased access' to a common structure with which Keil's (1981b) differentiation model is concerned (see chapter 7), in the sense of a mechanism component being 'reused' for a new function. The movement patterns initially controlled by a preadapted procedure for refixation have acquired a new role under the control of a procedure for 'anticipation'. Additionally, the infant's program for handling visible moving objects can be seen as differentiating to take account of their disappearance and reappearance.

Two questions must be posed concerning the interpretation of this sequence so far. First, the notion of detecting a mismatch and seeking alternative information to redefine the situation seems a basic requirement of a learning program, as the previous discussion suggested. Why, then, does backtracking not occur in 3-month-old subjects? One possibility is that a change from predominantly saccadic to smooth visual following is involved. This brings up the distinction between visually elicited and visually guided forms of movement control in perceptual-behavioural coordination, which was introduced in the discussion of manual search. Saccadic following – like early reaching – involves a visually elicited form of perceptual-behavioural coordination. The location of an object in the periphery of the visual field invokes an orienting response of turning the eyes to that location. In contrast with this, smooth following involves an ongoing coordination of behaviour with information on the speed and trajectory of object movement. Unlike saccadic pursuit, it is sensitive during its execution to ongoing or *current* motion parameters.

That change is not inevitable when available procedures fail to be applicable is clearly illustrated by Hess's (1956) prism study with young chicks. They do not alter their direction of pecking, despite failing to reach any grain (cf. the role of food rewards in young barn-owls' auditory-visual coordination (Knudsen, 1988)). The kind of action organization on which such creatures are reliant involves visually elicited coordination of perception and behaviour. For the human infant, however, we can suggest that we are seeing the effects of a learning program mechanism that provides the opportunity for relating *new information* to action. In the context of the representational redescription model discussed above, the infant of this level might best be envisaged as on the threshold of level-one performance, as far as visual search is concerned – about to begin the process of acquiring new, successfully functioning procedures.

The second question concerns why recoordination to Type-II anticipation should be common by 9 months of age but scarce among 6-month-olds? The phenomena certainly suggest that reinforcement of turning is not sufficient for it to develop. This age may constitute a lower boundary for the operation of a learning program in this way. It is notable that this is the age at which otherwise widely varying perspectives locate a new level of infant organization; for example, Kagan (1972) suggests that *maturational* changes within the nervous system make *hypothesis testing* possible. Mounoud (1982, 1986) also proposes maturational changes after 6 months. Zelazo (1984), concerned with the development of walking, considers 9 to 12 months as a period of *cognitive metamorphosis*, which he attributes to 'increasing processing speed'. Flavell (1985) suggests that the onset of 'fear' on sight of the visual cliff, towards the end of the first year, is linked with *conceptual* mediation based on maturational changes. The recoordination notion suggests that the infant is still very tied to the current context – not 'thinking' as Kagan argues – but it gives some idea of the type of change in psychological mechanism which may be viewed, at a neurophysiological level of explanation, as maturational.

We know very little about the neurophysiological and neuroanatomical development of human infants in relation to such issues. The frontal and prefrontal lobes of the brain are involved with voluntary planning functions, and Luria suggests that it is impossible to diagnose tumours of those regions until after 7 months of age (cited by Bruner, 1969). It is also worth noting that (in non-human primates) the prefrontal lobes are implicated in the ability to achieve novel connections between perception and behaviour as indexed by adaptation to displacing prisms (Bossom, 1965). At the behavioural level, what we can conclude is that the development of Type-II anticipation brings infant performance on the tracking task into line with the age norms that have already been noted for the ability to use new information as a guide for action in a range of tasks: the manual search task; behaviour of the support; use of landmark features; size–weight covariation in grasping; anticipatory hand shaping for concave/convex objects; and 'fear' on sight of the visual cliff.

The earlier discussion of manual search suggested how difficult it is to articulate the goal of activity from the subject's viewpoint. In this context, initial refixations of the re-emerged object involve visually elicited saccades. The infant is turning to a location to see *what* is there. Thus, the behaviour is goal-terminated. Type-II anticipation can be considered as goal-directed to the extent that it occurs in the absence of information about the location of the object that it will enable the infant to see. At this level, however, the infant's representation of the situation is probably a long way from being a general strategy, such as that suggested by Prazdny, in which witnessing an object's disappearance activates procedures that check for its presence in a new location. The characteristics of Type-II anticipation suggest that the infant's understanding of the visual search task has developed to the first level of the representational redescription model. Interplay between their innate abilities and environmental structure has supported their construction of 'successful procedures', but they are still at the Implicit-level and have yet to make explicit and develop an appreciation of the general features of the situation.

The Type-II anticipation of 9-month-olds is very situation specific. It requires consistent experience for its development and it shows little transfer. Aspects of this acquisition appear closely related to *contingency learning*. For instance, 9-month-olds fail to develop Type-II anticipation beyond a certain degree of spatial separation of the disappearance and reappearance locations (cf. Baillargeon's (1986) observations on tracking studies within this age range). Such factors suggest an important restriction on the type of new action coordinations that can be developed within a perceptual-behavioural representational system. However, the pattern of change outlined here cannot be explained effectively in the learning-theory language of association and generalization, nor straightforwardly in first-order connectionist terms.

At this point, the infant's representation of the situation is 'successful', but in need of *comparison* and *variablization* processes, which would make explicit the relative roles of components such as the object, its disappearance, and the location at which it disappears. This level of development is suited to the organization-oriented comparison procedures emphasized by Karmiloff-Smith, with the comparison process operating over records of Type-II anticipations as these are established in different contexts. This learning program comparison can be considered to make explicit fixed, variable and irrelevant components of the search context. For instance, kinetic occlusion of an object will occur in all cases, but the particular occluder involved and its location will not.

By 12 months of age, some infants show a completely general pattern of search. It can occur on the first trial in a new situation, and it is not disrupted if the object's movement path or disappearance location is changed. With regard to the parallels between change in this and other tasks, Acredolo (1988) observes that it is towards the end of the first year that infants reveal an ability to associate a landmark with a temporary hiding place in a *single trial*. This level of organization can be thought of as marking the explicit representation of, or attunement to, a new *constraint* by the infant: that of the invariant relationship between disappearance and reappearance of moving objects. The initial Type-II anticipations were characterized as a form of empirical contingency – many such contingencies may be independently acquired. A constraint, on the other hand, establishes a general relationship between variable elements. For the infant, it is represented procedurally, through the mediation of perceptual and behavioural processes and their coordination within the task environment by a novel action program structure. The 'implicit belief' that this supports is based in a form of *situated inference*, not (*contra* Prazdny, 1980) in any explicit representation of the object continuing to exist and move behind the occluder.

With the establishment of this level of organization, one would predict the increasingly persistent search that comes to characterize the infant moving into the second year. The traditional AI view of learning to solve a problem considers it as a process of representing a problem-space and then choosing appropriate operators to get from a current state to the sought goal. By way of contrast, we could say that the naturally intelligent infant *constructs* the problem of finding the object by learning to find the object. That is, naturally intelligent systems are equipped to construct the (novel) problem-space and the (novel) goal through experiencing solving the problem. In ecological psychology's terms, newly differentiated kinetic occlusion has become salient and now affords voluntary search.

This is what could be called a *conceptual level* of understanding in action. But the mechanism proposed here does not support the view that the infant needs an object concept in the sense of evocational imagery successfully or generally to anticipate object movement in this task. The development of general procedures for rediscovering invisibly moving objects must certainly be seen as *knowing* about the object's movement, and such procedures come to confer a genuine type of foresight on the subject who possesses them. Nevertheless the infant does not need to have an internal picture of the object's movement that controls search behaviour. We might hypothesize that recall of the invisible object in Mandler's (1984) sense of 'being reminded' may operate; as the infant's new procedure of turning to the object's reappearance location is invoked by its kinetic occlusion at the disappearance point, perhaps some 'awareness' of the object at the occluder exit is eventually 'brought to mind' as procedures to check the perceptual component for evidence of its reappearance are set in motion. This, however, would be an emergent property of the operation of the infant's procedures; it would not play a role in determining which procedures operate or how.

Nor has the infant transcended perception and the environment, although s/he could be said to be causally embedded in a broader segment of environmental structure. The infant has extended the range of action by making explicit new action-relevant information and constructing new goals. Considering Piaget's theory, one can agree that his emphasis on the role of action in developing object understanding is correct. But the way in which action is implicated is misstated due to his inappropriate conceptualization of action. Generalized and effective strategies for interacting with objects do not reveal a reconstruction of knowledge structures on a new level. They reveal a reorganization of perceptual-behavioural action systems through processes that appear to be shared with other naturalistic representational systems.

Final state – a postscript

This chapter has sketched a computational perspective on the kinds of mechanisms that would be required to underwrite an epigenetic framework for development. That perspective is characterized by its exploration of classical computational concepts as a route to understanding representation, and to understanding infants' perceptual-behavioural action as representation. Where and how does infancy end? I argued that action can be taken farther than is usually thought, to account for changes generally associated with radical qualitative change in mechanisms to a re-presentational level. The key developmental issue is: how much farther? Any developmental explanation can and should initially seek the mechanisms underlying novel representational phenomena in intrasystemic change. Only when that strategy reaches its limits must mechanisms responsible for discontinuity be sought in intersystemic developments.

Given our current state of knowledge, it seems feasible that infant abilities that are assumed to show a discontinuity towards the end of the sensorimotor period may eventually find an intra-systemic explanation in terms of a richly conceptualized model of action. This would challenge Piaget's (1953, 1955) claim that the stage VI infant's new capacity for *insight* in the manipulation of *spatial* relations among objects indicates a reconstruction of action, evidenced by their solutions to human equivalents of the well known 'stick-banana problem'. Such task solutions are not as magical as is sometimes assumed. Chimpanzees who have not had specific experience with sticks cannot solve the Kohler tasks. Kohler's analysis of his animals' performance in terms of its functional adequacy ignored many aspects relevant to developing a process model (Chance, 1960). The same is likely of infant behaviour on such tasks. Although Piaget claims there is now *mental combination* rather than *sensorimotor apprenticeship* (1953, p. 337), we rarely find evidence for a novel solution out of the blue. What we do see is the infant making some error that is not totally disastrous, but whose undesired consequence is anticipated before it actually occurs. It is this that leads to a change in strategy to something novel. For example, one of Piaget's children attempts to kneel on a stool and bend forward, the stool wobbles, she 'sees' the implication of this and moves the stool to rest against a sofa, only then resuming her kneeling attempt.

Such observations suggest that an ongoing transaction between subject and environment remains significant in the execution of such *insightful* solutions. As far as first-year infants' planfulness goes, their pragmatic knowledge comes to incorporate an understanding of the role of preconditions in activity. In the second year, action restructuring proceeds to an understanding of protection violations (where the order and consequences of multiple procedures must be considered). This is evidenced by advance ordering of procedures to avoid potentially conflicting subgoals, after considerable practice (Willatts, 1992). For example, they can remove a barrier before pulling a support-cloth to reach an object that would otherwise be dislodged.

It has been suggested that infants' action-based ability is restructured in the direction of an increased conceptualization of activity and of the situations to which it applies. In that sense, the constraints on what can be known through action are considerably more lax than the traditional picture makes it seem. But this does not entail that we should expect action to equip infants with the range of mental flexibility and combination of elements that has been associated with 'fully fledged' concepts. A key task for infancy must be to consider the possibilities for *addition to* (as opposed to a reconstruction of) action, viewed as a wholly unconventional system of representation, of partly and/or wholly conventional systems in Dretske's (1988) sense; and to clarify the novel abilities these systems and their interaction with perceptual-behavioural action mechanisms may support.

Contenders for such additions include the verbal system (Karmiloff-Smith, 1991; cf. the flexibility attributed to E-3 level representation grounded in intersystemic links with language); Leslie's (1983, 1987) metarepresentational operator (cf. the properties attributed to primary and secondary representations); and Mounoud's (1986) suggestion for a post-perceptual conceptual code. None of these yet offers a detailed account that specifies both the mechanisms underlying a new system and those responsible for prior pragmatic understanding in a way that could clarify fully how they interact with one another. A route into this is the kind of detailed computational account that has been advocated here.

This kind of computational approach has already shown parallels in organization and adaptive change in action and other systems. This does not imply that different naturalistic representational systems do not equip subjects with distinctive capacities for interaction with the world. It does imply a set of questions rather different from those generally posed within the infancy literature: what computational mechanisms underlie different types of naturalistic representation? How do they interact with one another? And what are the specific (spatial, temporal and causal) constraints on each type of representation? This final question, in particular, moves beyond the crude notion that action is tied to the world but other forms of representation are not. The question becomes: what range of selective correspondences does each system permit the subject to establish?

Comparing structure and function across domains of knowledge and systems of representation was one of the original aims of Bruner's functionalist perspective:

> I cannot believe that it suffices to argue simply that thought is action internalized or rendered into symbolic form. The task, rather, is to explore the range of development from the attainment of skill in action to the attainment of acumen: in problem-solving to explore in detail the kinship between them and the discontinuities as well. (Bruner, 1970, p. 91)

Developing a computational approach begins to make the goal of specifying mechanism, continuity and discontinuity at last feasible.

Notes

1. Anderson's (1978) 'indeterminacy of representation' thesis holds that it is impossible, in principle, to distinguish between alternative models of underlying processes so long as the models are behaviourally equivalent. However, establishing behavioural equivalence is not as straightforward as it is frequently taken to be by those who focus on its outcome.
2. Many interesting issues in machine learning are beyond the scope of this book. For a psychologically relevant overview see Draper (1987).
3. Winograd and Flores's (1986) alternative framework, which draws on the philosophy of Heidegger, faces difficulties surprisingly like those faced by Piaget's equilibration theory (Rutkowska, 1986).
4. Many machine learning systems can be characterized as 'storing' a dual representation of partially learned rules, concepts or methods: specific instances that are known to be examples of the rule, concept or method; and the most general description of instances encountered so far. The aim is to close the 'gap' between them (Bundy, Silver and Plummer, 1985).
5. Cf. Sleeman's (1987) discussion of the structure of training examples and the problem of algebra students' 'mis-generalization'.

References

Acredolo, L. (1978), 'Development of spatial orientation in infancy', *Developmental Psychology*, 14, pp. 224–34.

Acredolo, L. (1988), 'From signal to 'symbol': the development of landmark knowledge from 9–13 months', *British Journal of Developmental Psychology*, 6, pp. 369–71.

Agre, P. and Chapman, D. (1987), 'Pengi: an implementation of a theory of activity', *Proceedings of the Sixth National Conference on Artificial Intelligence, Seattle, Wash.*, vol. 1, Los Altos, Calif.: William Kaufmann.

Ames, A. (1968), *The Ames Demonstrations in Perception*, ed. W. H. Ittelson, New York: Hafner.

Anderson, J. R. (1978), 'Arguments concerning representations for mental imagery', *Psychological Review*, 85, pp. 249–77.

Aslin, R. N. (1987), 'Visual and auditory development in infancy', in J. D. Osofsky (ed.) *Handbook of Infant Development*, New York: Wiley.

Atkinson, J. and Braddick, O. (1989), 'Development of basic functions', in A. Slater and J. G. Bremner (eds), *Infant Development*, Hove and London: Lawrence Erlbaum.

Attneave, F. (1971), 'Multistability in perception', *Scientific American*, 225, pp. 62–71.

Baillargeon, R. (1986), 'Representing the existence and the location of hidden objects: object permanence in 6- and 8-month-old infants', *Cognition*, 23, pp. 21–41.

Baillargeon, R., Graber, M., Devos, J. and Black, J. (1990), 'Why do young infants fail to search for hidden objects?', *Cognition*, 36, pp. 255–84.

Baillargeon, R., Spelke, E. S. and Wasserman, S. (1985), 'Object permanence in five-month-old infants', *Cognition*, 20, pp. 191–208.

Ball, W. and Tronick, E. (1971), 'Infant responses to impending collision: optical and real', *Science*, 171, pp. 818–20.

Ball, W. and Vurpillot, E. (1976), 'La perception du mouvement en profoundeur chez le nourrisson', *Année Psychologique*, 76, pp. 383–400.

Ball, W. and Vurpillot, E. (1981), 'Action and perception of displacements in infancy', in G. E. Butterworth (ed.), *Infancy and Epistemology: An evaluation of Piaget's theory*, Hemel Hempstead: Harvester Wheatsheaf.

Ballard, D. H. (1989), 'Reference frames for animate vision', in N. S. Sridharan (ed.), *Proceedings of the Eleventh International Joint Conference on Artificial Intelligence*, vol. 2, San Mateo, Calif.: William Kaufmann.

Ballard, D. H., Hinton, G. E. and Sejnowski, T.J. (1983), 'Parallel visual computation', *Science*, 306, pp. 21–6.

Banks, M. S. and Ginsburg, A. P. (1985), 'Early visual preferences: a review and new theoretical treatment', in H. W. Reese (ed.), *Advances in Child Development and Behaviour*, New York: Academic Press.

Banks, M. S. and Salapatek, P. (1981), 'Infant pattern vision: a new approach based on the contrast sensitivity function', *Journal of Experimental Child Psychology*, 31, pp. 1–45.

Banks, M. S. and Salapatek, P. (1983), 'Infant visual perception', in M. M. Haith and J. J. Campos (eds), *Infancy and Developmental Psychology. Handbook of Child Psychology*, 4th edn, vol. 2, New York: Wiley.

Barr, A. and Feigenbaum, E. A. (1982), *The Handbook of Artificial Intelligence,* vol. 2, Los Altos, Calif.: William Kaufmann.

Barwise, J. (1983), 'Information and semantics', *Behavioral and Brain Sciences*, 6, p. 65.

Barwise, J. (1987), 'Unburdening the language of thought', *Mind and Language*, 2, pp. 82–96.

Barwise, J. and Perry, J. (1983), *Situations and Attitudes*, Cambridge, Mass.: MIT Press.

Barwise, J. and Perry, J. (1984), 'Shifting situations and shaken attitudes', Centre for the Study of Language and Information, Stanford University, Technical Report No. CSLI–84–13.

Benson, J. B. and Uzgiris, I. C. (1985), 'Effect of self-initiated locomotion on infant search-activity', *Developmental Psychology*, 21, pp. 923–31.

Bertenthal, B. I., Proffitt, D. R. and Kramer, S. J. (1987), 'Perception of biomechanical motions by infants: implementation of various processing constraints', *Journal of Experimental Psychology: Human perception and performance*, 13, pp. 577–85.

Block, N. (ed.) (1980), *Readings in Philosophy of Psychology*, vol. 1, London: Methuen.

Bobrow, D. and Raphael, B. (1974), 'New programming languages for artificial intelligence', *ACM Computing Surveys*, 6, pp. 155–74.

Boden, M. A. (1972), *Purposive Explanation in Psychology*, Cambridge, Mass.: Harvard University Press.

Boden, M. A. (1981), *Minds and Mechanisms: Philosophical psychology and computational models*, Hemel Hempstead: Harvester Wheatsheaf.

Boden, M. A. (1982), 'Is equilibration important? – A view from artificial intelligence', *British Journal of Psychology*, 73, pp. 165–73.

Boden, M. A. (1983), 'Artificial intelligence and animal psychology', *New Ideas in Psychology*, 1, pp. 1–33.

Boden, M. A. (1985), 'Is computational psychology constructivist?', *Archives de Psychologie*, 53, pp. 103–12.

Boden, M. A. (1988), *Computer Models of Mind*, Cambridge: Cambridge University Press.

Bossom, J. (1965), 'The effect of brain lesions on prism-adaptation in monkey', *Psychonomic Science*, 2, pp. 45–6.

Bower, T. G. R. (1966), 'The visual world of infants', *Scientific American*, 251, pp. 80–92.

Bower, T. G. R. (1967), 'The development of object-permanence: some studies of existence constancy', *Perception and Psychophysics*, 2, pp. 411–18.

Bower, T. G. R. (1972), 'Object perception in infants', *Perception*, 1, pp. 15–30.

Bower, T. G. R. (1974), *Development in Infancy*, 1st edn, San Francisco: Freeman.

Bower, T. G. R. (1977), *A Primer of Infant Development*, San Francisco: Freeman.

Bower, T. G. R. (1979), *Human Development*, San Francisco: Freeman.

Bower, T. G. R. (1982), *Development in Infancy*, 2nd edn, San Francisco: Freeman.

Bower, T. G. R. and Patterson, J. G. (1973), 'The separation of place, movement and object in the world of the infant', *Journal of Experimental Child Psychology*, 15, pp. 161–8.

Bower, T. G. R., Broughton, J. M. and Moore, M. K. (1970a), 'Infant responses to approaching objects: an indicator of response to distal variables', *Perception and Psychophysics*, 9, pp. 193–6.

Bower, T. G. R., Broughton, J. M. and Moore, M. K. (1970b), 'Demonstration of intention in the reaching behaviour of neonate humans', *Nature*, 228, pp. 679–81.

Bower, T. G. R., Broughton, J. M. and Moore, M. K. (1971), 'Development of the object concept as manifested in changes in the visual tracking behaviour of infants between 7 and 20 weeks of age. *Journal of Experimental Child Psychology*, 11, pp. 182–93.

Braddick, O. J. (1973), 'The masking of apparent motion in random dot patterns', *Vision Research*, 13, pp. 355–69.

Braddick, O. J. (1974), 'A short-range process for apparent motion', *Vision Research*, 14, pp. 519–27.

Brady, M. (1985), 'Artificial intelligence and robotics', *Artificial Intelligence*, 26, pp. 79–121.

Brady, M. and Daniel, R. (1986), 'Robotics', tutorial presented at Eighth European Conference on Artificial Intelligence, Brighton, 22 July.

Brainerd, C. J. (1978), *Piaget's Theory of Intelligence*, Englewood Cliffs, N.J.: Prentice Hall.

Brazelton, T. B., Koslowski, B. and Main, M. (1974), 'The origin of reciprocity: the early mother–infant interaction', in M. Lewis and L. A. Rosenblum (eds), *The Effect of the Infant on its Caregiver*. New York: Wiley.

Bremner, J. G. (1978), 'Egocentric versus allocentric coding in nine-month-old infants: factors influencing the choice of code', *Developmental Psychology*, 14, pp. 346–55.

Bremner, J. G. (1980), 'The infant's understanding of space', in M. Cox (ed.) *Are Young Children Egocentric?*, London: Concord.

Bremner, J. G. (1985), 'Object tracking and search in infancy', *Developmental Review*, 5, pp. 371–96.

Bremner, J. G. (1988), *Infancy*, Oxford: Blackwell.

Bremner, J. G. (1989), 'Development of spatial awareness in infancy', in A. Slater, and J. G. Bremner (eds), *Infant Development*, Hove and London: Lawrence Erlbaum.

Bronson, G. W. (1974), 'Postnatal visual development', *Child Development*, 45, pp. 873–90.

Bronson, G. W. (1982), *The Scanning Patterns of Human Infants: Implications for visual learning*, Norwood, N.J.: Ablex.

Brooks, R. A. (1991a), 'Challenges for complete creature architectures', In J.-A. Meyer and S. W. Wilson (eds), *From Animals to Animats: Proceedings of the first international conference on the simulation of adaptive behaviour*, Cambridge, Mass: MIT Press.

Brooks, R. A. (1991b), 'Intelligence without representation', *Artificial Intelligence*, 47, pp. 139–60.

Brown, A. (1979), 'Theories of memory and the problems of development', in L. S. Cermak and F. I. M. Craik (eds), *Levels of Processing in Human Memory*, Hillsdale, N.J.: Lawrence Erlbaum.

Brown, J. S. and van Lehn, K. (1980), 'Repair theory: a generative theory of bugs in procedural skills', *Cognitive Science*, 4, pp. 379–426.

Bruce, V. (1988), *Recognising Faces*, Hove and London: Lawrence Erlbaum.

Bruce, V. and Green, P. (1985), *Visual Perception: Physiology, psychology and ecology*, London: Lawrence Erlbaum.

Bruner, J. S. (1959), 'Inhelder and Piaget's *The Growth of Logical Thinking*', *British Journal of Psychology*, 50, p. 365.

Bruner, J. S. (1966), 'On cognitive growth', in J. S. Bruner, R. R. Olver and P. M. Greenfield (eds), *Studies in Cognitive Growth*, New York: Wiley.

Bruner, J. S. (1968), *Processes in Cognitive Growth: Infancy*, Barre, Mass.: Clark University Press with Barre Publishers.

Bruner, J. S. (1969), 'Origins of problem solving strategies in skill acquisition', paper presented at the XIX International Congress of Psychology, London, July.

Bruner, J. S. (1970), 'The growth and structure of skill', in K. J. Connolly (ed.) *Mechanisms of Motor Skill Development*, London: Academic Press.

Bruner, J. S. (1973), 'Organization of early skilled action', *Child Development*, 44, pp. 1–11.

Bruner, J. S. (1975), 'The ontogenesis of speech acts', *Journal of Child Language*, 2, pp. 1–19.

Bruner, J. S. and Koslowski, B. (1972), 'Visually preadapted constituents of manipulatory action', *Perception*, 1, pp. 3–14.

Bryant, P. E. and Trabasso, T. (1971), 'Transitive inferences and memory in young children', *Nature*, 232, pp. 456–8.

Bullinger, A. (1981), 'Cognitive elaboration of sensorimotor behaviour', in G. E. Butterworth (ed.), *Infancy and Epistemology: An evaluation of Piaget's theory*, Hemel Hempstead: Harvester Wheatsheaf.

Bundy, A., Silver, B. and Plummer, D. (1985), 'An analytic comparison of some rule–learning programs', *Artificial Intelligence*, 27, pp. 137–81.

Burge, J. (1974), 'A.I. and sensori–motor intelligence', *Proceedings of the First Conference of the Society for Artificial Intelligence and Simulation of Behaviour*, Brighton.

Bushnell, I. W. R., Sai, F. and Mullin, J. T. (1989), 'Neonatal recognition of the mother's face', *British Journal of Developmental Psychology*, 7, pp. 3–16.

Butterworth, G. E. (1975), 'Object identity in infancy: the interaction of spatial location codes in determining search errors', *Child Development*, 46, pp. 866–70.

Butterworth, G. E. (ed.) (1977), 'Object disappearance and error in Piaget's stage IV task', *Journal of Experimental Child Psychology*, 23, pp. 391–401.

Butterworth, G. E. (1981a), *Infancy and Epistemology: An evaluation of Piaget's theory*, Hemel Hempstead: Harvester Wheatsheaf.

Butterworth, G. E. (1981b), 'Object permanence and identity in Piaget's theory of infant cognition', in G. E. Butterworth (ed.), *Infancy and Epistemology: An evaluation of Piaget's theory*, Hemel Hempstead: Harvester Wheatsheaf.

Butterworth, G. E. (1981c), 'The origins of auditory-visual perception and visual proprioception in human development', in H. I. Pick and R. Walk (eds), *Perception and Experience*, vol. 2, New York: Plenum.

Butterworth, G. E. (1989), 'Events and encounters in infant perception', in A. Slater and G. Bremner (eds), *Infant Development*, Hove and London: Lawrence Erlbaum.

Butterworth, G. E. and Castillo, M. (1976), 'Coordination of auditory and visual space in newborn infants', *Perception*, 2, pp. 155–60.

Butterworth, G. E. and Cicchetti, D. (1978), 'Visual calibration of posture in normal and motor retarded Downs syndrome infants', *Perception*, 7, pp. 513–25.

Butterworth, G. E. and Hicks, L. (1977), 'Visual proprioception and postural stability in infancy: a developmental study', *Perception*, 6, pp. 255–62.

Butterworth, G. E. and Hopkins, B. (1988), 'Hand–mouth coordination in the new-born baby', *British Journal of Developmental Psychology*, 6, pp. 303–14.

Butterworth, G. E. and Jarrett, N. (1991), 'What minds have in common is space: spatial mechanisms serving joint visual attention in infancy', *British Journal of Developmental Psychology*, 9, pp. 55–72.

Buxton, B. F. and Buxton, H. (1983), 'Monocular depth perception from optical flow by space–time signal processing', *Proceedings of the Royal Society*, Series B, 218, pp. 27–47.

Campos, J. J., Hiatt, S., Ramsay, D., Henderson, C. and Svejda, M., (1978), 'The emergence of fear on the visual cliff', in M. Lewis and R. A. Rosenblum (eds), *The Development of Affect*, vol. 1, New York: Plenum Press.

Campos. J. J., Langer, A. and Krowitz, A. (1970), 'Cardiac responses on the visual cliff in prelocomotor human infants', *Science*, 170, pp. 196–7.

Caplan, D. (1987), *Neurolinguistics and Linguistic Aphasiology: An introduction*, Cambridge: Cambridge University Press.

Carey, S. (1985), 'Constraints on semantic development', in J. Mehler and R. Fox (eds), *Neonate Cognition: Beyond the blooming, buzzing confusion*, Hillsdale, N.J.: Lawrence Erlbaum.

Carey, S. (1991), 'Knowledge acquisition: enrichment or conceptual change?' in S. Carey and R. Gelman (eds), *The Epigenesis of Mind: Essays on biology and cognition*, Hillsdale, N.J.: Lawrence Erlbaum.

Carpenter, G. C. (1974), 'Mother's face and the newborn', *New Scientist*, 61, pp. 742–4.

Chance, M. R. A. (1960), 'Kohler's chimpanzees: how did they perform?' *Man*, 60, pp. 130–5.

Charlesworth, W. R. (1968), 'Cognition in infancy: where do we stand in the mid–sixties?', *Merrill-Palmer Quarterly*, 14, pp. 25–46.

Charlesworth, W. R. (1969), 'The role of surprise in cognitive development', in D. Elkind and J. H. Flavell (eds), *Studies in Cognitive Development*, New York: Oxford University Press.

Chomsky, N. (1965), *Aspects of the Theory of Syntax*, Cambridge, Mass.: MIT Press.

Chomsky, N. (1980), 'Rules and representations', *Behavioral and Brain Sciences*, 3, pp. 1–61.

Churchland, P. M. (1986a). 'Some reductive strategies in cognitive neurobiology', *Mind*, 95, pp. 279–309.

Churchland, P. M. (1986b), 'The continuity of philosophy and the sciences', *Mind and Language*, 1, pp. 5–15.

Churchland, P. M. (1987), 'Connectionism and neurocomputational explanation', lecture presented at the University of Sussex, June.

Churchland, P. M. (1989a), 'On the nature of explanation: a PDP approach', in P. M. Churchland, *The Neurocomputational Perspective: The nature of mind and the structure of science*, Cambridge, Mass.: MIT Press.

Churchland, P. M. (1989b), 'On the nature of theories: a neurocomputational perspective', in C. W. Savage (ed.), *The Structure and Function of Scientific Theories*, Minnesota Studies in the Philosophy of Science, Minneapolis: University of Minnesota Press.

Churchland, P. S. (1980), 'A perspective on mind–brain research', *Journal of Philosophy*, 128, pp. 185–207.
Clark, A. (1989), *Microcognition: Philosophy, cognitive science and parallel distributed processing*, Cambridge, Mass.: Bradford/MIT Press.
Clark, A. and Lutz, R. (eds) (1992), *Connectionism in Context*, London: Springer-Verlag.
Clark, A. and Karmiloff-Smith, A. (in press), 'The cognizer's innards: a psychological and philosophical perspective on the development of thought', *Mind and Language*.
Clark, E. (1983), 'Meanings and concepts', in J. Flavell and E. Markman (eds), *Cognitive Development. Handbook of Child Psychology*, 4th edn, vol. 3, New York: Wiley.
Cliff, D. (1990), 'Computational neuroethology: a provisional manifesto', University of Sussex, Cognitive Science Research Paper, Serial No. CSRP 162.
Clocksin, W. F. (1980), 'Perception of suface slant and edge labels from optical flow: a computational approach', *Perception*, 9, pp. 253–69.
Clowes, M. B. (1971), 'On seeing things', *Artificial Intelligence*, 6, pp. 79–116.
Cohen, L. B. (1979), 'Our developing knowledge of infant cognition', *American Psychologist*, 34, pp. 894–99.
Cohen, L. B. (1988), 'The relationship between infant habituation and infant information processing', *European Bulletin of Cognitive Psychology*, 8, pp. 445–54.
Cohen, L. B. and Younger, B. A. (1984), 'Infant perception of angular relations', *Infant Behaviour and Development*, 7, pp. 37–47.
Cole, M., Hood, L. and McDermott, R. (1978), 'Ecological niche picking', in U. Neisser (ed.), *Memory Observed: Remembering in natural contexts*, San Francisco: Freeman, 1982.
Collett, T. S. and Land, M. F. (1978), 'How hoverflies compute interception courses', *Journal of Comparative Physiology A*, 125, pp. 191–204.
Crick, F. (1989), 'The recent excitement about neural networks', *Nature*, 337, pp. 129–32.
Crick, F. and Asanuma, S. (1986), 'Certain aspects of the anatomy and physiology of the cerebral cortex', in J. L. McClelland and D. E. Rumelhart (eds), *Parallel Distributed Processing*, vol. 2: *Psychological and Biological Models*, Cambridge, Mass.: Bradford/MIT Press.
Cromer, R. F. (1980), 'Empirical evidence in support of non-empiricist theories of mind', *Behavioral and Brain Sciences*, 3, pp. 16–18.
Davies, M. (1992), 'Thinking persons and cognitive science', in A. Clark and R. Lutz (eds), *Connectionism in Context*, London: Springer-Verlag.
Day, R. H. and McKenzie, B. E. (1977), 'Constancies in the perceptual world of the infant', in W. Epstein (ed.), *Stability and Constancy in Visual Perception: Mechanisms and processes*, New York: Wiley.
Dennett, D. (1971), 'Intentional systems', *Journal of Philosophy*, 68, pp. 87–106.
Dennett, D. (1978a), *Brainstorms*, Cambridge, Mass.: Bradford/MIT Press.
Dennett, D. (1978b), 'Why not the whole iguana?', *Behavioral and Brain Sciences*, 1, pp. 103–4.
Dennett, D. (1982), 'On thoughts about words', *Cognition*, 12, pp. 211–17.
Dennett, D. (1983), 'Intentional systems in cognitive ethology: the "Panglossian paradigm" defended', *Behavioral and Brain Sciences*, 6, pp. 343–90.
Dennett, D. (1988), 'Science, philosophy and interpretation', *Behavioral and Brain Sciences*, 11, pp. 535–46.

Dennett, D. (1991), *Consciousness Explained*, London: Allen Lane, Penguin Press.

Dennis, W. (1960), 'Causes of retardation among institutional children: Iran', *Journal of Genetic Psychology*, 96, pp. 47–59.

Dennis, W. and Dennis, M. G. (1940), 'The effect of cradling practices upon the onset of walking in Hopi children', *Journal of Genetic Psychology*, 56, pp. 77–86.

Deregowski, J. B. (1978), 'In search of a wider perspective: cross-cultural studies', in A. Burton and J. Radford (eds), *Thinking in Perspective*, London: Methuen.

Diamond, A. (1981), 'Retrieval of an object from an open box: the development of visual-tactile control of reaching in the first year of life', paper presented at the Society for Research in Child Development, Boston, April.

Diamond, A. (1985), 'Development of the ability to use recall to guide action, as indicated by infants' performance on AB', *Developmental Psychology*, 56, pp. 868–83.

Diamond, A. (1988), 'Differences between adult and infant cognition: is the crucial variable presence or absence of language?', in L. Weiskrantz (ed.), *Thought Without Language*. Oxford: Oxford University Press.

Draper, S. W. (1980), *Reasoning About Depth in Line-Drawing Interpretation*, unpublished D.Phil thesis, University of Sussex.

Draper, S. W. (1987), 'Machine learning and cognitive development', in J. C. Rutkowska and C. Crook (eds), *Computers, Cognition and Development*, Chichester: Wiley.

Dresher, E. and Hornstein, N. (1976), 'On some supposed contributions of artificial intelligence to the scientific study of language', *Cognition*, 4, pp. 321–98.

Drescher, G. (1986), 'Genetic AI: translating Piaget into LISP', *Instructional Science*, 14, pp. 357–80.

Dretske F.I. (1981), *Knowledge and the Flow of Information*, Oxford: Blackwell.

Dretske F. I. (1983), 'Précis of *Knowledge and the Flow of Information*', *Behavioral and Brain Sciences*, 6, pp. 55–90.

Dretske F. I. (1988), *Explaining Behaviour: Reasons in a world of causes*, Cambridge, Mass.: Bradford/MIT Press.

Dreyfus H. L. (1981), 'From micro-worlds to knowledge representation: AI at an impasse', in J. Haugeland (ed.), *Mind Design: Philosophy, psychology, artificial intelligence*, Cambridge, Mass.: Bradford/MIT Press.

Enroth-Cugell, C. and Robson, J. G. (1966), 'The constrast sensitivity of retinal ganglion cells of the cat', *Journal of Physiology*, 187, pp. 517–52.

Ericsson, K. A. and Simon, H. A. (1984), *Protocol Analysis: Verbal reports as data*, Cambridge, Mass.: MIT Press.

Evans, G. (1982), *The Varieties of Reference*, Oxford: Oxford University Press.

Evans, W. F. (1973), 'The stage IV error in Piaget's theory of object concept development: an investigation of the role of activity', unpublished dissertation, University of Houston.

Evans, W. F. and Gratch, G. (1972), 'The stage IV error in Piaget's theory of object concept development: difficulty in object conceptualization or spatial localization?', *Child Development*, 43, pp. 682–8.

Fagen, J. W., Morrongiello, B. A., Rovee-Collier, C. and Gekoski, M. J. (1984), 'Expectancies and memory retrieval in three-month-old infants', *Child Development*, 55, pp. 936–43.

Fantz, R. L. (1958), 'Pattern vision in young infants', *Psychological Record*, 8, pp. 43–47.

Fantz, R. L. (1965), 'Visual perception from birth as shown by pattern selectivity', repr. in L. J. Stone, H. T. Smith and L. B. Murphy (eds), (1974), *The Competent Infant*, London: Tavistock.

Fantz, R. L. (1966), 'Pattern discrimination and selective attention as determinants of perceptual development from birth', in A. H. Kidd and J. L. Rivoire (eds), *Perceptual Development in Children*, New York: International Universities Press.

Fisher, R. B. (1986), 'Identity independent object segmentation in 2½-D sketch data', *Proceedings of the Seventh European Conference on Artificial Intelligence*, vol. 2.

Flavell, J. (1985), *Cognitive Development*, Englewood Cliffs, N.J.: Prentice Hall.

Fodor, J. A. (1968), *Psychological Explanation*, New York: Random House.

Fodor, J. A. (1972), 'Some reflections on L. S. Vygotsky's "Thought and Language"', *Cognition*, 1, pp. 83–95.

Fodor, J. A. (1975), *The Language of Thought*, New York: Cromwell.

Fodor, J. A. (1980), 'Methodological solipsism considered as a research strategy in cognitive science', *Behavioral and Brain Sciences*, 3, pp. 63–110.

Fodor, J. A. (1983), *The Modularity of Mind: An essay on faculty psychology*, Cambridge, Mass.: Bradford/MIT Press.

Fodor, J. A. (1987), 'A situated grandmother? Some remarks on proposals by Barwise and Perry', *Mind and Language*, 2, pp. 123–40.

Fodor, J. A. and Pylyshyn, Z. W. (1981), 'How direct is visual perception?: some reflections on Gibson's ecological approach', *Cognition*, 9, pp. 139–96.

Fodor, J. A. and Pylyshyn, Z. W. (1988), 'Connectionism and cognitive architecture: a critical analysis', *Cognition*, 28, pp. 3–71.

Friedman, S. (1972), 'Habituation and recovery of visual response in the alert human newborn', *Journal of Experimental Child Psychology*, 13, pp. 339–49.

Frisby, J. P. (1979), *Seeing: Mind, brain and illusion*, Oxford: Oxford University Press.

Furth, H. G. (1969), *Piaget and Knowledge: Theoretical foundations*, Englewood Cliffs, N.J.: Prentice Hall.

Gardner, J. K. (1971), 'The development of object identity in the first six months of human infancy', paper presented at the Meeting of the Society for Research in Child Development, Minneapolis, April.

Gauker, C. (in press), 'An extraterrestrial perspective on conceptual development', *Mind and Language*.

Gayl, I. E., Roberts, J. O. and Werner, J. S. (1983), 'Linear systems analysis in infant visual pattern preferences', *Journal of Experimental Child Psychology*, 35, pp. 30–45.

Gelman, R. (1990), 'Structural constraints on cognitive development', *Cognitive Science*, 14, pp. 3–9.

Gibson, E. J. (1979), 'Commentary', in M. H. Bornstein and W. Kessen (eds), *Psychological Development From Infancy*, Hillsdale, N.J.: Erlbaum.

Gibson, E. J. (1987), 'Introductory essay: what does infant perception tell us about theories of perception?', *Journal of Experimental Psychology: Human perception and performance*, 13, pp. 515–23.

Gibson, E. J. and Rader, N. (1979), 'Attention: the perceiver as performer', in G. A. Hale and M. Lewis (eds), *Attention and Cognitive Development*, New York: Plenum Press.

Gibson, E. J. and Spelke, E. S. (1983), 'The development of perception', in J. H. Flavell and E. M. Markman (eds), *Cognitive Development. Handbook of Child Psychology*, 4th edn, vol. 3, New York: Wiley.

Gibson, E. J. and Walk, R. D. (1960), 'The visual cliff', *Scientific American*, 202, pp. 64–71.

Gibson, J. J. (1950), *The Perception of the Visual World*, Boston, Mass.: Houghton-Mifflin.

Gibson, J. J. (1968), *The Senses Considered as Perceptual Systems*, London: George Allen and Unwin.

Gibson, J. J. (1979), *The Ecological Approach to Visual Perception*, Boston, Mass.: Houghton-Mifflin.

Gibson, J. J. and Gibson, E. J. (1955), 'Perceptual learning: differentiation or enrichment?', *Psychological Review*, 68, pp. 32–41.

Gibson, K. R. (1981), 'Comparative neuro-ontogeny: its implications for the development of human intelligence', in G. E. Butterworth (ed.), *Infancy and Epistemology: An evaluation of Piaget's theory*, Hemel Hempstead: Harvester Wheatsheaf.

Gould, J. L. (1982), *Ethology: The mechanisms and evolution of behaviour*, New York: Norton.

Gratch, G. (1972), 'A study of the relative dominance of vision and touch in 6 month old infants', *Child Development*, 43, pp. 615–23.

Gratch, G. (1975), 'Recent studies based on Piaget's view of object concept development', in L. B. Cohen and P. Salapatek (eds), *Infant Perception: From sensation to cognition*, vol. 2, *Perception of Space, Speech, and Sound*, London: Academic Press.

Gratch, G. (1982), 'Responses to hidden persons and things by 5-, 9-, and 16-month-old infants in a visual tracking situation', *Developmental Psychology*, 18, pp. 232–7.

Gratch, G., Appel, K. J., Evans, W. F., Le Compte, G. K. and Wright, N. A. (1974), 'Piaget's stage IV object concept error: evidence for forgetting or object conception?', *Child Development*, 45, pp. 71–7.

Gregory, R. (1970), *The Intelligent Eye*, London: Weidenfeld and Nicholson.

Haber, R. (1983), 'Can information be objectivized?', *Behavioral and Brain Sciences*, 6, pp. 70–1.

Haith, M. M. (1980), *Rules That Babies Look By*, Hillsdale, N.J.: Lawrence Erlbaum.

Haith, M. M. and Campos, J. J. (1977), 'Human infancy', *Annual Review of Psychology*, 28, pp. 251–93.

Hamlyn, D. W. (1978), *Experience and the Growth of Understanding*, London: Routledge and Kegan Paul.

Hamlyn, D. W. (1981), 'Cognitive systems, "folk psychology", and knowledge', *Cognition*, 10, pp. 115–18.

Harnad, S. (1992), 'Connecting object to symbol in modelling cognition', in A. Clark and R. Lutz, (eds), *Connectionism in Context*, London: Springer-Verlag.

Harris, P. L. (1974), 'Perseverative search at a visibly empty place by young infants', *Journal of Experimental Child Psychology*, 45, pp. 535–42.

Harris, P. L. (1975), 'The development of search and object permanence during infancy', *Psychological Bulletin*, 82, pp. 332–44.

Harris, P. L. (1976), 'Subject, object and framework: a theory of spatial development', paper presented at the Annual Conference of the Developmental Section of the British Psychological Society, Guildford.

Harris, P. L. (1983), 'Infant cognition', in M. M. Haith and J. J. Campos (eds), *Infancy and Developmental Psychology. Handbook of Child Psychology*, 4th edn, vol. 2, New York: Wiley.

Harris, P. L. (1989), 'Object permanence in infancy', in A. Slater and G. Bremner (eds), *Infant Development*, Hove and London: Lawrence Erlbaum.

Haugeland, J. (1978), 'The nature and plausibility of cognitivism', *Behavioral and Brain Sciences*, 1, pp. 215–60.

Hebb, D. O. (1946), 'On the nature of fear', *Psychological Review*, 53, pp. 259–76.

Hebb, D. O. (1949), *The Organization of Behaviour*, New York: Wiley.

Hess, E. H. (1956), 'Space perception in the chick', *Scientific American*, 195, pp. 71–80.

Hinton, G. (1979), 'Some demonstrations of the effects of structural descriptions in mental imagery', *Cognitive Science*, 3, pp. 231–50.

Hinton, G. (1984), 'Some computational solutions to Bernstein's problems, in H. T. A. Whiting (ed.), *Human Motor Actions: Bernstein reassessed*, Amsterdam: North Holland.

Hobson, P. (1991), 'Against the theory of "Theory of Mind"', *British Journal of Developmental Psychology*, 9, pp. 33–53.

Hood, B. and Willatts, P. (1986), 'Reaching in the dark to an object's remembered position: evidence for object permanence in 5-month-old infants', *British Journal of Developmental Psychology*, 4, pp. 57–65.

Horn, B. K. P. (1975), 'Image intensity understanding', Massachusetts Institute of Technology, Artificial Intelligence Laboratory, Artificial Intelligence Memorandum, No. 335.

Hubel, D. H. and Wiesel, T. N. (1959), 'Receptive fields of single neurons in the cat's striate cortex'. *Journal of Physiology*, 148, pp. 574–91.

Huffman, D. A. (1971), 'Impossible objects as nonsense sentences', *Machine Intelligence*, 6, pp. 295–323.

Humphrey, N. K. (1972), 'Seeing and nothingness', *New Scientist*, 53, pp. 241–55.

Humphrey, N. K. (1974), 'Vision in a monkey without striate cortex', *Perception*, 3, pp. 241–55.

Humphrey, N. K. (1976), 'The social function of intellect', in P. P. G. Bateson and R. A. Hinde (eds), *Growing Points in Ethology*, Cambridge: Cambridge University Press.

Humphreys, G. W. (1987), 'Objects, words, brains and computers: framing the correspondence problem in object and word recognition', *Bulletin of the British Psychological Society*, 40, pp. 207–10.

Humphreys, G. W. and Riddoch, J. (1986), 'Information processing systems which embody computational rules: the connectionist approach', *Mind and Language*, 1, pp. 201–12.

Hunt, J. McVicker (1961), *Intelligence and Experience*, New York: Ronald.

Hunt, J. McVicker (1965), 'Intrinsic motivation and its role in cognitive development', in D. Levine (ed.), *Nebraska Symposium on Motivation*, Lincoln: University of Nebraska Press.

Israel, D. (1988), 'Bogdan on information', *Mind and Language*, 3, pp. 123–40.

Jacobson, S. (1979), 'Matching behaviour in the young infant', *Child Development*, 50, pp. 425–30.

Jeffrey, W. (1968), 'The orienting reflex and attention in cognitive development', *Psychological Review*, 75, pp. 323–34.

Jeffrey, W. (1976), 'Habituation as a mechanism for perceptual development', in T. J. Tighe and R. N. Leaton (eds), *Habituation*, Hillsdale, N.J.: Lawrence Erlbaum.

Johnson, M. (1988), 'Memories of mother', *New Scientist*, 18, pp. 60–2.

Johnson, M. (1990), 'Cortical maturation and the development of visual attention in early infancy', *Journal of Cognitive Neuroscience*, 2, pp. 81–95.

Johnson-Laird, P. N. (1983), *Mental Models: Towards a cognitive science of language, inference and consciousness*, Cambridge: Cambridge University Press.

Johnson-Laird, P. N. (1988), *The Computer and the Mind: An introduction to cognitive science*, London: Fontana.

Jones, R., Spelke, E. S. and Alley, T. (1985), 'Work group on perceptual development', in W. H. Warren and R. E. Shaw (eds), *Persistence and Change: Proceedings of the first international conference on event perception*, Hillsdale, N.J.: Lawrence Erlbaum.

Jordan, M. I. (1986), 'An introduction to linear algebra in parallel distributed processing', in D. E. Rumelhart, J. L. McClelland and the P.D.P. Research Group, *Parallel Distributed Processing*, vol. 1, *Foundations*, Cambridge, Mass.: Bradford/ MIT Press.

Kagan, J. (1972), 'Do infants think?', *Scientific American*, 226, pp. 74–82.

Kagan, J. (1979), 'Structure and process in the human infant: the ontogeny of mental representation', in M. H. Bornstein and W. Kessen (eds), *Psychological Development from Infancy: Image to intention*, Hillsdale, N.J.: Lawrence Erlbaum.

Karmel, B. Z. and Maisel, E. B. (1975), 'A neuronal activity model for infant attention', in L. B. Cohen and P. Salapatek (eds), *Infant Perception: From sensation to cognition*, New York: Academic Press.

Karmiloff-Smith, A. (1979), 'Micro- and macrodevelopmental changes in language acquisition and other representational systems', *Cognitive Science*, 3, pp. 91–118.

Karmiloff-Smith, A. (1984), 'Children's problem solving', in M. E. Lamb, A. L. Brown and B. Rogoff (eds), *Advances in Developmental Psychology*, vol. 3, Hillsdale, N.J.: Lawrence Erlbaum.

Karmiloff-Smith, A. (1986), 'From meta-processes to conscious access: evidence from children's metalinguistic and repair data', *Cognition*, 23, pp. 95–147.

Karmiloff-Smith, A. (1987), 'Beyond modularity: a perspective on human consciousness', paper presented at the Annual Conference of the British Psychological Society, Brighton, April.

Karmiloff-Smith, A. (1990), 'Constraints on representational change: evidence from children's drawings'. *Cognition*, 34, pp. 57–83.

Karmiloff-Smith, A. (1991), 'Beyond modularity: innate constraints and developmental change', in S. Carey and R. Gelman (eds), *The Epigenesis of Mind: Essays on biology and cognition*, Hillsdale, N.J.: Lawrence Erlbaum.

Karmiloff-Smith, A. and Inhelder, B. (1974), 'If you want to get ahead, get a theory', *Cognition*, 3, pp. 195–212.

Kaye, K. (1982), *The Mental and Social Life of Babies: How parents create persons*, Hemel Hempstead: Harvester Wheatsheaf.

Keele, (1982), 'Learning and control of coordinated motor patterns: the programming perspective', in J. A. S. Kelso (ed.), *Human Motor Behaviour: An introduction*, Hillsdale, N.J.: Lawrence Erlbaum.

Keil, F. C. (1979), *Semantic and Conceptual Development: An ontological perspective*, Cambridge, Mass.: Harvard University Press.

Keil, F. C. (1981a), 'Children's thinking: what never develops?', *Cognition*, 10, pp. 159–66.

Keil, F. C. (1981b), 'Constraints on knowledge and cognitive development', *Psychological Review*, 88, pp. 197–227.

Keil, F. C. (1985), 'Mechanisms in cognitive development and the structure of knowledge', in R. J. Sternberg (ed.), *Mechanisms of Cognitive Development*, San Francisco: Freeman.

Keil, F. C. (1990), 'Constraints on constraints: surveying the epigenetic landscape', *Cognitive Science*, 14, pp. 135–68.

Keil, F. C. (1991), 'Godzilla vs. Mothra and the Sydney opera house: boundary conditions on functional architecture in infant visual perception and beyond', *Mind and Language*, 6, pp. 239–51.

Kellman, P. J. (1984), 'Perception of three-dimensional form by human infants', *Perception and Psychophysics*, 36, pp. 353–8.

Kellman, P. J. and Short, K. R. (1987), 'Development of three-dimensional form perception', *Journal of Experimental Psychology: Human perception and performance*, 13, pp. 545–57.

Kellman, P. J. and Spelke, E. S. (1983), 'Perception of partly occluded objects in infancy', *Cognitive Psychology*, 15, pp. 483–524.

Kellman, P. J., Gleitman, H. and Spelke, E. S. (1987), 'Object and observer motion in the perception of objects by infants', *Journal of Experimental Psychology: Human perception and performance*, 13, 586–93.

Kellman, P. J., Spelke, E. S. and Short, K. R. (1986), 'Infant perception of object unity from translatory motion in depth and vertical translation', *Child Development*, 57, pp. 72–86.

Kelso, J. A. S. (ed.) (1982), *Human Motor Behaviour: An introduction*, Hillsdale, N.J.: Lawrence Erlbaum.

Kessen, W., Haith, M. M. and Salapatek, P. (1970), 'Human infancy: a bibliography and guide', in P. H. Mussen (ed.), *Carmichael's Manual of Child Psychology*, New York: Wiley.

Kestenbaum, R., Termine, N. and Spelke, E. S. (1987), 'Perception of object boundaries by 3-month-old infants', *British Journal of Developmental Psychology*, 5, pp. 367–83.

Kirsh, D. (1991), 'Today the earwig, tomorrow man?', *Artificial Intelligence*, 47, pp. 161–84.

Kleiner, K. A. and Banks, M. S. (1987), 'Stimulus energy does not account for 2-month-olds' face preferences', *Journal of Experimental Psychology*, 13, pp. 594–600.

Knudsen, E. I. (1988), 'Sensitive and critical periods in the development of sound localization', in S. Easter, R. F. Barald, and B. M. Carlson (eds), *From Message to Mind: Directions in developmental neurobiology*, Sunderland, Mass.: Sinauer Associates.

Koenderink, J. J. (1980), 'Why argue about direct perception?', *Behavioral and Brain Sciences*, 3, pp. 390–1.

Kugler, P. N., Kelso, J. A. S. and Turvey, M. T. (1982), 'On the control and coordination of naturally developing systems', in J. A. S. Kelso and J. E. Clark (eds), *The Development of Movement Control and Coordination*, Chichester: Wiley.

Lawler, R. (1981), 'The progressive construction of mind', *Cognitive Science*, 5, pp. 1–30.

Lawler, R. (1985), *Computer Experience and Cognitive Development: One child's learning in a computer culture*, Chichester: Ellis Horwood

Le Compte, G. K. and Gratch, G. (1972), 'Violation of a rule as a method of diagnosing infants' levels of object concept', *Child Development*, 43, pp. 385–96.

Lee, D. N. and Aronson, E. (1974), 'Visual proprioceptive control of standing in human infants', *Perception and Psychophysics*, 15, pp. 529–32.

Legerstee, M. (1992), 'A review of the animate–inanimate distinction in infancy: implications for models of social and cognitive knowing', *Early Development and Parenting*, 1, pp. 59–67.

Legerstee, M., Pomerleau, A., Malcuit, G. and Feider, H. (1987), 'The development of responses to people and a doll: implications for research in communication', *Infant Behaviour and Development*, 10, pp. 81–95.

Lenat, D. (1979), 'On automated scientific theory formation: a case study using the AM program', in J. E. Hayes, D. Michie and O. I. Mikulich (eds), *Machine Intelligence*, vol. 9, Chichester: Ellis Horwood.

Leslie, A. M. (1983), 'Pretend play and representation in the second year of life', poster presented at the Annual Conference of the British Psychological Society Developmental Section, Oxford, September.

Leslie, A. M. (1984), 'Spatiotemporal continuity and the perception of causality in infants', *Perception*, 13, pp. 287–305.

Leslie, A. M. (1987), 'Pretense and representation: the origins of a "theory of mind"', *Psychological Review*, 94, pp. 412–26.

Leslie, A. M. (1990), 'Pretence, autism and the basis of "theory of mind"', *The Psychologist*, 3, pp. 120–3.

Leslie, A. M. and Keeble, S. (1987), 'Do six-month-old infants perceive causality?', *Cognition*, 25, pp. 265–88.

Lettvin, J., Maturana, H., McCulloch, W. and Pitts, W. H. (1959), 'What the frog's eye tells the frog's brain', *Proceedings IRE*, 47, pp. 1940–51.

Levin, J. A. and Kareev, Y. (1980), 'Problem solving in everyday situations', *The Quarterly Newsletter of the Laboratory of Comparative Human Cognition*, 2, pp. 47–52.

Lewis, M. (1969), 'Infants' responses to facial stimuli during the first year of life', *Developmental Psychology*, 1, pp. 75–86.

Longuet-Higgins, H. C. (1978), 'On describing cognitive processes', *Behavioral and Brain Sciences*, 1, p. 110.

Longuet-Higgins, H. C. and Prazdny, S. (1980), 'The interpretation of a moving retinal image', *Proceedings of the Royal Society*, Series B, 208, pp. 385–97.

Luger, G., Bower, T. G. R. and Wishart, J. G. (1983), 'A model of the development of the early infant object concept', *Perception*, 12, pp. 21–34.

Luria, A. R. (1961), *The Role of Speech in the Regulation of Normal and Abnormal Behaviour*, Oxford: Pergamon Press.

Luria, A. R. (1973), *The Working Brain*, Harmondsworth: Penguin.

McDermott, D. (1983), 'Under what conditions can a machine attribute meanings to symbols?', in A. Bundy, (ed.), *Proceedings of the Eighth International Joint Conference on Artificial Intelligence*, Karlsruhe, Germany, vol. 1, Los Altos, Calif.: William Kaufmann.

McDonnell, P. M. (1975), 'The development of visually guided reaching', *Perception and Psychophysics*, 18, pp. 181–5.

McGarrigle, J. R. and Donaldson, M. (1974), 'Conservation accidents', *Cognition*, 3, pp. 341–50.

McGarrigle, J. R., Grieve, R. and Hughes, M. (1978), 'Interpreting inclusion: a contribution to the study of the child's cognitive and linguistic development', *Journal of Experimental Child Psychology*, 26, pp. 528–50.

McKenzie, B. E. and Day, R. H. (1973), 'Perceptual shape constancy in early infancy', *Perception*, 3, pp. 315–26.

McKenzie, B. E., Day, R. H. and Ihsen, E. (1984), 'Localization of events in space: young infants are not always egocentric', *British Journal of Developmental Psychology*, 2, pp. 1–9.
Mach, E. (1885), *The Analysis of Sensations*, translated (1959) C. M. Williams, New York: Dover.
Mandler, G. (1975), *Mind and Emotion*, New York: Wiley.
Mandler, J. M. (1983), 'Representation', in J. H. Flavell and E. M. Markman (eds), *Cognitive Development. Handbook of Child Psychology*, 4th edn, vol. 3. New York: Wiley.
Mandler, J. M. (1984), 'Representation and recall in infancy', in M. Moscovitch (ed.), *Advances in the Study of Communication and Affect*, vol. 9, New York: Plenum Press.
Mandler, J. M. (1992), 'Concept formation in infancy', paper presented to the British Infancy Research Group, Exeter, July.
Mandler, J. M. (in press) 'How to build a baby II: conceptual primitives', *Psychological Review*.
Maratos, O. (1982), 'Trends in the development of early imitation in infancy', in T. G. Bever (ed.), *Regressions in Development: Basic phenomena and theories*, Hillsdale, N.J.: Lawrence Erlbaum.
Marr, D. (1976), 'Early processing of visual information', *Philosophical Transactions of the Royal Society*, Series B, 275, pp. 483–519.
Marr, D. (1977), 'Artificial intelligence – a personal view', *Artificial Intelligence*, 9, pp. 37–48.
Marr, D. (1980), 'Visual information processing: the structure and creation of visual representations', *Philosophical Transactions of the Royal Society*, Series B, 290, pp. 199–218.
Marr, D. (1982), *Vision: A computational investigation into the human representation and processing of visual information*, San Francisco: Freeman.
Marr, D. and Hildreth E. (1980), "Theory of edge detection', *Proceedings of the Royal Society*, Series B, 207, pp. 187–217.
Marr, D. and Nishihara, H. K. (1978), 'Representation and recognition of the spatial organization of three-dimensional shapes', *Proceedings of the Royal Society*, Series B, 200, pp. 269–94.
Marr, D. and Poggio, T. (1976), 'Cooperative computation of stereo disparity', *Science*, 94, pp. 283–7.
Marr, D. and Ullman, S. (1981), 'Directional selectivity and its use in early visual processing', *Proceedings of the Royal Society*, Series B, 211, pp. 151–80.
Maturana, H. R. (1970), 'Neurophysiology of cognition', in P. Garvin (ed.), *Cognition: A multiple view*, New York: Spartan Books.
Maturana, H. R. and Varela, F. J. (1980), *Autopoiesis and Cognition: The realization of the living*, Dordrecht: Reidel.
Maurer, D. and Lewis, T. L. (1979), 'Peripheral discrimination by 3-month-old infants', *Child Development*, 50, pp. 276–9.
Mehler, J. (1985), 'Introduction', in J. Mehler and R. Fox (eds), *Neonate Cognition: Beyond the blooming, buzzing confusion*, Hillsdale, N.J.: Lawrence Erlbaum.
Meltzoff, A. N. (1981), 'Imitation, inter-modal coordination and representation in early infancy', in G. E. Butterworth (ed.), *Infancy and Epistemology: An evaluation of Piaget's theory*, Hemel Hempstead: Harvester Wheatsheaf.
Meltzoff, A. N. and Moore, M. K. (1977), 'Imitation of facial and manual gestures by human neonates', *Science*, 198, pp. 75–8.

Meltzoff, A. N. and Moore, M. K. (1983), 'Newborns imitate adult facial gestures', *Child Development*, 54, pp. 702–9.

Meltzoff, A. N. and Moore, M. K. (1985), 'Cognitive foundations and social functions of imitation and intermodal representation in infancy', in J. Mehler and R. Fox (eds), *Neonate Cognition: Beyond the blooming, buzzing confusion*, Hillsdale, N.J.: Lawrence Erlbaum.

Mendelson, M. J. and Haith, M. M. (1976), 'The relation between audition and vision in the human newborn', Monographs Society for Research in Child Development, 41, No. 4, Chicago: University of Chicago Press.

Mervis, C. and Rosch, E. (1981), 'Categorization of natural objects', *Annual Review of Psychology*, 32, pp. 89–115.

Meyer, J.-A. and Guillot, A. (1991), 'Simulation of adaptive behaviour in animats: review and prospect', in J.-A. Meyer and S. W. Wilson (eds), *From Animals to Animats: Proceedings of the first international conference on the simulation of adaptive behaviour*, Cambridge, Mass.: Bradford/MIT Press.

Michaels, C. F. and Carello, C. (1981), *Direct Perception*, New York: Prentice Hall.

Miller, G. A., Galanter, E. and Pribram, K. (1960), *Plans and the Structure of Behavior*, New York: Holt.

Miller, L. (1978), 'Has artificial intelligence contributed to our understanding of the human mind? A critique of arguments for and against', *Cognitive Science*, 2, pp. 111–27.

Minsky, M. L. (1980), 'Decentralized minds', *Behavioral and Brain Sciences*, 3, pp. 439–40.

Minsky, M. L. and Papert, S. A. (1988), *Perceptrons*, 2nd edn, Cambridge, Mass.: MIT Press.

Mitchell, T. M. (1983), 'Learning and problem-solving', in A, Bundy (ed.), *Proceedings of the Eighth International Joint Conference on Artificial Intelligence*, Karlsruhe, Germany, vol. 2, Los Altos, Calif.: William Kaufmann.

Morton, J. and Johnson, M. (1991), 'CONSPEC and CONLERN: A two process theory of infant face recognition', *Psychological Review*, 98, pp. 164–81.

Mounoud, P. (1981), 'Cognitive development: construction of new structures or construction of internal organizations?', in I. E. Sigel, D. M. Brodzinsky and R. M. Golinkoff (eds), *New Directions in Piagetian Theory and Practice*, Hillsdale, N.J.: Lawrence Erlbaum.

Mounoud, P. (1982), 'Revolutionary periods in early development', in T. G. Bever (ed.), *Regressions in Mental Development: Basic phenomena and theories*, Hillsdale, N.J.: Lawrence Erlbaum.

Mounoud, P. (1986), 'Similarities between developmental sequences at different age periods', in I. Levin (ed.), *Stage and Structure – Reopening the Debate*, Norwood, N.J.: Ablex.

Mounoud, P. and Bower, T. G. R. (1974), 'Conservation of weight in infants', *Cognition*, 3, pp. 29–40.

Mounoud, P. and Hauert, C. A. (1982a), 'Development of sensorimotor organization in young children: grasping and lifting objects', in G. E. Forman (ed.), *Action and Thought: From sensorimotor schemes to thought operations*, New York: Academic Press.

Mounoud, P. and Hauert, C. A. (1982b), 'Sensorimotor and postural behaviours: their relation to cognitive development', in W. W. Hartup (ed.), *Review of Child Development Research*, vol. 6, Chicago and London: University of Chicago Press.

Mounoud, P. and Vinter, A. (1981), 'Representation and sensorimotor development', in G. E. Butterworth (ed.), *Infancy and Epistemology: An evaluation of Piaget's theory*, Hemel Hempstead: Harvester Wheatsheaf.

Mozer, M and Smolensky, P. (1989), 'Using relevance to reduce network size automatically', *Connection Science*, 1, pp. 3–17.

Muller, A. A. and Aslin, R. N. (1978), 'Visual tracking as an index of the object concept', *Infant Behavior and Development*, 1, pp. 309–19.

Neilson, J. and Dockrell, J. (1982), 'Cognitive tasks as interactional settings', in G. E. Butterworth and P. H. Light (eds), *Social Cognition: Essays in the development of understanding*, Hemel Hempstead: Harvester Wheatsheaf.

Neisser, U. (ed.), (1982), *Memory Observed: Remembering in natural contexts*, San Francisco: Freeman.

Neisser, U. (1988), 'Five kinds of self-knowledge', *Philosophical Psychology*, 1, pp. 35–60.

Nelson, K. (1977), 'Cognitive development and the acquisition of concepts', in R. C. Anderson, R. J. Spiro and W. E. Montague (eds), *Schooling and the Acquisition of Knowledge*, Hillsdale, N.J.: Lawrence Erlbaum.

Newell, A. (1970), 'Remarks on the relationship between artificial intelligence and cognitive psychology', in R. Banerji and M. D. Mesarovic (eds), *Theoretical Approaches to Non-Numerical Problem-Solving*, Berlin: Springer-Verlag.

Newell, A. (1980), 'Physical symbol systems', *Cognitive Science*, 4, pp. 135–83.

Newell, A. and Simon, H. A. (1963), 'GPS – a program that simulates human thought', in E. A. Feigenbaum and J. Feldman (eds), *Computers and Thought*, New York: McGraw-Hill.

Newell, A. and Simon, H. A. (1972), *Human Problem Solving*, Englewood Cliffs, N.J.: Prentice Hall.

Newell, A. and Simon, H. A. (1976), 'Computer science as empirical inquiry: symbols and search', *Communications of the ACM*, 19, pp. 113–26.

Nickerson, R. S. and Adams, M. J. (1979), 'Long term memory for a common object', *Cognitive Psychology*, 11, pp. 287–307.

Norman, D. A. (1980), 'Twelve issues for cognitive science', *Cognitive Science*, 4, pp. 1–32.

Norman, D. A. (1986), 'Reflections on cognition and parallel distributed processing', in J. L. McClelland and D. E. Rumelhart (eds), *Parallel Distributed Processing*, vol. 2, *Psychological and Biological Models*, Cambridge, Mass.: Bradford/MIT Press.

Norris, D. (1991), 'The constraints on connectionism', *The Psychologist*, 4, pp. 293–6.

Oatley, K. (1972), *Brain Mechanisms and Mind*, London: Thames and Hudson.

Oatley, K. (1978), *Perceptions and Representations: The theoretical bases of brain research and psychology*, London: Methuen.

Olson, R., Yonas, A. and Cooper, R. G. (1980), 'Development of pictorial perception', in M. A. Hagen (ed.), *The Perception of Pictures*, vol. 2, New York: Academic Press.

Osofsky J. D. (ed.) (1987), *Handbook of Infant Development*, New York: Wiley.

Oyama, S. (1985), *The Ontogeny of Information: Developmental systems and evolution*, Cambridge: Cambridge University Press.

Oyama, S. (1988), 'How do you transmit a template?', *Behavioral and Brain Sciences*, 11, pp. 644–5.

Palmer, S. E. (1978), 'Fundamental aspects of cognitive representation', in E. Rosch and B. B. Lloyd (eds), *Cognition and Categorization*, Hillsdale, N.J.: Lawrence Erlbaum.

Partridge, D. (1987), 'Human decision making and the symbolic search space paradigm', *AI & Society*, 1, pp. 103–14.

Partridge, D. and Paap, K. (1988), 'An introduction to learning', *Artificial Intelligence Review*, 2, pp. 79–102.

Peacocke, C. (1986), 'Explanation in computational psychology', *Mind and Language*, 1, pp. 101–23.

Pentland, A. (1985), 'Perceptual organization and the representation of natural form', Stanford, Calif.: Stanford Research International Technical Note No. 357, July.

Piaget, J. (1953), *The Origin of Intelligence in the Child*, London: Routledge and Kegan Paul.

Piaget, J. (1955), *The Child's Construction of Reality*, London: Routledge and Kegan Paul.

Piaget, J. (1968a), *On the Development of Memory and Identity*, Barre, Mass.: Clark University Press with Barre Publishers.

Piaget, J. (1968b), *Structuralism*, London: Routledge and Kegan Paul.

Piaget, J. (1963), 'Le langage et les opérations intellectuelles', translated (1969) in H. G. Furth, *Piaget and Knowledge: Theoretical foundations*, Englewood Cliffs, N.J.: Prentice Hall.

Piaget, J. (1970), *Genetic Epistemology*, New York and London: Columbia University Press.

Piaget, J. (1971), *Biology and Knowledge*, Edinburgh: Edinburgh University Press.

Piaget, J. (1972), 'Intellectual evolution from adolescence to adulthood', *Human Development*, 15, pp. 1–12.

Piaget, J. (1974), *The Child and Reality: Problems of genetic epistemology*, London: Frederic Muller.

Piaget, J. (1976), *The Grasp of Consciousness*, London: Routledge and Kegan Paul.

Piaget, J. (1978a), *The Development of Thought: Equilibration of cognitive structures*, Oxford: Basil Blackwell.

Piaget, J. (1978b), *Success and Understanding*, London: Routledge and Kegan Paul.

Piaget, J. and Inhelder, B. (1969), *The Psychology of the Child*, London: Routledge and Kegan Paul.

Piaget, J. and Inhelder, B. (1973), *Memory and Intelligence*, London: Routledge and Kegan Paul.

Pieraut-Le Bonniec, G. (1985), 'Hand–eye coordination and infants' construction of convexity and concavity', *British Journal of Developmental Psychology*, 3, pp. 273–80.

Pinker, S. and Prince, A. (1988), 'On language and connectionism', *Cognition*, 28, pp. 73–193.

Poggio, T. and Reichardt, W. (1976), 'Visual control of orientation behaviour in the fly. Part II: towards the underlying neural interactions', *Quarterly Review of Biophysics*, 9, pp. 377–438.

Poincaré, H. (1905), *The Value of Science*, translated (1958) G. B. Halstead, Paris: Flammarion; New York: Dover, 1958.

Prazdny, S. (1980), 'A computational study of a period of infant object-concept development', *Perception*, 9, pp. 125–50.

Prazdny, S. (1981), 'Determining the instantaneous direction of motion from optical flow generated by a curvilinearly moving observer', *Computer Graphics and Image Processing*, 17, pp. 238–48.

Premack, D. (1990), 'The infant's theory of self-propelled objects', *Cognition*, 36, pp. 1–16.
Premack, D. and Woodruff, G. (1978), 'Does the chimpanzee have a theory of mind?', *Behavioral and Brain Sciences*, 4, pp. 515–26.
Priest, A. and Young, R. M. (1986), 'Methods for evaluating micro-theory systems', in J. Self (ed.), *Artificial Intelligence and Human Learning: Intelligent computer-aided instruction*, London and New York: Chapman and Hall.
Pylyshyn, Z. W. (1978), 'Computational models and empirical constraints', *Behavioral and Brain Sciences*, 1, pp. 93–127.
Pylyshyn, Z. W. (1979), 'Complexity and the study of artificial and human intelligence', in M. Ringle, (ed.), *Philosophical Perspectives in Artificial Intelligence*, Atlantic Highlands, N.J.: Humanities Press.
Pylyshyn, Z. W. (1980), 'Computation and cognition: issues in the foundations of cognitive science', *Behavioral and Brain Sciences*, 3, pp. 111–69.
Pylyshyn, Z. W. (1981), 'Complexity and the study of natural and artificial intelligence', in J. Haugeland (ed.), *Mind Design*, Cambridge, Mass.: Bradford.
Pylyshyn, Z. W. (1984), *Computation and Cognition: Toward a foundation for cognitive science*, Cambridge, Mass.: MIT Press.
Rader, N., Bausano, M., and Richards, M. E. (1980), 'On the nature of the visual cliff avoidance response in human infants', *Child Development*, 51, pp. 61–8.
Reddy, V. (1992), 'Protodeclarative pointing: gaze control or shared attention?', paper presented at the Annual Conference of the British Psychological Society Developmental Section, Edinburgh, September.
Redshaw, M. (1978), 'Cognitive development in gorilla and human infants', *Human Evolution*, 7, pp. 133–41.
Reed, E. S. (1986), 'Review of V. Bruce and P. Green, *Visual Perception: Physiology, psychology and ecology*' (1985), London: Lawrence Erlbaum, *International Society for Ecological Psychology Newsletter*, 2, pp. 7–9.
Regan, D., Beverley, D. I. and Cynader, M. (1979), 'Stereoscopic subsystems for position in depth and for motion in depth', *Proceedings of the Royal Society*, Series B, 204, pp. 485–501.
Richards, M. P. M. (1974), *Integration of a Child into a Social World*, Cambridge: Cambridge University Press.
Ritchie, G. and Hannah, M. (1984), 'AM: a case study in AI methodology', *Artificial Intelligence*, 23, pp. 249–68.
Rogers, B. and Graham, M. (1979), 'Motion parallax as an independent cue for depth perception', *Perception*, 8, pp. 125–34.
Rogoff, B. and Lave, J. (eds), (1984), *Everyday Cognition: Its development in social context*, Cambridge, Mass.: Harvard University Press.
Rogoff, B., Malkin, C. and Gilbride, K. (1984), 'Interaction with babies as guidance in development', in B. Rogoff and J. V. Wertsch (eds), *Children's Learning in the Zone of Proximal Development*, San Francisco: Jossey-Bass.
Rozin, P. (1976), 'The evolution of intelligence and access to the cognitive unconscious', in J. M. Sprague and A. A. Epstein (eds), *Progress in Psychobiology and Physiological Psychology*, New York: Academic Press.
Rumelhart, D. E. and McClelland, J. L. (1986), 'PDP models and general issues in cognitive science', in D. E. Rumelhart and J. L. McClelland (eds), *Parallel Distributed Processing*, vol. 1, *Foundations*, Cambridge, Mass.: Bradford/MIT Press.

Rumelhart, D. E., McClelland, J. L. and the PDP Research Group (1986), *Parallel Distributed Processing*, vol. 1, *Foundations*, Cambridge, Mass.: Bradford/MIT Press.

Rumelhart, D. E., Smolensky, P., McClelland, J. L. and Hinton, G. E. (1986), 'Schemata and sequential thought processes in PDP models', in J. L. McClelland and D. E. Rumelhart (eds), *Parallel Distributed Processing*, vol. 2, *Psychological and Biological Models*, Cambridge, Mass.: Bradford/MIT Press.

Runeson, S. (1977), 'On the possibility of "smart" perceptual mechanisms', *Scandinavian Journal of Psychology*, 18, pp. 172–9.

Russell, J. (1984), *Explaining Mental Life: Some philosophical issues in psychology*, London: Macmillan.

Rutkowska, J. C. (1983), 'Patterns of behaviour on the visual tracking paradigm in the first year of life: increasing success in object search or construction of a new problem–space?', paper presented at the Annual Conference of the British Psychological Society Developmental Section, Oxford, September.

Rutkowska, J. C. (1984), 'Explaining infant perception: insights from artificial intelligence?', University of Sussex, Cognitive Science Research Paper, Serial No. CSRP 005.

Rutkowska, J. C. (1985), 'Does the phylogeny of conceptual development increase our understanding of concepts or of development?', in G. E. Butterworth, J. C. Rutkowska and M. Scaife (eds), *Evolution and Developmental Psychology*, Hemel Hempstead: Harvester Wheatsheaf.

Rutkowska, J. C. (1986), 'Developmental psychology's contribution to cognitive science', in K. S. Gill (ed.), *Artificial Intelligence for Society*, Chichester: Wiley.

Rutkowska, J. C. (1987), 'Computational models and developmental psychology', in J. C. Rutkowska and C. Crook (eds), *Computers, Cognition and Development*, Chichester: Wiley.

Rutkowska, J. C. (1988), 'Perception, action and representation in infancy: a computational approach', unpublished D.Phil thesis, University of Sussex.

Rutkowska, J. C. (1990), 'Action, connectionism and enaction: a developmental perspective', *AI & Society*, 4, pp. 96–114.

Rutkowska, J. C. (1991), 'Looking for "constraints" in infants' perceptual-cognitive development', *Mind and Language*, 6, pp. 215–38.

Rutkowska, J. C. (1992a), 'A computational alternative to the Piagetian infant', *Infancia y Apprendizaje*, 57, pp. 99–120.

Rutkowska, J. C. (1992b), 'Early prehension intention: Bruner and Koslowski (1972) revisited', paper presented at the Annual Conference of the British Psychological Society Developmental Section, Edinburgh, September.

Salapatek, P. (1975), 'Pattern perception in early infancy', in L. B. Cohen and P. Salapatek (eds), *Infant Perception: From sensation to cognition*, New York: Academic Press.

Satinoff, E. (1978), 'Neural organization and evolution of thermal regulation in animals', *Science*, 201, pp. 16–22.

Schiff, W., Caviness, J. A. and Gibson, J. J. (1962), 'Persistent fear responses in rhesus monkeys to the optical stimulus of looming', *Science*, 136, pp. 982–3.

Schmidt, R. A. (1982a), 'More on motor programs', in J. A. S. Kelso (ed.), *Human Motor Behaviour: An introduction*, Hillsdale, N.J.: Lawrence Erlbaum.

Schmidt, R. A. (1982b), 'The schema theory', in J. A. S. Kelso (ed.), *Human Motor Behaviour: An introduction*, Hillsdale, N.J.: Lawrence Erlbaum.

Schultz, D. (1981), *A History of Modern Psychology*, 3rd edn, New York and London: Academic Press.

Scott, P. D. and Vogt, R. C. (1983), 'Knowledge oriented learning', in A. Bundy (ed.), *Proceedings of the Eighth International Joint Conference on Artificial Intelligence*, Karlsruhe, Germany, vol. 1, Los Altos, Calif.: William Kaufmann.

Scriven, M. (1977), 'Commentary on: Schooling and the facilitation of knowing. By J. D. Bransford, K. E. Nitsch and J. J. Franks', in R. C. Anderson, R. J. Spiro and W. E. Montague (eds), *Schooling and the Acquisition of Knowledge*, Hillsdale, N.J.: Lawrence Erlbaum.

Shapiro, B. A. (1973), *A Survey of Problem Solving Languages and Systems*, National Science Foundation, Technical Report No. TR 235.

Shapiro, D. C. and Schmidt, R. A. (1982), 'The schema theory: recent evidence and developmental implications', in J. A. S. Kelso and J. E. Clark (eds), *The Development of Movement Control and Coordination*, Chichester: Wiley.

Shaw, R. E. and McIntyre, M. (1974), 'Algoristic foundations of cognitive psychology', in W. Weimer and D. S. Palermo (eds), *Cognition and the Symbolic Processes*, Hillsdale, N.J.: Lawrence Erlbaum.

Shaw R. E. and Bransford, J. (eds) (1977), *Perceiving, acting and knowing: Toward an ecological psychology*, Hillsdale, N.J.: Lawrence Erlbaum.

Shaw R. E. and Turvey, M. T. (1980), 'Methodological realism', *Behavioral and Brain Sciences*, 3, pp. 94–7.

Shepard, R. N. (1984), 'Ecological constraints on internal representation: resonant kinematics of perceiving, imagining, thinking and dreaming', *Psychological Review*, 91, pp. 417–47.

Simon, H. A. (1957), *Models of Man*, New York: Wiley.

Simon, H. A. (1962), 'An information processing theory of cognitive development', in W. Kessen and C. Kuhlman (eds), *Thought in the Young Child*, Chicago: University of Chicago Press.

Simon, H. A. (1969), *The Sciences of the Artificial*, Cambridge, Mass.: MIT Press.

Sinha, C. (1985), 'A socio-naturalistic approach to human development', in G. E. Butterworth, J. C. Rutkowska and M. Scaife (eds), *Evolution and Developmental Psychology*, Hemel Hempstead: Harvester Wheatsheaf.

Slater, A. and Bremner, J. G. (eds), (1989), *Infant Development*, Hove and London: Lawrence Erlbaum.

Slater, A., Mattock, A. and Brown, E. (1991), 'Form perception at birth: Cohen and Younger (1984) revisited', *Journal of Experimental Child Psychology*, 51, pp. 395–406.

Slater, A., Morison V. and Rose, D. (1983), 'Perception of shape by the new-born baby', *British Journal of Developmental Psychology*, 1, pp. 135–42.

Slater, A., Morison, V. and Somers, M. (1988), 'Orientation discrimination and cortical function in the human newborn', *Perception*, 17, pp. 597–602.

Slater, A., Morison, V., Town, C. and Rose, D. (1985), 'Movement perception and identity constancy in the new-born baby', *British Journal of Developmental Psychology*, 3, pp. 211–20.

Slater, A., Morison, V., Somers, M., Mattock, A., Brown E., and Taylor, D. (1990), 'Newborn and older infants' perception of partly occluded objects', *Infant Behaviour and Development*, 13, pp. 33–49.

Sleeman, D. (1987), 'Cognitive science, AI and developmental psychology. Are there links? Could there be links?', in J. C. Rutkowska and C. Crook (eds), *Computers, Cognition and Development*, Chichester: Wiley.

Sloman, A. (1978), *The Computer Revolution in Philosophy*, Hemel Hempstead: Harvester Wheatsheaf.

Sloman, A. (1980), 'What kind of indirect process is perception?', *Behavioral and Brain Sciences*, 3, pp. 401–4.

Sloman, A. (1982), 'Computational epistemology', *Cahiers de la Fondation Archives Jean Piaget*, Geneva, June, No. 2–3, pp. 51–93.

Sloman, A. (1983a), 'Commentary on M. A. Boden, "Artificial intelligence and animal psychology"', *New Ideas in Psychology*, 1, pp. 41–50.

Sloman, A. (1983b), 'Image interpretation: the way ahead?', in O. J. Braddick and A. C. Sleigh (eds), *Physical and Biological Processing of Images*, Berlin: Springer-Verlag.

Sloman, A. (1984), 'The structure of the space of possible minds', in S. Torrance (ed.), *The Mind and the Machine: Philosophical aspects of artificial intelligence*, Chichester: Ellis Horwood.

Sloman, A. (1987a), 'The explanatory status of mentalistic concepts', Institute of Cognitive and Information Sciences Seminar, University of Sussex, May.

Sloman, A. (1987b), 'What are the purposes of vision?' University of Sussex, Cognitive Science Research Paper, Serial No. CSRP 066.

Sloman, A. (1989), 'On designing a visual system. (Towards a Gibsonian computational model of vision)', *Journal of Experimental and Theoretical Artificial Intelligence*, 1, pp. 289–337.

Smith, B. C. (1985), 'Is computation formal?', unpublished manuscript, Stanford, Calif.: Centre for the Study of Language and Information, Stanford University.

Smith, L. (1982), 'Class inclusion and conclusions about Piaget's theory', *British Journal of Psychology*, 73, pp. 267–76.

Smolensky, P. (1986), 'Neural and conceptual interpretations of parallel distributed processing models', in J. L. McClelland and D. E. Rumelhart, (eds), *Parallel Distributed Processing*, vol. 2, *Psychological and Biological Models*, Cambridge, Mass.: Bradford/MIT Press.

Smolensky, P. (1987), 'Connectionist AI, symbolic AI and the brain', *Artificial Intelligence Review*, 1, pp. 95–110.

Smolensky, P. (1988), 'On the proper treatment of connectionism', *Behavioral and Brain Sciences*, 11, pp. 1–23.

Spelke, E. S. (1982), 'Perceptual knowledge of objects in infancy', in J. Mehler, E. C. T. Walker and M. Garrett (eds), *Perspectives on Mental Representation: Experimental and theoretical studies of cognitive processes and capacities*, Hillsdale, N.J.: Lawrence Erlbaum.

Spelke, E. S. (1985), 'Perception of unity, persistence and identity: thoughts on infants' conceptions of objects', in J. Mehler and R. Fox (eds), *Neonate Cognition: Beyond the blooming, buzzing confusion*, Hillsdale, N.J.: Lawrence Erlbaum.

Spelke, E. S. (1988), 'Where perceiving ends and thinking begins: the apprehension of objects in infancy', in A. Yonas (ed.), *Perceptual Development in Infancy: The Minnesota symposium on child psychology*, vol. 20, Hillsdale, N.J.: Lawrence Erlbaum.

Spelke, E. S. (1990), 'Principles of object perception', *Cognitive Science*, 14, pp. 29–56.

Spelke, E. S. (1991), 'Physical knowledge in infancy: Reflections on Piaget's theory', in S. Carey and R. Gelman (eds), *The Epigenesis of Mind: Essays on biology and cognition*, Hillsdale, N.J.: Lawrence Erlbaum.

Spelke, E. S. and Born, W. S. (1982), 'Perception of visible objects by three-month-old infants', unpublished paper, cited in Spelke 1982.

Steels, L. (1982), 'Constraints as consultants', *Proceedings of the Fifth European Conference on Artificial Intelligence*, Orsay, France.

Steels, L. (1990), 'Towards a theory of emergent functionality', in J.-A. Meyer and S. W. Wilson (eds), *From Animals to Animats: Proceedings of the first international conference on the simulation of adaptive behaviour*, Cambridge, Mass.: MIT Press.

Stelmach, G. E. (1982), 'Motor control and motor learning: the closed loop perspective', in J. A. S. Kelso (ed.), *Human Motor Behaviour: An introduction*, Hillsdale, N.J.: Lawrence Erlbaum.

Stern, D. (1985), *The First Relationship*, London: Fontana.

Stewart, J. (1989), 'A biologist's view of connectionism', paper presented at the Collège International de Philosophie and Université de Technologie de Compiègne Seminar, 'La Cognition', Chantilly, France, July.

Stich, S. (1978), 'Beliefs and subdoxastic states', *Philosophy of Science*, 45, pp. 499–518.

Stillings, N. A., Feinstein, M. H., Garfield, J. L., Rissland, E. L., Rosenbaum, D. A., Weisler, S. E. and Baker-Ward, L. (1986), *Cognitive Science: An introduction*, Cambridge, Mass.: Bradford/MIT Press.

Stone, L. J., Smith, H. T. and Murphy, L. B. (eds) (1974), *The Competent Infant*, London: Tavistock.

Suchman, L. (1987), *Plans and Situated Actions: The problem of human–machine communication*, Cambridge: Cambridge University Press.

Sussman, G. J. (1975), *A Computer Model of Skill Acquisition*, New York: American Elsevier.

Sylvester-Bradley, B. (1985), 'Failure to distinguish between people and things in early infancy', *British Journal of Developmental Psychology*, 3, pp. 281–92.

Thelen, E. (1984), 'Learning to walk: ecological demands and phylogenetic constraints', in L. P. Lipsitt and C. Rovee-Collier (eds), *Advances in Infancy Research*, vol. 3, Norwood, N.J.: Ablex.

Thornton, S. (1987), 'Adaptive production systems as models of cognitive development', in J. C. Rutkowska and C. Crook (eds), *Computers, Cognition and Development*, Chichester: Wiley.

Trevarthen, C. (1968), 'Two mechanisms of vision in primates', *Psychologische Forschung*, 31, pp. 299–337.

Trevarthen, C. (1974), 'Conversations with a two-month-old', *New Scientist*, 62, pp. 230–5.

Trevarthen, C. (1975), 'Early attempts at speech', in R. Lewin (ed.), *Child Alive*, London: Temple-Smith.

Trevarthen, C. (1980), 'Neurological development and the growth of psychological functions', in H. J. Sants (ed.), *Developmental Psychology and Society*, London: Macmillan.

Trevarthen, C. (1984), 'How control of movement develops', in H. T. A. Whiting (ed.), *Human Motor Actions: Bernstein reassessed*, Amsterdam: North Holland.

Trevarthen, C. (1992), 'Consciousness of persons in infancy: its motives and functions', paper presented to the British Infancy Research Group, Exeter, July.

Trevarthen, C., Hubley, P. and Sheeran, L. (1975), 'Les activités innées du nourrisson', *La recherche*, 6, pp. 447–58.

Turvey, M. T. (1977), 'Preliminaries to a theory of action with reference to vision', in

R. E. Shaw and J. Bransford (eds), *Perceiving, Acting and Knowing: Towards an ecological psychology*, Hillsdale, N.J.: Lawrence Erlbaum.
Turvey, M. T. and Kugler, P. N. (1984), 'An ecological approach to perception and action', in H. T. A. Whiting (ed.), *Human Motor Actions: Bernstein reassessed*, Amsterdam: North Holland.
Turvey, M. T., Shaw, R. E., Reed, E. S. and Mace, W. M. (1981), 'Ecological laws of perceiving and acting', *Cognition*, 9, pp. 237–301.
Ullman, S. (1979), *The Interpretation of Visual Motion*, Cambridge, Mass.: MIT Press.
Ullman, S. (1980), 'Against direct perception', *Behavioral and Brain Sciences*, 3, pp. 373–81.
Valsiner, J. (1987), *Culture and the Development of Children's Action*, Chichester: Wiley.
van Brakel, J. (1991), 'Meaning, prototypes and the future of cognitive science', *Minds and Machines*, 1, pp. 233–57.
Varela, F. J. (1988), 'Cognitive science: a cartography of current ideas', unpublished English manuscript, author's translation of *Connaître – Les Sciences Cognitives: Tendances et perspectives*, Paris: Editions du Seuil (1989).
von Frisch, K. (1966), *The Dancing Bees*, London: Methuen.
von Hofsten, C. (1977), 'Binocular convergence as a determinant of reaching behaviour in infancy', *Perception*, 6, pp. 139–44.
von Hofsten, C. (1980), 'Predictive reaching for moving objects by human infants', *Journal of Experimental Child Psychology*, 30, pp. 369–82.
von Hofsten, C. (1982), 'Eye–hand coordination in the newborn', *Developmental Psychology*, 18, pp. 450–67.
von Hofsten, C. (1984), 'Developmental changes in the organization of prereaching movements', *Developmental Psychology*, 20, pp. 378–88.
von Hofsten, C. (1985), 'Perception and action', in M. Frese and J. Sabini (eds), *Goal Directed Behaviour: The concept of action in psychology*, Hillsdale, N.J.: Lawrence Erlbaum.
von Hofsten, C. and Lee, D. N. (1985), 'Dialogue on perception and action', in W. H. Warren and R. E. Shaw (eds), *Persistence and Change: Proceedings of the first international conference on event perception*, Hillsdale, N.J.: Lawrence Erlbaum.
von Hofsten, C. and Spelke, E. S. (1985), 'Object perception and object-directed reaching in infancy', *Journal of Experimental Child Psychology*, 114, pp. 192–212.
Vuyk, R. (1981), *Overview and Critique of Piaget's Genetic Epistemology 1965–1980*, 2 vols, London: Academic Press.
Vygotsky, L. S. (1978), *Mind in Society*, Cambridge, Mass.: Harvard University Press.
Vygotsky, L. S. (1981), 'The genesis of higher mental functions', in J. Wertsch (ed.), *The Concept of Activity in Soviet Psychology*, Armonk, N.Y.: M. E. Sharpe.
Waltz, A. (1975), 'Generating semantic descriptions from drawings of scenes with shadows', in P. Winston (ed.), *The Psychology of Computer Vision*, New York: McGraw Hill.
Warrington, E. K. and Taylor, A. M. (1973), 'The contribution of the right parietal lobe to object recognition', *Cortex*, 9, pp. 152–64.
Wattam-Bell, J. (1992), 'The development of sensitivity to visual motion', paper presented to the British Infancy Research Group, Exeter, July.
Whiting, H. T. A. (ed.) (1984), *Human Motor Actions: Bernstein reassessed*, Amsterdam: North Holland.

Wilcox, S. and Katz, S. (1981), 'The ecological approach to development: an alternative to cognitivism', *Journal of Experimental Child Psychology*, 32, pp. 247–63.

Willatts, P. (1979), 'Adjustment of reaching to change in object position by young infants', *Child Development*, 50, pp. 911–13.

Willatts, P. (1984), 'Stages in the development of intentional search by young infants', *Developmental Psychology*, 20, pp. 389–96.

Willatts, P. (1989), 'Development of problem-solving in infancy', in A. Slater and G. Bremner (eds), *Infant Development*, Hove and London: Lawrence Erlbaum.

Willatts, P. (1992), 'Development of planning strategies in infancy', paper presented to the British Infancy Research Group, Exeter, July.

Winograd, T. (1973), 'The processes of language understanding', in J. Benthall (ed.), *The Limits of Human Nature*, London: Allen Lane.

Winograd, T. (1977), 'On some contested suppositions of generative linguistics about the scientific study of language', *Cognition*, 5, pp. 151–79.

Winograd, T. (1987), 'Cognition, attunement and modularity', *Mind and Language*, 2, pp. 97–103.

Winograd, T. and Flores, C. F. (1986), *Understanding Computers and Cognition: A new foundation for design*, Norwood, N.J.: Ablex.

Wishart, J. G. and Bower, T. G. R. (1985), 'A longitudinal study of the development of the object concept', *British Journal of Developmental Psychology*, 3, pp. 243–59.

Wood, D., Bruner, J. S. and Ross, G. (1976), 'The role of tutoring in problem solving', *Journal of Child Psychology and Psychiatry*, 17, pp. 89–100.

Woodfield, A. (1988), 'On the very idea of acquiring a concept', in J. Russell (ed.), *Philosophical Perspectives on Developmental Psychology*, Oxford: Basil Blackwell.

Yonas, A. and Granrud, C. E. (1985), 'The development of sensitivity to kinetic, binocular and pictorial depth information in human infants', in D. I. Ingle, M. Jeannerod and D. N. Lee (eds), *Brain Mechanisms and Spatial Vision*, Dordrecht: Martinus Nijhoff.

Yonas, A., Granrud, C. E. and Pettersen, L. (1985), 'Infants' sensitivity to relative size information for distance', *Developmental Psychology*, 21, pp. 161–7.

Young, D. (1988), 'Describing the information for action', in O. G. Meijer and K. Roth (eds), *Complex Movement Behaviour: 'The' motor-action controversy*, Amsterdam: North Holland.

Young, R. M. (1976), *Seriation by Children: An artificial intelligence analysis of a Piagetian task*, Basel: Birkhauser Verlag.

Youniss, J. and Furth H. G. (1973), 'Reply to Bryant and Trabasso, 1971', *Nature*, 244, pp. 314–15.

Zelazo, P. R. (1976), 'From reflexive to instrumental behaviour', in L. P. Lippsitt (ed.), *Developmental Psychobiology*, Hillsdale, N.J.: Lawrence Erlbaum.

Zelazo, P. R. (1984), '"Learning to walk": recognition of higher order influences', in L. P. Lipsitt and C. Rovee-Collier (eds), *Advances in Infancy Research*, vol. 3, Norwood, N.J.: Ablex.

Name index

Subject index